FROM THE WIRELESS TO THE WEB

FROM THE WIRELESS

The evolution o

Dedication

To all those writers who have set down their glimpses of the future of world communications and to those dedicated enthusiasts who have turned those visions into reality: Jules Verne, HG Wells, Isaac Asimov, Robert Heinlein, Alfred Bester, Arthur C Clarke, Guglielmo Marconi, Reginald Fessenden, Edwin Armstrong, John Logie Baird, Philo Farnsworth, William Shockley, Von Neumann and those still too young to be mentioned.

To my family who have remained supportive in the face of an author's obsession.

By the same author

In Marconi's Footsteps 1894 to 1920: Early Radio

TO THE WEB

elecommunications 1901–2001

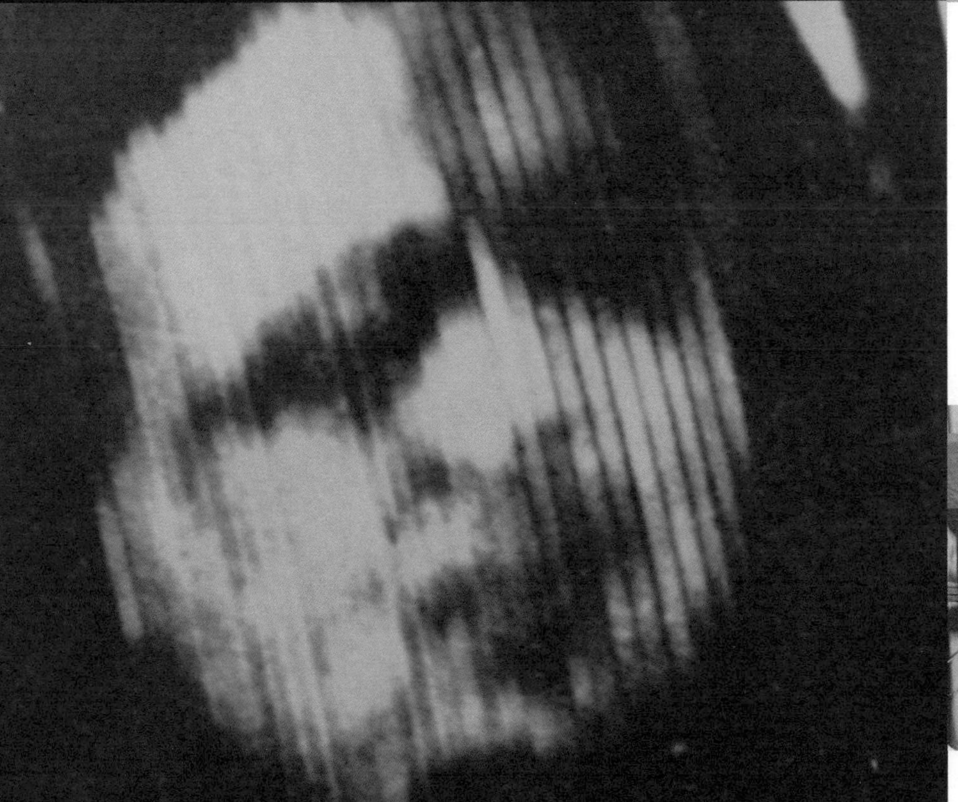

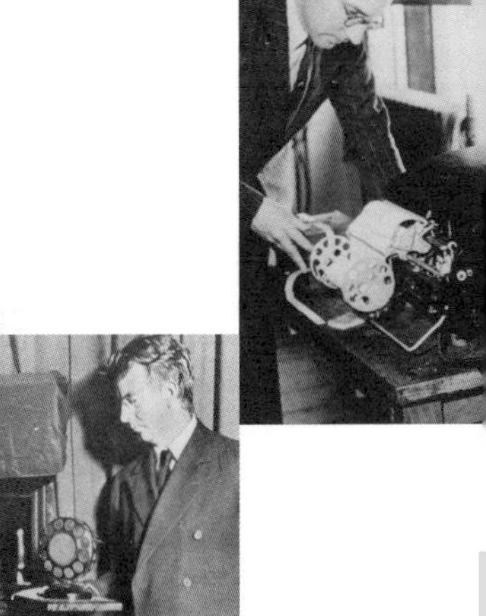

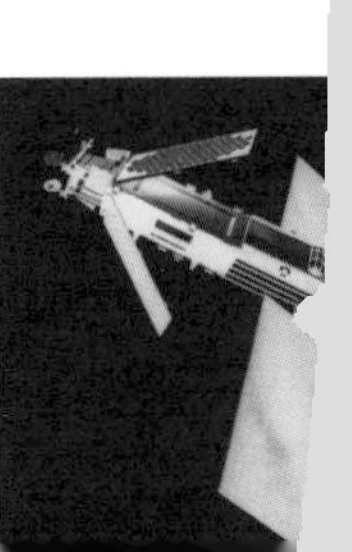

PETER R. JENSEN

Contents

A UNSW Press book

Published by
University of New South Wales Press Ltd
University of New South Wales
Sydney 2052 Australia
www.unswpress.com.au

First published 2000

National Library of Australia
Cataloguing-in-Publication entry:

Jensen, Peter R.
From the wireless to the web: the evolution of telecommunications 1901–2001

Bibliography
Includes index.
ISBN 0 86840 458 6.

1. World Wide Web (Information retrieval system) — History. 2. Telecommunication — History — 20th century. 3. Computers — History. I. Title.

025.04

Text design Dana Lundmark
Printer BPA, Melbourne

Acknowledgments

This book would not have been possible without the generous assistance of many people in Australia, Britain, the United States and Canada. Referring to everyone who has helped would be impossible task and, in any case, it may well be that someone who should be acknowledged would be inadvertently omitted. In those cases, then, it may be assumed that the debt of gratitude is forthcoming even if not specifically noted.

In Australia, I should like to acknowledge the assistance and encouragement of Ian O'Toole, Colin McKinnon and other members of the Historic Radio Society of Australia. In addition, John Geremin and John Deane of the Australasian Computer Museum Society Inc. have helped with information. The encouragement of my friends in the Wireless Institute of Australia and the Australasian Communications and Computer Association Ltd was also of value in keeping my 'nose to the grindstone' over a protracted period. In particular, the help of Alaric Havyatt, Peter Naish and Max Burnet, who read the text looking for technical and historical 'howlers' and also provided formidable editorial input, is very much appreciated.

In Great Britain, over a protracted period, I have been assisted by Roy Rodwell, the former Archivist of the Marconi Company. In addition, David Rudram of the Amberley Chalkpits Museum has given much-appreciated assistance in providing access to obscure but important equipment for photography, as has Major Pickard and his associates at the Royal Signals Museum at Blandford Forum. In this same context of providing access to equipment, the help of Mr Bill Journeaux, whose private collection is to be envied, is acknowledged. On the Isle of Wight, Douglas Byrne gave me access to material at Puckpoole Park and Arreton Manor, and in London the collection of Gerald Wells was inspiring. In the west of Cornwall, access to the Cables and Wireless Museum at Porthcurnow was rather like stepping back in time and provided interesting material on the early attempts to create a girdle around the earth, however primitive. Lieutenant Commander Legge of the radio museum at Fareham was also a friendly and helpful guide to the Royal Navy collection and provided photographs of relevant early equipment.

In Canada, members of the Newfoundland Amateur Radio Society were extremely helpful, in particular Joe and Michelle Craig, who looked after strangers from a far land. A visit to Heart's Content helped to complete the picture of early efforts to provide communications across the North Atlantic.

I should like to mention the following organisations and persons who provided advice or illustrations: Mr A Sale of the Bletchley Park Trust, United Kingdom; Mr E Malmgren and Mr G Roberts of the Qantas Historical Archives; and Mr M Pharaoh of the Mawson Trust, Adelaide.

And finally, I would like to mention the editor of this book, whose meticulous and determined amendments to the text have produced a substantial improvement and polish for which I wish to express my gratitude.

Key to sources

The following organisations and individuals are acknowledged as the source of the photographs contained in this book and are thanked for allowing the author to include their material in this publication:

AWA	Amalgamated Wireless Australasia Limited
ALC	Alcatel Photographs
ANAR	Australian National Antarctic Research Images library
BBC	British Broadcasting Corporation
BPM	Bletchley Park Museum
CB	Carl Benz Collection
DEA	John Dean photographs
EA	*Electronics Australia*
ET	Enrico Tedeschi
FAR	Royal Naval Communications Museum, Fareham
IBM	IBM — Thomas J Watson Research Centre
IRID	Iridium South Pacific Consortium
LUC	Lucent Technology, Bell Laboratories and Texas Instruments
MAR	Marconi Archives, Chelmsford
MAW	Mawson Trust, Adelaide University
MER	Merseyside Maritime Museum
NASA	NASA Web site photojournal.jpl.nasa.gov/
OTC	Overseas Telecommunications Commission (Australia) — Archives
PRJ	Peter Jensen Photographic Archives
QAN	QANTAS Archives
RFDS	Royal Flying Doctor Service, Australia
SCI	Science Museum London
TEL	Telstra, Archives, Australia
UNK	Unknown source, public purchase
WIN	Winsome McCallum Photographs
YM	Ypres Museum, public display

Sources of copies made by the author Permission has been sought to use this material and where received this is expressly acknowledged with appreciation:

25Y	Odham's Press
40B	Hutchinson Publishers
AOB	Sampson Low Marston
BAF	Illustrated London News and Sketch
BAW	Hutchinson Publishers
BBC	British Broadcasting Corporation
BOS	Thomas Nelson and Sons
DKI	Lovat Dickson
GAL	Heineman
MMS	George G Harrap
MWT	Naval Institute Press
PRR	Funk and Wagnall
SAW	Her Majesty's Stationery Office
SWA	CR Leutz
USM	*Radio News* (US magazine, 1932)
WIT	Oxford University Press

Time line

1822	Babbage's paper on the difference engine no. 1 is published
1844	Morse telegraph link is established
1865	Maxwell's equations for electromagnetic waves are written
1866	Trans-Atlantic cable is laid
1876	Bell patents the telephone
1887	Hertz demonstrates existence of electromagnetic waves
1895	Marconi experiments with wireless telegraphy
1901	Trans-Atlantic wireless is established
1906	Triode valve/tube is developed
1912	Sinking of the RMS *Titanic*
1914	First World War
1922	Broadcasting commences
1936	Electronic television commences
1939	Second World War
1947	The transistor is invented
1948	Electronic stored program computer runs at Manchester
1957	Russia launches Sputnik 1
1958	First microchip is patented
1962	United States launches Telstra 1 telecommunications satellite
1969	ARPANET is initiated
1974	INTEL makes 8080 microprocessor
1977	Apple microcomputer starts selling
1981	IBM produces the PC
1984	NFS Net commences operation
1992	Microsoft produces Windows 3.1
1994	Mosaic Communications produces Netscape Navigator 1
1995	The World Wide Web becomes publicly available
1999	Trans-world satellite communication system — Iridium

Introduction

The Internet, as it is known today, is the child of two parents — telecommunications and the electronic computer. This book aims to show how, from the earliest days, the present complexity of the World Wide Web was progressively created. The birth date of the Internet is seen to have been set by the bridging of the Atlantic with radio signals for the first time in 1901. Its coming of age was marked by the later creation of the first electronic computers after 1940. Even before 1901 there was a long period of development that laid the groundwork for the system of global communications that was to follow.

Here is the story of the creation of the system of international communication based on the cable, radio and, more recently, the satellite. It is also the story of how the threat of world war and the need to transfer data from computer to computer in a secure manner initiated a fundamental change in the technology of communications. This was a change that would lead directly to the form of international communication that is now available.

The World Wide Web has been in a state of rapid development over the last few years and has caught public imagination as no other development has done in recent times. It seems probable that this new mode of communication and trade is destined to have a huge impact on the way all the members of the world community interact in the future. Suddenly the traditional boundaries of the international order have been traversed. The development of increasingly intimate modes of communications such as video conferencing has brought people closer together. Only language continues as a barrier to world communication and even this will go with the use of the computer for machine translation in the future.

The world has changed enormously in the last few years in ways that are, as yet, not entirely obvious. This process is likely to continue with increasing momentum into the future. Humanity is riding a wave of development, the ramifications of which have barely begun to be comprehended, and the human race faces a future as profoundly different as that faced by the people living at the start of the Industrial Revolution in the 18th century.

The year 2001 is seen as marking the start of a communications revolution with a potential to change the way we live. What follows is the story of how our present system of communication began and developed.

CHAPTER

1

Before 1901

Early communications

Two hundred years ago, many people lived in isolated villages scattered across the face of the earth. The major exception to this was the new class of town dwellers located in the rapidly growing cities and towns of the industrial revolution. Away from such places, country folk usually spent the whole of their lives close to the places in which they had been born and mostly where their ancestors had lived for generations before that.

In a time before universal literacy, communication was largely by word of mouth and news travelled slowly in country areas and was frequently much distorted in the process. However, the demands of business in the cities and towns saw the development of a network of communication linkages. Initially, messages were carried by runners but later by couriers on horseback. Before long, elaborate signalling systems were developed using mechanical semaphores allowing communications over greater distances and shortening the time for a message to travel from its origin to its destination.

Such a system was introduced in France in 1793, and a refinement involving Venetian blind-like shutters was installed between London and Dover in Great Britain in 1795. Close by, at the little town of Deal and somewhat later, a time signal was provided every day to ships moored offshore in the lee of the Goodwin Sands. This was achieved by the dropping of a large, black ball set on the roof of a tower. In Victorian times, this building was located right at the top edge of the shingle beach. Normally the black ball rested at the lower end of a tall pole on the roof but as the hour approached it was slowly winched to the top. This action provided a warning that the hour was near, and precisely at one o'clock the ball was allowed to drop to the bottom of the pole.

Deal time ball tower, Kent, England *(PRJ)*

At a later date, an electric signal sent from the observatory at Greenwich was used to activate a trip mechanism and this in turn released the ball from its elevated position.

By such means as the semaphores and time balls, the demand for trading news and essential time information for navigational purposes was partially met. However, for all the advantages of an arrangement of semaphore signals over a horseborne messenger, a person was still required at each intermediate semaphore station. Apart from inevitable errors that crept into messages conveyed in this fashion, the system was still painfully slow.

COOKE AND WHEATSTONE TELEGRAPH

The work of Cooke, Wheatstone and, later, Morse and Vail was to change the method of passing messages for all time. At a stroke, with the introduction of the electric telegraph, the frequent intermediate stations were largely eliminated and with them the inevitable errors. Compared with the horse-and-rider method of message distribution and the mechanical semaphore, the messages travelled along the overhead wires at the speed of light.

For all its huge advantage over a system of messages transmitted by hand and packhorse, the electric telegraph was still tied to the ground. The wires through which the messages passed were still bound inexorably to the earth by the poles on which they were hung. Until the latter part of the 19th century, this was to be the only method of passing messages by electricity and, for anyone moving about on the land or out at sea, an infallible means of communication was still not available. The electric signalling lamp, flags and the heliograph remained the basis of communication in such circumstances and inevitably the range over which messages could be passed was severely limited.

The early systems of electrical communication provided a stepping stone to the development of radio communication, not only in the apparatus that was produced but also in the method of signalling that was developed. As early as the middle of the 18th century, there were proposals to use the newly discovered electrostatic force as a basis for signalling and, over the next 70 years, a number of novel ideas emerged.

Significant in the evolution of electrical signalling was the work of a Russian, Baron Schilling. His experiments had been conducted in 1832 and the telegraph lines established between certain buildings in St Petersburg for the Czar were noticed by a medical student named William Cooke who was working in Germany at that time. Cooke was later to join forces with Professor Charles Wheatstone, inventor at a later time of the code ascribed to Baron Playfair and known as the 'Playfair' code.

From the collaboration of Cooke and Wheatstone came the first electric telegraph patent of 1837 and its application in the linking of two London railway stations. The refined version of the Cooke–Wheatstone system used a needle which swung between two electromagnets activated by separate telegraph lines. It was therefore possible to have the needle swing to the left or to the right depending on which magnet was activated. An associated code was developed which initially was to be read visually. However, it was soon realised that the sound of the needle striking the stops on either side could be readily understood directly by the ear. This was then elaborated into a system in which the striker needle produced a markedly different tone depending on which way it swung.

Although it provided a useful and reasonably rapid method of sending signals between remote locations, the Cooke–Wheatstone system was very soon overtaken by another electrical signalling method which was elegant in its simplicity and reliability. This was the work of an American, Samuel Morse, and his collaborator, Alfred Vail. These two inventors produced the system which can be seen as the direct antecedent of the first method of radio communication using morse code.

MORSE AND VAIL TELEGRAPH

In 1844 a telegraph link was established between Washington and Baltimore using the Morse–Vail system. The first message sent via this new communication link was a quotation from the Bible — 'What hath God wrought' — a somewhat enigmatic question that may have at last received an answer in the burgeoning of the Internet.

As ultimately developed, the Morse–Vail telegraph used a sounder which produced a 'click-clack' noise, depending on whether a key-down or key-up condition applied. The system had been intended initially to operate with a mechanical embossing machine to convert the key-down, key-up periods into marks and spaces that could be read directly by eye. But as with the Cooke–Wheatstone system, the telegraphists quickly learned to read the sound of the relay. In this fashion the system of sound signalling, now known universally as morse code, was born.

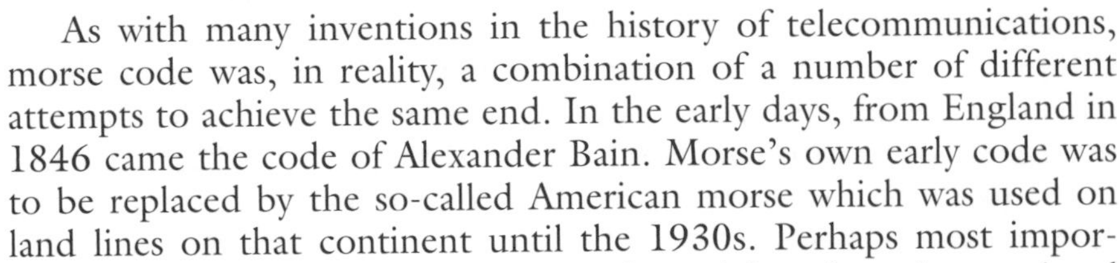

As with many inventions in the history of telecommunications, morse code was, in reality, a combination of a number of different attempts to achieve the same end. In the early days, from England in 1846 came the code of Alexander Bain. Morse's own early code was to be replaced by the so-called American morse which was used on land lines on that continent until the 1930s. Perhaps most importantly and based on international agreement in 1851, a new code combining elements from all its predecessors, the continental code, was introduced. This was to be the code adopted universally for radio when it arrived in the latter part of the 19th century. The various codes are shown in the illustration, and the derivation of the continental code from the earlier codes is evident.

Morse key and sounder, c. 1870 *(PRJ)*

Following the establishment of the continental code as the standard mode of communication by radio, mechanised systems were later introduced to increase the rapidity of signalling and, with the introduction of radio teletype, the Baudot and Murray codes were used. These codes did not use the variable length elements of Morse and Bain but a code based on simple on/off or mark and space elements. Interestingly, the derivation of this system could be seen in the much earlier problem of sending morse dots and dashes across the undersea cables that proliferated in the latter part of the 19th century.

In an era before electronic amplification and the inclusion of 'repeaters' in these undersea cables, dots and dashes became hopelessly distorted so as to be indecipherable. This resulted from the electrical characteristics of very long cables and the behaviour of electrons passing through them. The problem was ultimately solved by introducing a

Above Sounder unit in boxed reflector for use in a repeater station *(PRJ)*

Right Morse code paper tape perforator *(PRJ)*

mark–space code, where the elements were created by sending positive and negative pulses relative to earth through the cables and thus removing any ambiguity.

By the latter part of the 19th century, the invention of Morse and Vail had become the basis of an ever-growing network of linkages around the world, and even the remote commercial centres of Australia were connected to the northern hemisphere. The extraordinary work of Sir Charles Todd and the creation of the overland telegraph cable from Adelaide to Darwin and thence to Java by undersea cable in 1872 was to see the isolation of Australia banished forever with the dawn of the world communication age.

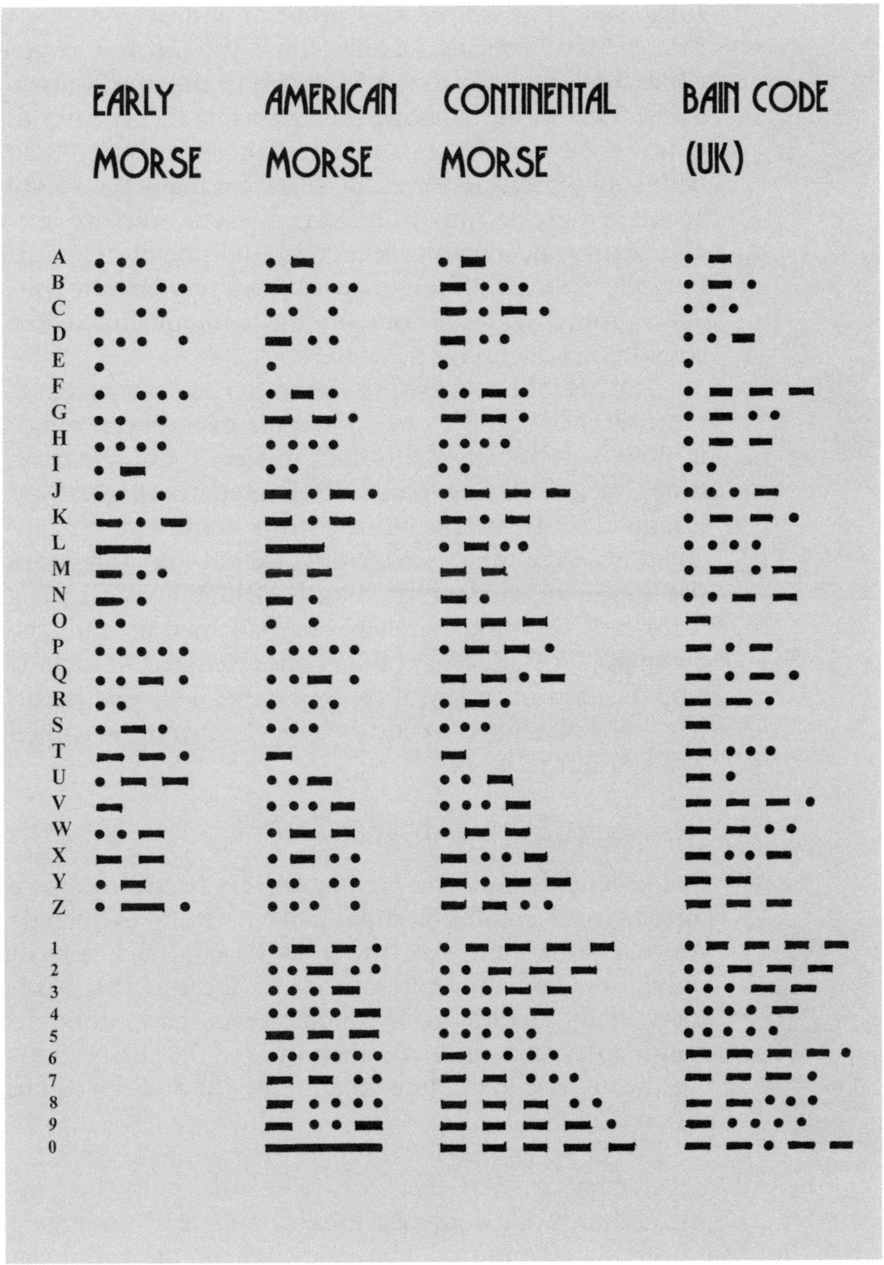

EARLY MORSE | AMERICAN MORSE | CONTINENTAL MORSE | BAIN CODE (UK)

A B C D E F G H I J K L M N O P Q R S T U V W X Y Z

1 2 3 4 5 6 7 8 9 0

Various telegraphic signalling codes including morse and continental morse code *(PRJ)*

REIS, BELL AND EDISON TELEPHONE

With the establishment of telegraphic communications within the continental areas and across the oceans, attention soon turned to the possibility of conveying the voice by electricity. The pioneering work of Philip Reis in 1861 was completely overtaken by the inventions of Professor Alexander Graham Bell, who created the first effective receiving device in 1876. The electromagnetic telephone receiver that Bell developed was at first used as a transmitter as well as for reception. However, its low output made it unsuitable for the unamplified system in which it was to be used.

Skeleton telephone, c. 1900 *(PRJ)*

The lamp black microphone invented by Thomas Edison in 1878 solved this problem and, assisted by a local battery, provided a substantial voltage and a capability to drive a signal over a substantial distance. Known as the 'button microphone', the principle that it embodied of a carbon element, the resistance of which varied with sound pressure waves, persisted for many years until replaced quite recently with electrodynamic microphones able to provide a greater level of sound fidelity. The carbon microphone, although providing a very healthy voltage output, was notorious for the poor quality of the sound reproduction it provided.

The telephone network based on the electromagnetic receiver and the carbon granule microphone rapidly grew to serve the cities and towns of late Victorian times, but for many years was limited to the seaward boundaries of Europe and the other continents. A capability to cross the oceans with voice did not come until the advent of wireless in 1896, but the fidelity that is a feature of the modern telephone had to wait until the technology of undersea cables had developed to include amplification in the form of 'repeaters'. This was not to happen until almost a century after the invention of Bell was first put to work.

TRANS-AUSTRALIAN TELEGRAPH

Coincidentally and as a prelude to the extension of the international communication cable network to include Australia, the time ball tower at Deal played a most significant part. Indeed, without it, there is the possibility that Charles Todd might never have gone to South Australia and the history of the first trans-Australian telegraph line might also have been quite different.

Receiver unit from Bell telephone system *(PRJ)*

In 1854 Charles Todd was in charge of the galvanic department at the Greenwich Observatory on the outskirts of central London. Todd's interest in telegraphy had developed when he had been at Cambridge University where he had been employed as Assistant Astronomer at the observatory. At the Greenwich Observatory, Todd was able to expand this interest in telegraphy through contact with the superintendent of the Telegraph Department, Charles Walker.

In 1854, the South Australian government wrote to the Colonial Office in London requesting assistance in the construction of a telegraph line to give early warning of a possible Russian naval attack — the Crimean War was then at its peak. In characteristic bureaucratic fashion, the Colonial Office duly referred the matter to Sir George Airy, the Astronomer Royal. When Airy came to decide who should be offered the position in South Australia, it was a problem with the telegraph line to the Deal time ball that was to help him reach a decision. He had sent young Charles Todd to rectify the problem, and this was achieved within a few days. Having established his capacity for both independent and effective action, Todd was then appointed to the position of Government Astronomer and Superintendent of Telegraphs in Adelaide. At the age of 28, Todd, together with his new wife, departed for what then must have seemed like the other end of the universe, Australia.

In November 1855, at the time of the arrival of Charles and Alice Todd in Adelaide, the centre of Australia was almost completely unexplored territory, and Darwin, on the northern edge of the continent, seeming quite remote and inaccessible. All this changed with the remarkable exploration of John McDouall Stuart, who successfully traversed the inhospitable inland region in July 1862 after a number of previously abortive attempts. With the completion of that epic journey, Todd urged that a telegraph line be established along the route followed by Stuart, connecting Adelaide to Darwin.

Not long after, in 1866, the first trans-Atlantic cable was successfully laid by the steamship *Great Eastern*. This ship had been designed by the famous engineer Isambard Kingdom Brunel, who subsequently

Landing the first undersea telegraph cable at Darwin in 1872 *(TEL)*

went on to devise and complete many other notable engineering projects. With the success of the trans-Atlantic crossing, it was apparent that a telegraph connection from Great Britain to Adelaide on the other side of the globe was likely to be possible in the not too distant future. As it transpired, the establishment of the undersea link from Java to the northern tip of Australia and the opening of the trans-continental line from Darwin to Adelaide were to consume another 8 years and involve the installation of approximately 36 000 poles across the mostly arid but sometimes flooded inland of Australia. In an extraordinary exercise of tenacity characteristic of the Victorian age, Charles Todd, later knighted, was to oversee the construction of the new telegraph line through an almost unknown and very hostile land. No doubt the cost of the Australian section of the link to London, in an estimate of remarkable precision suggested to have been an amount of £479 174 18s 3d, was an accurate reflection of the difficulties encountered.

The main telegraphs receiving room at Darwin *(TEL)*

Notwithstanding the potential value of the new telegraphic connection to Britain, the cost of the enterprise was very substantial and represented a huge investment for the small colony of South Australia, particularly given the small size of its population at that time, 80 000 people. When that estimated amount is converted to modern-day figures, one arrives at a sum of approximately £100 million sterling or, say, A$230 million — a figure quite sufficient to give the contemporary treasurer of the State of South Australia pause for deep reflection.

HEART'S CONTENT

Unfortunately, a visit to Darwin today does not allow one to see any of the artefacts of the telegraph system that was established there in 1872. The main beach of the old town seems to have been overlayed by progressive filling and certainly any sign of the old cable is absent. By comparison, the trans-Atlantic cable referred to earlier is still visible at its landing place at Heart's Content, Newfoundland, and is likely to have been very similar in both a physical and technical sense to what was built at Darwin some years later.

As the photographs of the Heart's Content telegraph station show, the cables consisted of an exterior bundle of wrought iron wires which encased a core of copper wires, surrounded by an insulating layer of gutta-percha. This material, at that time only recently discovered and having some similarities to latex rubber, provided a very effective separation between the copper wires and the sea water in which the cable was laid. The cable was covered externally by cloth impregnated with tar and wax, and this would also have significantly helped to separate the inner copper core from the surrounding salt water in electrical terms.

However, apart from the need to provide adequate insulation between sea water and copper core, what proved a far more serious problem in the initial attempts to lay a trans-Atlantic cable was the strength of the iron cable reinforcing. It was not until this had been increased by almost three times was it possible to produce a cable that would not break under its own weight when extended to the sea floor some kilometres or so below in the deepest parts of the route. In addition, the length of the cable, over 2300 nautical miles, and its unbroken form of construction made for a huge and heavy load which had contributed to the lack of success of the 1857 attempt. At this time, two American ships, the USS *Niagara* and USS *Agamemnon*, had been used as the cable-laying vessels. The use of Brunel's extraordinary and revolutionary *Great Eastern* had been required to cope with this problem and led to the ultimate success of the 1866 expedition. It is noteworthy that this ship, with a length of 213 metres (700 feet) and a displacement of 22 950 tonnes (22 500 tons), was five times bigger than any other contemporary vessel afloat and would not be equalled in size for the best part of 40 years after 1866.

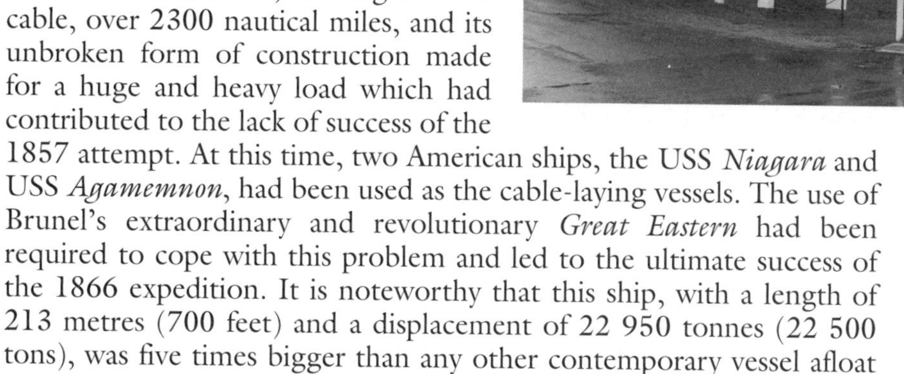

On land, the terminal station at Darwin is also likely to have been somewhat similar to the earlier example at Heart's Content and to have used similar apparatus. Much of this telegraphic equipment was to

The telegraph station at Heart's Content, Newfoundland *(PRJ)*

Below left The telegraph repeater room at Heart's Content showing the Murray multiplex rotors from the 1920s *(PRJ)*

Below right Early telegraph cable on the foreshore at Heart's Content *(PRJ)*

remain in operation for many years, being replaced progressively by devices intended to increase both the reliability and volume of data being passed through the station. Again, the photographs of the Heart's Content station demonstrate the changes that occurred between the earliest installation in 1866 and the technology that was in use when the station was closed down in the early 1960s.

HERTZ AND THE ETHER

Heinrich Hertz at the time of the ether experiments of 1887 *(EA)*

By 1887 the electric telegraph was already a marvel that had changed the world and its system of communication forever. However, for all the effectiveness of the new system, out at sea or on the move communications still had to rely on primitive means with an extremely limited range. The world was ready for something revolutionary and the work of a rather obscure German scientist was to provide the initial impetus. However, before this work as the basis of a quite new method of communication was fully realised, some years would have to elapse.

By 1888, the first glimmerings of a revolutionary new method of sending signals could be seen in the work of that German scientist, Heinrich Hertz. Initially, however, few saw the implications of the minuscule beam of light that was cast by his research and by the discharge across the 'spark gap' of his simple receiver. It would be another 8 years before the technology would be developed to exploit the new medium of the ether and Hertzian waves. Further, it would not be the acknowledged scientists of that time, at the end of the 19th century, who would turn that feeble spark-generated beam of energy into a shaft of brilliant illumination. On the contrary, it would be a young man uninhibited by a comprehensive knowledge of the science of physics of the late 19th century, knowledge which could have easily dissuaded him from proceeding with his own experiments. The young man was, of course, Guglielmo Marconi and his work is now generally well known.

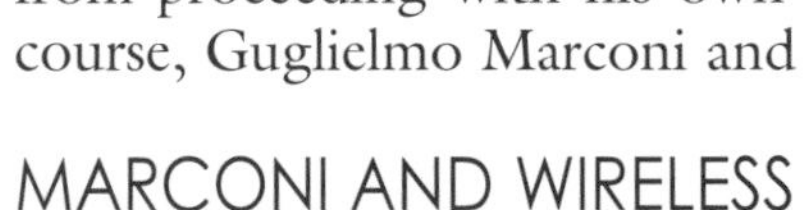

MARCONI AND WIRELESS

At a very early stage, Marconi developed a driving ambition to set the ether waves of Hertz to work and to create a communication system

Marconi's first receiver of 1895 at the Villa Grifone *(PRJ)*

The country home of the Marconi family, Villa Grifone, near Bologna *(PRJ)*

suitable for sending messages over a great distance. When the Italian postal authorities were apparently indifferent to his ideas, his mother proposed a visit to England where they had family relatives to assist and there was an obvious need for 'wireless' communication. With the surrounding sea to disconnect this island domain from the main body of Europe, isolation was certainly a useful deterrent to potential aggressors. However, this isolation also had the effect of limiting trade-related communications, making the passing of messages a very expensive and time-consuming activity.

Great Britain in 1895 had an enormous need for a system of wireless communication. She had a large and complex navy and, as a part of the fleet, newly developed vessels that moved about far too quickly to be easily controlled by light signalling or semaphore using flags. Perhaps even more importantly for the future success of the wireless or radio system of communication that was just about to be born, there was a 'midwife' available in the person of Captain Henry Jackson.

For a number of years this naval officer had been thinking about the problem of controlling fast motor-torpedo boats. Not only had he envisaged the use of Hertzian waves as the basis of a system of communication but he had started to put his thoughts into action. His early efforts all involved the same method of transmission as used by Hertz and as later applied by Oliver Lodge and demonstrated in 1894. However, as a busy serving naval officer, he had little time to devote to this interesting area of research and it tended to languish. By the time a working method of transmission and reception had been developed

A replica of Marconi's first transmitter at the Villa Grifone *(PRJ)*

by Jackson, the youthful Marconi had already sprung onto the world stage with a refined and reliable system of his own which was patented in June 1896.

After his arrival in Great Britain, Marconi received considerable encouragement from a number of eminent people, perhaps most notably the Chief Engineer of the Post Office, William Preece. However, it was probably just as much the support of Captain Jackson that was significant in promoting the work of Marconi in its initial period at the end of the 19th century. Certainly Jackson was instrumental in having the Royal Navy adopt Marconi apparatus in 1901. It is apparent that a considerable rapport developed between the young man and his much older supporter in Jackson. Because of the secret nature of reports produced by Jackson for consumption by the naval authorities,

Sir Oliver Lodge, inventor of tuning, or syntony as it was then known *(MMS)*

it was not until many years later that the true extent of his part in introducing radio became apparent.

In terms of the story of the development of world communications, the introduction of wireless into the ships of the Royal Navy represents one of the most important early milestones. However, since 1901 there have been many such milestone events. As later chapters will demonstrate, these milestones have been both technical and practical in nature and have tended to occur with increasing frequency and greater technical complexity as time has passed.

In the last 20 years, with the development of the microcomputer as a domestic and business appliance, this escalation of development and complexity has been particularly marked. We can now see that over a period of 100 years there has been an implicit goal for all this development. This has been the achievement of a system of universal communication available to all and now a reality in the World Wide Web.

CHAPTER

2

The mid-1800s

Babbage —frustrated genius

Many people involved with computers with only the scantiest knowledge of their history may be aware of the name Babbage and associate it with early developments in this field. What they may not know is that the story of Babbage is not at all a happy one. With modern computers as a basis for a contemporary perspective, although his life could be seen as involving great intellectual brilliance and inventiveness, it was in the end one of frustration, loneliness and monumental failure. It can now be seen with great clarity that here was a man wildly ahead of his time 'intellectually' — so much so that his work, despite the immense promise that it offered, was effectively forgotten for nearly 100 years.

Simply put, what this extraordinary Victorian gentleman set out to produce was a mechanical device that would compute in much the same manner as the modern electronic computer, but in a time long before the discovery of the electron. In the past, when the name Babbage was spoken, it was usually in the context of suggestions that his ideas involved tolerances and technology that exceeded what was practical to achieve in 1830. It has now been shown that it was not the technology of the Victorian era that was at fault and prevented his ideas from being realised, but failure of support by the government and inadequate funding. In the end it was the lack of resources to finish his projected analytical engine and the later difference engine no. 2 that effectively destroyed this irascible genius.

Charles Babbage in his later years *(EA)*

EARLY COMPUTATION MACHINES

When Charles Babbage was born in Devonshire, England, in 1791, some of the initial steps towards mechanised computation had already been taken. Based on the work of John Napier and his invention of logarithms in 1614, the slide rule was soon invented by William Oughtred in 1621. The next significant step was the work of Blaise Pascal who, in 1642, created a counting machine to help his father in his numerical work.

Pascal's calculating engine, known as the 'Pascaline', had interconnected cogs and gear wheels which could add and subtract numbers. It could also multiply and divide in a rather slow and clumsy fashion by successive addition and subtraction. Necessary improvements in the multiply and divide function led to it being superseded by a machine invented by the mathematician Gottfried Leibnitz, the great rival of Isaac Newton. This device used rods with a series of elongated teeth of varying length which were used to carry out the multiply and divide functions of the calculator. Remarkably, this was to remain a fundamental design feature of mechanical calculating machines up until the 1960s when electronic calculators made the electromechanical apparatus obsolete.

In Victorian times it was commonplace that the tables of logarithms developed by Napier in 1614 would be filled with computational and transcription errors. Numerical calculations based on such tables were inevitably inaccurate and numerical accuracy was of particular significance to the science of navigation. In Britain in the 1800s, navigation was of paramount importance to trade which supported the British Empire. Inaccurate tables of logarithms and other functions were the bane of Victorian navigators and seamen and the cause of untold numbers of shipwrecks. Clearly something far better was needed urgently, and Babbage and his friends were well aware of that need.

COMPUTATION ENGINES

How great ideas arise is frequently an obscure matter and that certainly applies to the origins of the solution to the accuracy problem now proposed by Babbage. It appears that his radical solution stemmed from a discussion that occurred at Trinity College, Cambridge, in 1812. One of his friends at university was John Herschel, who later became the

Astronomer Royal. The problem of numerical accuracy of the logarithm tables was being discussed and, during the course of the conversation with Herschel, the question 'Why can't it be done by steam?' was asked. This idea of mechanising the production of numerical tables became Babbage's obsession and something that he worked on almost continuously for the next 60 years until his death in 1871.

During the next 10 years, Babbage, as a man of independent wealth, was able to work on the problem of mechanical solutions to numerical computation. In 1822 he presented a paper to the Royal Astronomical Society and this was the basis of what later became known as difference engine no. 1. For this work he was later awarded the Society's first gold medal. Work on building this machine, which was based on a rather arcane way of calculating known as the 'method of finite differences', commenced in that same year and, by 1832, a demonstration model had been produced and was shown to be able to calculate satisfactorily.

DIFFERENCE ENGINE NO. 1

Carrying the pilot project through to the complete machine proved an intractable problem. At this stage it appears that the problem of mechanical tolerance and repeatability did indeed arise, and this led to disputes with the workshop of Joseph Clements where the difference engine no. 1 was being constructed. To avoid 'jamming' of the mechanism which involved thousands of moving parts and cogs, great accuracy in fabrication of the component parts was called for. No doubt this also led to the financial disputes that finally saw the work on the machine come to a halt. However, what was to prove the insuperable barrier to success of this project was the decision of the government, in the absence of tangible results, not to provide further funding.

Of his many personality traits, one ultimately proved most destructive of Babbage's work — his inability to be satisfied with the final form of any of his inventions. Time and again in the history of his work one can read of fresh ideas arising and the decision taken to make modifications, however time consuming. Always for Babbage such changes proved essential and irresistible.

As difference engine no. 1 was being developed and its many problems were responded to but ultimately to no avail, Babbage turned his mind to a far more complex machine. This is now known as the analytical engine. This machine was later described in a large number of elaborate and extremely beautiful engineering/architectural drawings prepared by the draftsman CG Jarvis. Careful study of what was proposed in this material indicates that here, in principle, was the basis of the modern computer.

THE ANALYTICAL ENGINE

In his autobiographical work, Babbage described the analytical engine as consisting of two parts, the store and the mill. However, looked at with the eye of a modern computer user and applying contemporary terminology, one can see the resemblances to the computers of today.

The components of the analytical engine can be seen to consist of:

- input area for entering the numbers to be processed — input/output (IO)
- numerical processor (the mill) — central processing unit (CPU)
- process control — programs held in external hard drive
- numerical memory — random access memory (RAM)
- output area for printing results — input/output (IO).

As suggested by Jarvis's drawings, the operation of the machine was extremely complex as it involved some 25 000 parts including cogs, wheels, levers and cams. To see it in operation would no doubt be a far more satisfactory experience than attempting to understand how it could have worked from the written word or even from drawings. What certainly helps comprehension of the operation of this strange class of counting machines are recent computer graphics developed to explain the operation of a later device, difference engine no. 2.

A section of the analytical engine as constructed by Babbage's son, Colonel Henry Babbage *(PRJ)*

A very pretty piece of animated drawing has been created to show how the number wheels in this machine turn and advance to carry out the necessary additions for the method of differences. This may be seen at the Science Museum in London together with the reproduction of difference engine no. 2 and a variety of other machines, including one that was built by Babbage's son, Henry Provost Babbage, between 1890 and 1911. This is a small, demonstration unit of the analytical engine, consisting of a part only of the mill and the printing mechanism. Access to the models also helps in understanding how the machines were intended to operate.

In system terms, the similarity of the analytical engine to the modern computer is obvious. Apart from the processing of numbers undertaken by the mechanical 'mill', process control and memory was provided through the use of punched cards, similar to those used at that time in the weaving industry and invented by Joseph Marie Jacquard. However, what clearly distinguishes this machine from a contemporary computer are the cogs, levers and printer which Babbage intended to drive with a steam engine. Further, because of the mechanical system proposed, Babbage asserted that the machine would either calculate correctly or it would jam, but it could never lie! Compared with Babbage's analytical engine, not telling the truth is an occasional problem of modern computers that has still not been entirely eliminated.

The new machine was offered to the government but, not entirely surprisingly, given the earlier expenditure, no enthusiasm was shown and the project languished. In this regard, Babbage's acerbic tongue and capacity to make enemies had not assisted his cause, despite the support of his staunch friends such as John Herschel.

DIFFERENCE ENGINE NO. 2

With the refusal of the government to provide further financial support, it having by now contributed just over £17 000 sterling, Babbage set the drawings of the analytical engine to one side and turned his attention to a new form of difference engine. This machine was of a simpler design and had far fewer parts than the original difference engine but was able to calculate to a much higher order of accuracy. It was to be known as difference engine no. 2, and work on it commenced in 1849.

Again, financial difficulties and dissatisfaction with the design led to it never being constructed. However, of the three machines of Babbage, it is by far the most extensively documented, being described in great detail in over 20 drawings prepared to a very high graphical standard — one of the reasons that it was later chosen to be the basis for a commemorative replica. This project was commissioned by the Science Museum in 1987.

Analysis of these drawings and Babbage's writings allows a comprehension of its intended mode of operation. It can be deduced that, compared with a modern computer which uses a binary system of numbers, the difference engine no. 2 used decimal numbers. These numbers were engraved on toothed brass wheels set in eight vertical stacks in which each stack could represent a multi-digit number. There were 31 number wheels in each stack of numbers and hence potentially the same number of digits. Unlike a modern computer, where numbers are either 1 or 0, in a decimal digital system values can fall between integral numbers. The machine must be able to negotiate such transitional values without jamming up. In the difference engine no. 2, at the end of each cycle of addition and numerical transfer from one stack to the next, the wheels were locked into position at integral number values or zero. This locking also operated when number wheels were not in use.

ADA LOVELACE

Apart from extreme intellectual brilliance, Babbage was described as a man of 'great charm, vitality and good health'. Apparently he got on well with women and, in particular, the only daughter of the poet Byron, Augusta Ada, later the Countess of Lovelace. This extremely bright young woman, who at an early age developed an interest and expertise in music, machinery and mathematics, was clearly drawn to the much older Babbage in a platonic meeting of great minds. Although the extent of her contribution to the development of the analytical engine is not entirely clear, what this young woman certainly did do for Babbage was translate and extend a comprehensive discussion of how the analytical engine operated. This had been prepared initially by a young Italian military engineer, Captain Luigi Menabrea, in 1842.

Apart from this extremely useful activity, it appears that Ada Lovelace was significantly involved in setting down the method by which the analytical engine might be 'programmed', to use the modern description. Among other things, it is apparent that she came to realise that whatever the brilliance of the conception, such a machine for computing was able to do only what its human operator instructed it to do. In this respect she clearly anticipated the fundamental law of modern computing which is expressed in the acronym GIGO. This says that if what you tell the machine to do is nonsensical, the response will be equally nonsensical — 'garbage in, garbage out'.

Unfortunately, and perhaps unsurprisingly as the daughter of Lord Byron, Ada Lovelace lived an extremely turbulent and controversial life and subsequently died at the age of 36 in 1852, struck down by a particularly vicious form of cancer. This was another body blow for Babbage who had already seen his young wife, Georgina, die in childbirth in 1827 and only three of his eight children survive their childhood. Later he was to lose all but the eldest of his three sons to the impossibly distant Antipodes, and lived out his old age in an intellectual vacuum, perhaps to some extent of his own making.

TRAGEDY AND TRIUMPH

In many respects, the story of Babbage involves elements of tragedy, given the extraordinary intellectual heights that his work had achieved. The failure of the incredibly advanced analytical engine represents the most poignant aspect of this tragedy because, in so many respects, although mechanical in its operation, it worked in exactly the same way as do modern electronic computers.

Replica of the difference engine no. 2 as built at the Science Museum, London, with Doron Swade (Senior Curator of Computers and Control) in the foreground and engineers Reg Crick and Barry Holloway *(SCI)*

Nearly 100 years were to pass before an opportunity was made to expose the full genius of Babbage and, somewhat surprisingly, it came from the collaboration between an Australian mathematician, Dr Alan Bromley, and the newly appointed curator of computing at the Science Museum, Doron Swade. In 1979, Bromley carried out a detailed analysis of Babbage's machines as a research study while on sabbatical leave in the United Kingdom. In particular he was concerned to deal with the frequently discussed issue of whether an inability to achieve appropriate engineering tolerances for this type of machine, given the technology of early Victorian times, had caused the failure of Babbage's project.

Bromley ultimately came to the conclusion that the 1 to 2/1000th of an inch required in the machinery that Babbage had specified was quite capable of being achieved in the 1820s. Bromley then carried out further research at the Science Museum and subsequently, in 1985, presented a proposal to the museum that a full-scale reproduction of difference engine no. 2 be undertaken, based on the perception that it was within the skill of Victorian engineers to have created it. The rest, as they say, is history.

Initially, a trial section of the calculating machinery of the difference engine no. 2 was produced and was shown to operate properly. Subsequently, over a 4-year period, an extraordinary full-size reproduction was built at the Science Museum. The construction was carried out on the ground floor in full view of the public. However, upon completion and successful operation of the replica, it was moved upstairs and placed inside a glass case away from the prying hands of museum visitors.

The difference engine no. 2 being built on the ground floor of the Science Museum *(PRJ)*

The difference engine no. 2 completed and housed in a glass case on an upper floor *(PRJ)*

The winding handle for the difference engine no. 2 which replaced a steam engine intended to drive the machine *(PRJ)*

Shortly after the completion of the difference engine no. 2, an opportunity to operate this machine, which involved turning the crank handle, revealed that the operation was akin to the use of an old-fashioned clothes mangle. A video film made at that time shows the peculiar rippling and serpentine motion of the number wheels as they rotate, rise and carry to add, before coming to rest at the beginning of the next cycle of addition.

CALCULATE OR CRASH

As previously noted, Babbage boasted that his machines could never lie and would jam up first — unfortunately it appears that this claim has now been shown to be entirely correct. In October 1996, the author visited the Science Museum to inspect the Babbage machine again and discuss its method of operation with members of the staff. It appeared that the machine was not currently in operation, having been unsympathetically used by a person involved in some public relations photography. Apparently she had turned the handle as if she were trying to start up a vintage motor car and this had resulted in the jamming of some of the registers. The machine required complete adjustment and fine-tuning by the original engineering staff. However, assurance was received that at a previous time, when fully complete and adjusted, it had calculated numbers to 30 places of decimals as originally intended. To that extent it was a great success.

At first sight, a machine designed in the 19th century might seem to have little relevance to modern communications. In architectural terms, here was a mechanical device that shared common structural characteristics with machines powered by electricity, having relatively few moving parts and using technology that would have been totally incomprehensible to scientists of Babbage's time. However, when its intended mode of operation is appreciated, the link to modern electronic computation is clear and hence its relation to later systems of communication.

CHAPTER 3

1901

Marconi's gamble

By the early part of 1901, the new system of wireless had achieved a limited degree of success, particularly for ships at sea. Despite its novelty, having been patented only 5 years before in June 1896, the Royal Navy had been very impressed by its capabilities and had obtained a number of radio systems from the recently formed Marconi Wireless Telegraph Company. In 1900, the maritime ship owners started to follow the example set by the Admiralty. Most significant, and to some extent rather surprising given the conflagration and warfare that was to erupt not very far into the future, it was the German shipping company, Norddeutscher Marine, which was the first organisation to have Marconi apparatus installed on its flagship, the *Kaiser Wilhelm der Grosse*.

Now driven by a desire to achieve ever greater distance and in the face of competition with the well-established cable telegraph companies, Marconi cast around to find a means of confirming what he already sensed — that wireless was an economical and significant alternative to the vulnerable undersea telegraph links.

Marconi as a young man travelling to Canada in November 1901 *(MAR)*

As Marconi considered what might be done, earlier experiments came to mind. During 1899, Marconi had been involved in tests of wireless by the Royal Navy. Installations had been made on the HMS *Juno*, HMS *Alexandria* and HMS *Europa*, and signals had been received from the shore station at the Needles on the Isle of Wight, 100 kilometres (60 miles) away — well over the horizon and out of sight. Now a fantastic idea floated into his mind. If the radio signals could jump over the hill of land and water between England and a ship far out to sea over the horizon, perhaps they could also straddle the enormous mountain of water between England and America — a trans-Atlantic crossing, in fact.

By the early part of 1900, not only had the English Channel been traversed by radio signals from Dover to Wimereux but also by that time a reliable linkage had been established between the Niton Station on the Isle of Wight and one much further along the coast of England at the Haven Hotel, south of Bournemouth. Given these successes, which all seemed to fly in the face of accepted contemporary scientific knowledge, Marconi's enthusiasm for and confidence in such a wild project was perhaps understandable. Despite it having been shown that radio waves, by some unexplained mechanism, could reach well beyond the horizon, expecting them to traverse the gulf that separated England and America was another thing entirely. Evidently what he planned to do involved a considerable element of optimism and, given the inevitable cost involved, a significant level of risk.

Early Ruhmkorff coil, c. 1890 *(PRJ)*

PROFESSOR FLEMING

No doubt in an endeavour to minimise the risk, and characteristic of Marconi when confronted with a technical problem beyond his own solution with the application of sweat and tears, he now found an expert in the person of Professor JA Fleming. This was a very sound choice as events were to transpire, for within another 4 years Fleming was to invent the thermionic valve, or tube as it is called in America.

Fleming's speciality was high-voltage alternating current. He was, in many ways, an ideal choice to assist in creating a new technology about which there was very little basic data and everything was in the process of development. As his later books were to demonstrate, Fleming was an extremely methodical and systematic scientist, and once he had attacked and solved a problem, most of the underlying scientific basis was also laid bare for all to see. With a formidable capacity to render problems into the language of mathematics, Fleming's analysis of the new domain of

Diagram of the Poldhu transmitter in 1901 as used to send the first signal to St John's, Newfoundland *(PRJ)*

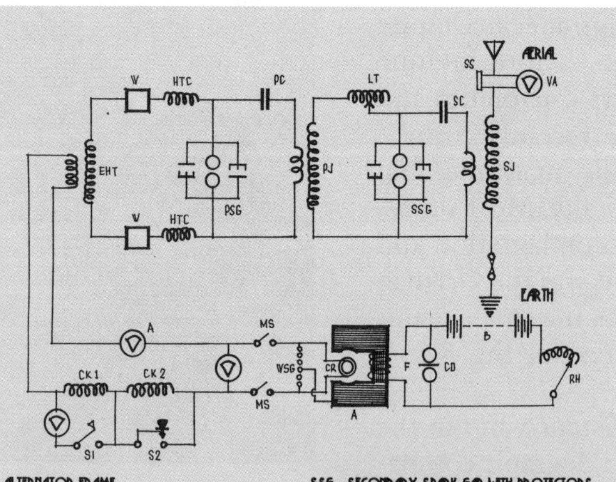

high frequency, alternating current and radio frequency energy is to be seen in his many publications and remains as a testimony to his scientific capability.

For a period of about 6 years, from the initial work of Marconi near Bologna in Italy, the most effective means of producing radio frequency energy was by creating a spark with an induction or Ruhmkorff coil. This device, despite consuming several hundred watts of electric power, was capable of an effective output power measured only in milliwatts. Based on his estimation of power requirements for the bridging of the Atlantic, Marconi specified an apparatus with an input power of about 12 000 watts and this was presented to Fleming as his first major design and construction task. He responded with a two-stage transmitter (see the diagram of the Poldhu transmitter). By having two stages coupled (as shown in the diagram), the 'back' electromotive force was prevented from finding its way into the generator and overstressing its windings and allowing destructive 'arc over' to occur. (See the section 'Spark Transmission Methods', pp. 54–61.)

Aerial photograph of Poldhu radio station taken in the 1920s *(MAR)*

This decision was to have some profound and completely unexpected consequences. As recently confirmed by experiment and analysis, the Fleming transmitter produced radio frequency energy very rich in harmonics. The higher frequency harmonics are now thought to have enabled the transmitter to take advantage of the as yet undiscovered ionosphere. In the light of contemporary knowledge, it seems unlikely that this transmitter, operating on the fundamental frequency set by its capacitive and inductive constants, could have produced a signal that would have crossed the Atlantic at the power level in use.

POLDHU WIRELESS STATION

The Poldhu radio station including early antenna structures *(MAR)*

Apart from a new and costly transmitter built to an entirely new principle, Marconi had concluded that a formidable antenna system was also required if his experiment was to be successful. Here again was a costly project and, in this instance, one that required major construction work on both sides of the Atlantic. Evidently the whole exercise was likely to be seen by sound businessmen as fraught with danger and very likely to fail. No doubt the directors of the newly formed Marconi Company were very nervous of the prospect of failure of the project that the young Marconi had proposed. As a result, the whole experiment was conducted in extreme secrecy, including the purchase of land in the far west of England at Poldhu. Here, during 1901, was constructed a radio station to house the new transmitter of Fleming and to serve as the site for an enormous drum-shaped array of wires which was to be the antenna.

As the photographs reveal, this antenna was in both a visual and an electrical sense impressive, but about as unsound an engineering structure as one could imagine. To the eye of the engineer or architect, it can be seen as intrinsically unstable. On this basis, then, and rather unsurprisingly, following a particularly savage gale the whole structure collapsed in a tangled heap of spars and ropes. Following this expensive disaster, it was hurriedly replaced with a much simpler and, to modern eyes, strangely familiar form of antenna. Apart from the array of wires running up from the transmitter house, the general form of this antenna foreshadowed the dipole antenna and double supporting structure that is still extensively used for transmission and reception a hundred years later.

Only a few weeks later, on 17 September 1901, the similar antenna structure that had been constructed at South Wellfleet near Cape Cod in America was also demolished. It too had succumbed to the attack of a quite abnormally vicious winter gale which came sweeping out of the North Atlantic.

Over the months of 1901, Marconi shuttled back and forth across the Atlantic by steamer, supervising the work and making the necessary decisions as problems arose. In particular, following the destruction of the antenna system in North America, he decided that it would be necessary to make do with a much simpler wire antenna on the western side of the Atlantic. This would be supported either on balloons or kites. The erection of the much-simplified antenna at Poldhu was another change which had to be made in the face of mounting cost and pressure from an ever more anxious board of directors, unable to be certain that what Marconi was attempting to achieve was likely to be successful. Again as a response to the new and temporary form of antenna to be used, Marconi decided to attempt the reception of the experimental transmission from Newfoundland rather than from the American site. Evidently the greater proximity of Poldhu to the Newfoundland site as compared with Cape Cod could make all the difference between success and failure.

Poldhu in 1990 with the radio amateur hut on the right *(PRJ)*

The transmitter at Poldhu as designed by Dr Fleming in 1900 *(MAR)*

ST JOHN'S, NEWFOUNDLAND

By the end of November, Marconi had established a listening post at St John's, Newfoundland, at the site of the old lookout and memorial to John Cabot at Signal Hill. In setting up this station in the old Fever Hospital close to the memorial, Marconi was given much assistance by the Governor of the province, Sir Cavendish Boyle, and the Premier, Sir Robert Bond.

After an agonising wait of some days, at last on 12 December 1901, faintly in the earpiece of the specially modified earphone, Marconi heard the three dots denoting the letter 'S' in morse code. The earphone was then passed to his assistant, George Kemp, who confirmed that he, too, had heard the faint signal from Poldhu. The following day and again for a brief period, the letter 'S' was heard, and Marconi then felt sufficiently convinced of the success of the experiment to advise the press what had occurred.

The Italian naval coherer used by Marconi to receive the signal from Poldhu in December 1901 *(MAR)*

Inevitably the news of the successful crossing of the Atlantic by a radio signal caused a furore. Although there were eminent men who believed and supported Marconi's work, there were also detractors who accused him of deception. Furthermore, for the best part of a century this has remained a controversial subject. Only quite recently, at the centenary conference of the Institute of Electrical Engineers in London in 1995, it erupted yet again — despite some apparently very sound analysis to substantiate the 'harmonic' theory of propagation of the radio waves discussed earlier. Even a replication of the original experiment, and assuming its success, seems unlikely to satisfy the doubters.

The author travelled to Newfoundland in a somewhat better time of the year for travel, September, compared with the expedition of Marconi in late November 1901. A written description of Newfoundland, even when supplemented by illustrations, fails to convey the flavour of the place. First impressions of St John's, well hidden by the fog, are that it is quite a tiny settlement and little changed from the time of Marconi. Along the edge of the sheltering harbour, that remains more or less true. But when the full extent of the place can be seen, it transpires that St John's is a substantial city. Apart from the heritage core of wooden houses and the old docks, it spreads out into a hinterland of modern timber and concrete buildings and vast shopping centres.

We stayed in the most famous heritage street of St John's, Gower Street, where in 1901 Marconi had stayed in Cochrane House, which had unfortunately been burned to the ground in 1993.

As the map shows, St John's harbour is a narrow appendix joined to the Atlantic by an even narrower aperture with high and forbidding headlands on either side. It must have been a terrible place to access by sailing vessel during a winter tempest. One false move and the rocks would claim an unwary ship.

In 1893, high above on the northern headland, the memorial to Cabot's discovery of Newfoundland in 1497 was built. A substantial blockhouse of what looks like granite, it has an octagonal turret on the south-eastern corner, and from the harbour seems a minute pimple on the top of the bare headland.

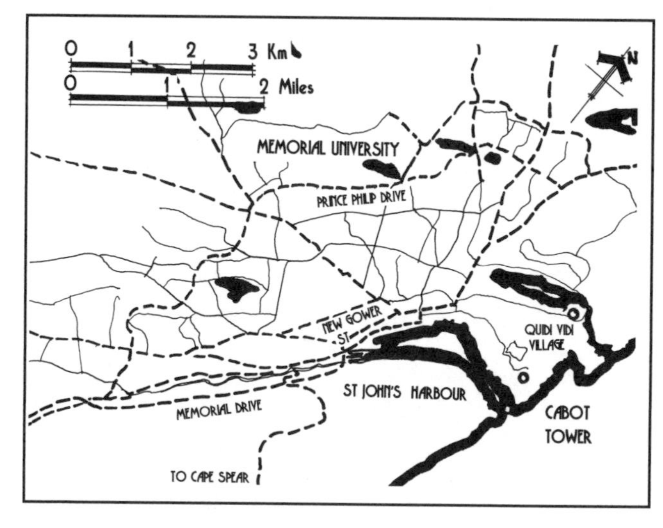

Map of the city of St John's, Newfoundland, and location of the Cabot Tower *(PRJ)*

Not far away from Cabot Tower, across the open parade ground which is now a car park, once stood a building which was used to isolate fever patients. An undistinguished building, it was demolished long ago. However, in 1901, within its walls was conducted an experiment that was to reverberate around the world and start a revolution in communication which continues to the present. Here it was that the 'S' signal sent from distant Poldhu was received after many agonising hours of waiting.

Gower Street in the old part of St John's, Newfoundland *(PRJ)*

Cape Race, south of St John's on the Avalon Peninsula, is a wild and lonely place, and was the site of the old shore station of the Marconi Marine Company, but as a result of an environmental clean-up by the Canadian government, virtually all the old structures — the antenna masts, the antenna bases, the guying blocks and the foundations of the station buildings — have gone. Only the partially demolished antenna base of an old metal tubular mast remains together with wrought-iron fragments of the mast left lying in the long grass. The tubular mast itself had been chopped up with oxyacetylene torches — all in the interests of ecological purity. This was sad, given the part that this wireless station had played in the tragedy of the *Titanic*.

Back in St John's, from the Society of Newfoundland Radio Amateurs' shack in the upper level of the Cabot Tower, it was possible to talk to radio amateurs at Poldhu on 14.160 megahertz with the greatest of ease. Using a solid-state transceiver, this quite trivial exercise served to emphasise the extraordinary advances that have been made in the last 100 years.

The one small irony of this routine amateur contact was that, in order to confirm arrangements, it was necessary to make a telephone call to Poldhu. Of course there was a very good precedent for this — Marconi had found it necessary to use the undersea cable telegraph to make arrangements with Dr J Ambrose Fleming at Poldhu to send the 12 000-watt signal across the Atlantic.

St John's, Newfoundland — the harbour and the Cabot Memorial Tower on the hilltop in the distance *(PRJ)*

HEART'S CONTENT

Access to the undersea cable is relatively easy from St John's, for it is only 100 or so kilometres (60 miles)

to the north-west at Heart's Content that it came ashore after a long journey across the Atlantic from Valencia Island off the southern tip of Ireland. It had been brought there on the British engineer Brunel's famous vessel, the *Great Eastern*, in 1866 after many tribulations and vicissitudes.

At the museum at Heart's Content, much of the original telegraphic apparatus is on show and, in addition, the undersea cables can be seen coming out of the Atlantic and up the beach to a conduit under the road. As the museum literature says, this was the first telegraphic link between the old and new worlds.

Top St John's in 1901 *(UNK)*

Above The lighthouse at Cape Race, Newfoundland *(PRJ)*

In Marconi's wireless experiment of 1901 lay the seeds of much later efforts and development. This was a project which relied on portable apparatus for its success. Even the antenna system was a temporary one, and the use of kites was to be replicated particularly for emergency signalling in the future. Certainly the base station at Poldhu was a substantial and effectively permanent installation. However, even at Poldhu, the antenna system was close to the form of dipole antennae of the future, as used particularly for military portable high-frequency communications, once the ionosphere was discovered and understood.

It seems fitting that Newfoundland should have been the place where the first wireless signal from Great Britain was received in 1901, given its earlier involvement with undersea cable telecommunications. What stands out in the experiment of 1901 is the use of the new 'wireless' medium to provide a connection to a very distant place, which previously had been accessed by the wires of the telegraph with enormous difficulty and expense. In the success of the project, one can see the first tentative link in what was to become, ultimately, a network of communication linkages that would girdle the earth by the centenary of that event.

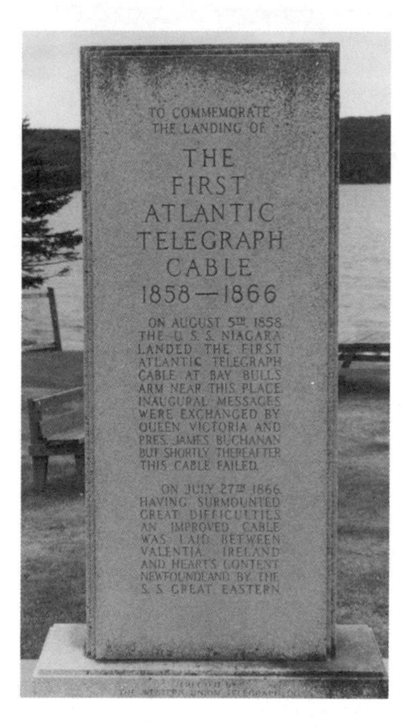

Right Memorial tablet at Heart's Content commemorating the landing of the first successful trans-Atlantic telegraph cable *(PRJ)*

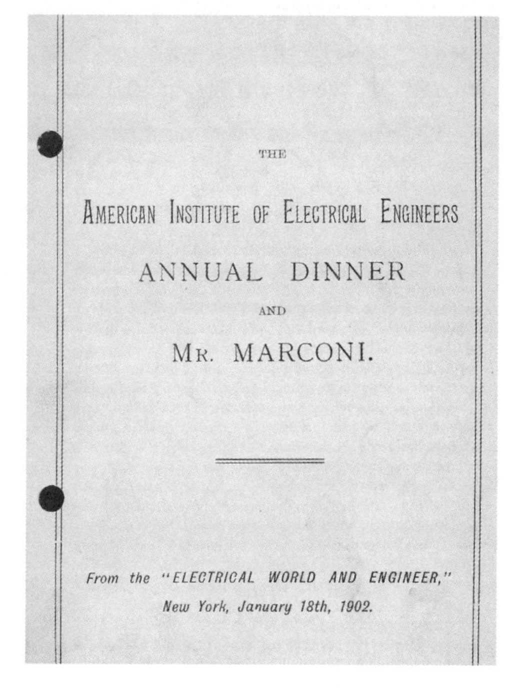

THE

AMERICAN INSTITUTE OF ELECTRICAL ENGINEERS

ANNUAL DINNER

AND

MR. MARCONI.

From the "ELECTRICAL WORLD AND ENGINEER,"
New York, January 18th, 1902.

Far right Menu for a commemorative dinner for Marconi in 1902 *(MAR)*

The bay at Heart's Content *(PRJ)*

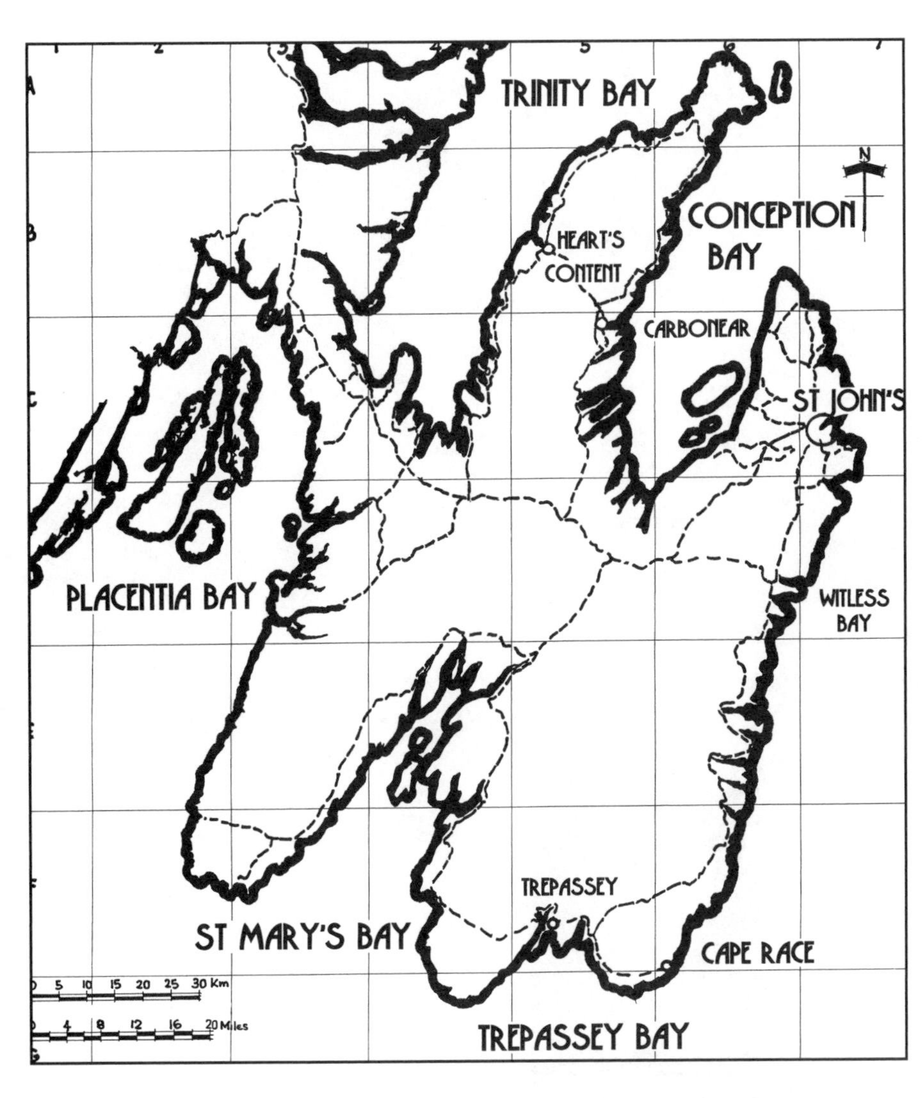

Map of the Avalon Peninsula, Newfoundland *(PRJ)*

CHAPTER
4
1901 to 1914

Telecommunication

The first tenuous filament of a new system of intercontinental communications might have been stretched across the cold grey waters of the Atlantic from Poldhu to St John's, but converting that first strand into a communications highway was to prove a very long and costly exercise. However, what was no longer in doubt was the reality and potential capability of radio communication or, to use the earlier term, wireless. In this instance, the distinction between cable and wireless communications was particularly significant given that Heart's Content, the landing place of the first trans-Atlantic telegraph cable, lay only 100 or so kilometres (60 miles) to the north-west of St John's.

Inevitably, cable-based communications would become the target in the competition to achieve supremacy in carrying messages between America and the Old World. The success of the trans-Atlantic experiment could be seen as propelling the early work of Marconi out of the laboratory as a half-finished technical trick into being a valid alternative to cable-based communications. From the beginning it had been a driving ambition of the young Italian inventor to beat the cable companies at their own game and on their own ground, although in this instance it was sea water rather than 'terra firma' that was the battlefield.

The years that followed, up until the commencement of the First World War, were a time of intense technological development in radio. In the first instance, Marconi and the company that bore his name could be seen to have achieved a significant lead over any other person or organisation in the field. Initially, this allowed the company to develop an associated commercial monopoly in the supply of radio systems, certainly in regard to the equipping of vessels at sea. In Britain itself the situation was somewhat less rosy because of laws relating to the primacy of the post office and the government's desire to protect its revenue base deriving from that instrumentality. However, overseas, the Marconi Company was not bound by any such restraint. This desirable position was soon to be challenged at the international level.

Naval signaller using a 'crows foot' morse key in 1903 *(FAR)*

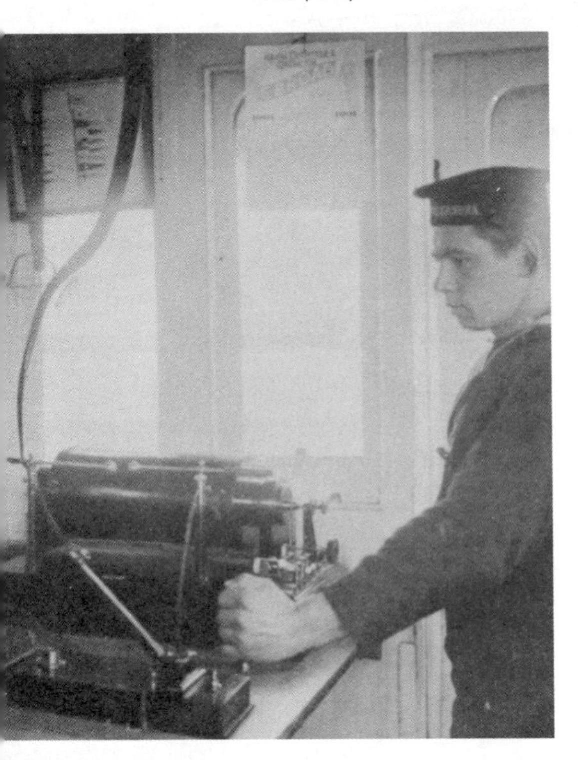

Realistically, it was not to be expected that other nations would allow an entrepreneur like Marconi to control a strategically important system of communication such as radio for any length of time. Given the turbulence of the political situation which preceded 1914, it was to be anticipated that Germany in particular would strive to develop its own system of wireless free of control of British Marconi interests.

TENSIONS IN EUROPE

Apart from anything else that might have caused friction between Great Britain and Germany, it is now clear that a well-developed jealousy on the part of the Kaiser of his cousin, the King of England, and of the Royal Navy was a particular element in the slide towards war that occurred over the period before 1914. In his own small way, it is apparent that Marconi managed to pour fuel on this particular fire in his dealings with Germany and its demands to have access to British shore stations from German ships at sea.

A 'crows foot' morse key at the Archives, Great Badow, near Chelmsford *(PRJ)*

In the first few years after 1901, German ships were under an embargo by Marconi shore stations if they did not carry Marconi apparatus — a situation that infuriated the Kaiser. Ultimately, this restriction was removed by international agreement, much to the chagrin of the Marconi Company. However, in terms of the greater good at an international level, it was probably quite appropriate that such access should have been permitted. The financial problem that this decision created for the Marconi Company was considerable. Evidently the cost of installation of such shore stations was entirely borne by the com-pany and free access to those not using Marconi apparatus did not produce any revenue for it.

Apart from such 'straws in the wind' anticipating the forthcoming conflict, the rate of development in the international community was quite remarkable, and over the period 1901 to 1914 dramatic advances were made on a wide front and in many places. Although in no way intended to be comprehensive, it is possible to review this emerging field of radio on the basis of some of the main participants at that time. Given the pre-eminence of the Marconi Company, it is logical to start by considering what happened in the United Kingdom during this period.

UNITED KINGDOM

In the great experiment conducted between St John's in Newfoundland and Poldhu in Cornwall, Marconi had taken both a coherer receiver and an Italian navy coherer. As has been suggested in more recent material, it is thought that the naval coherer was in reality operating as a true diode, that is, a device having true unidirectional current transfer characteristics. Whatever the mode of operation, it appears that it was this latter device with which Marconi heard the famous three-dot signal which translates to 'S' in morse code. By comparison, the conventional coherer set resolved nothing but static and no signal was able to be received with it.

Marconi came back from St John's determined to find a more satisfactory form of receiver and in remarkably short order had created the magnetic detector. Looking like an antique reel-to-reel tape recorder, this device was to prove remarkably efficient at detecting spark-generated pulses and was to be the principal receiving instrument of radio stations supplied by the Marconi Company for both marine and terrestrial use until well after 1914. (See the section 'Reception before the Valve', pp. 94–97.)

As energy and funds were poured into making the trans-Atlantic communications service a reality, two particular advances occurred to mirror the changes in technical effort to achieve long-distance communications. The quality of the spark signal and the level of power produced were increased by the introduction of rotary spark transmitters. (See the section 'Spark

Top Wireless apparatus on the SS *Arlanza* in 1904 *(MAR)*

Bottom The transmitter and receiver at the Lizard wireless station, Cornwall, in 1904 *(MAR)*

Transmission Methods', pp. 54–61.) Further, as wavelengths steadily increased, this was mirrored by an increase in the dimensions of antenna systems. Fortunately, the ultimate gargantuan proportions of the trans-Atlantic radio stations were to some extent contained with the discovery of the directional antenna system in which the array of wires was run away from the intended target in a line. This can be compared with the earlier antenna array in which a circular pattern of wires was brought together to a common point over the transmitting house.

In 1904, when carrying out installation and training of personnel in Russia, CS Franklin devised a new form of tuner for receiving purposes. This device, known as the multiple tuner, was patented in 1907 and in many respects could be seen as the longest living survivor of the earliest period of radio. In principle it remains to the present day as the fundamental device for achieving tuning and matching of a random-length antenna to a transmitter or receiver. Most radio amateurs have at some time or another built an antenna tuning unit (ATU) to a design fundamentally similar to that of Franklin.

All these various improvements were brought together in the Marconi trans-Atlantic stations at Clifden on the west coast of Ireland and at Glace Bay in Nova Scotia. By 1910 these stations were operating with rotary spark transmission at powers of 300 kilowatts running into incredible antenna systems having longitudinal dimensions of upwards of 2 kilometres (over 1 mile) and set to a frequency of 42 kilohertz. Ponderous and extravagant as such installations may look in contemporary terms, they allowed the transmission of enormous amounts of telegraphic material at relatively high data rates compared with the undersea cable system. In terms of Marconi's ambitions, they could be seen as successfully providing the competition to the undersea cables that he had hoped for before going to Newfoundland in 1901.

Apart from the land-based stations, marine installations were occurring at a considerable rate. From the first installation in a British ship, the SS *Lake Champlain* in 1901, by 1904 some 124 marine installations

Antenna towers at Glace Bay, Nova Scotia, in 1904 *(OTC)*

Below left A rotary spark generator in operation in its 'silencer' compartment as used at Clifden, on the west coast of Ireland, in 1907 *(OTC)*

Below right The wireless telegraphy station at Waenfawr, near Caernarvon, North Wales, built in 1912 and as seen in 1992 *(PRJ)*

had been leased to ship owners. Furthermore, with the particular commercial arrangements required to respond to the application of British law, these installations were operated by wireless operators trained and employed by the Marconi Company, who were provided to the ships' owners as part of a leasing arrangement.

Right Dr John Ambrose Fleming *(MAR)*

Left Dr Lee de Forest compares an audion with a power output valve *(PRR)*

One other extraordinarily important invention of this period should be mentioned — the diode valve or tube. Produced from the fertile brain of Professor Fleming, chief technical adviser to the Marconi Company and designer of the first transmitter installed at Poldhu used in the trans-Atlantic experiment, the thermionic valve was later further developed by an American, Dr Lee de Forest. (See the section 'Valves, Reception and Transmission', pp. 206–13.)

The diode depended on the movement of electrons across a vacuum in the presence of a voltage applied to a second electrode separated from the incandescent filament in a light bulb. In fact, the phenomenon of particulate movement away from the negative leg of a heated filament in an evacuated light bulb had been noticed initially in 1880 by the indefatigable experimenter Thomas Alva Edison during his quest for the elusive ideal electric light. Known thereafter as the 'Edison effect', the streak of clear glass on the side of the bulb away from the negative leg of the incandescent filament in the evacuated lamp bulb was an intriguing mystery for which scientists of the Victorian era could find no explanation. All that seemed clear was that the negative leg of the filament was throwing off some sort of fine material that was deposited uniformly on the inside of the bulb, except where the positive leg created an obstruction to the flow. The result was a clear line on the inside of the glass which constituted the shadow of the other leg of the filament. It would take the discovery of the electron before there was a basis for explaining this peculiar condition.

A very early thermionic diode valve (tube) as used by Dr Fleming for his detector experiments *(MAR)*

Despite the absence of an adequate scientific explanation, Edison carried out some experiments with a separate electrode set inside the evacuated space and soon discovered that a flow of current would occur across the vacuum if the electrode was connected to the positive leg. If the electrode was connected to the other or negative leg, there was no current flow. This would now be described as diode action in which unidirectional current flowed, depending on whether the electrode in the bulb was positive or negative with respect to the heated filament. In characteristic fashion, Edison promptly found a use for this new discovery and patented it. After that, the direct current diode was set aside and for a long time simply forgotten. Many years later, in the latter part of 1904, Dr J Ambrose Fleming was to put the Edison effect to very good use in the rectification of high-frequency radio signals.

Suffering from the onset of deafness, Fleming had been thinking of new ways to detect radio waves because of his inability to hear the acoustic output of the coherers or magnetic detectors that were in use at that time. At an earlier time he had experimented with the unidirectional flow of current in an evacuated bulb with a second electrode and had a number of test incandescent bulbs stored in his laboratory. One of these was tried as a detector and Fleming found that it did exactly what he had hoped, allowing a sensitive galvanometer to indicate when a signal had been received.

Apart from solving Fleming's hearing problem by allowing a visual display to indicate the presence of a radio signal, at a later stage the diode, both in its original form and as developed with additional electrodes, was to be the basis of a revolution in electronic radio reception and transmission.

Named the 'oscillation valve' by Fleming, the idea was very quickly taken up by de Forest who added the third electrode and created the triode. In this form it was to become the basis for the next generation of transmitters and receivers after the First World War. The triode was developed into a useful device at radio frequencies just in time to participate in the development of radio during the war years, and from then until about 1960 it would be the basis of the whole of the radio industry. In addition, it was to be the basis of the first of the electronic computers and, until solid-state devices were invented in 1947, it would reign supreme.

From April 1913 to May 1914, Meissner in Germany, CS Franklin in Great Britain, Edwin Armstrong in America and Captain HJ Round, also in Great Britain, converted the triode from a rather unpredictable and temperamental device into the basis of a new industry and a critical element in the next phase of radio development after the First World War.

GERMANY

In the experimental work conducted by Marconi on Salisbury Plain in 1896, one of the guests was Dr Ferdinand Braun of the Charlottenburg Institute. Braun brought to the experiments a scientist's appreciation of what Marconi was doing but, more importantly, an understanding of the basic principles of radio technology as they applied at that time. Braun immediately recognised a fundamental flaw in the arrangements used in Marconi's coherer receiver, and on his return to Germany he patented a solution. This fault involved the method of matching the antenna characteristics to the coherer. Its solution by Braun was to be the basis of the first radio system employed by a business that he entered into with the well-known firm Siemens and Halske. More seriously, it was to be the first break in the Marconi patents and allowed a major rival in the field of radio engineering to arise.

Another visitor to the demonstrations of Marconi in England was the electrical engineer Dr Adolf Slaby. On his return to Germany, and in association with Count von Arco, he developed a wireless system that was known as the Slaby–Arco system.

In principle, apart from different forms of hardware, the methods developed by both Braun and Slaby–Arco had a similar array of functions to the Marconi wireless as it had been developed to about 1901. However, what was highly significant about these two new German systems was that they would become the basis of creating a new organisation known as the Gesellschaft für Drahtlose Telegraphie. Later, this company was to be more generally known as Telefunken, the name that initially had been applied to the wireless apparatus that was produced by this organisation.

Given the extent to which Telefunken was to be come the arch-rival of the Marconi Company thereafter, Marconi must have later wondered at the wisdom of having allowed Braun and Slaby to see his work at close quarters as it had been developed to 1896.

Although undoubtedly sought-after as a means of avoiding the Marconi patents for spark transmission and reception, Telefunken was later to manufacture equipment with a fundamentally superior means of generating spark energy. This involved 'quenching' the spark-generated radio frequency energy which in turn produced something closer to a continuous wave in the antenna. (See the section 'Spark Transmission Methods', pp. 54–61.)

This system of quenching was initially developed by Max Wien but its simplicity and elegance, coupled with the desirable musical quality of the signal that it produced, led to it being adopted by Telefunken for virtually all of its early radio installations. It was also extremely efficient in terms of conversion of electrical power into radio frequency energy, having a conversion efficiency of about 75 per cent. Later, when legal disputes had been resolved just before the First World War, it was adopted in Marconi marine stations and could be seen to displace even rotary spark installations.

In the period just before the First World War, Marconi was embroiled in the so-called Marconi scandal concerning the purchase of shares by Ministers in the government of Prime Minister Lloyd George. Although Marconi was never suspected or accused of having played any part in this unfortunate affair, it did have a substantial and adverse impact on the company of which he was the founder. During this same period, a proposal to create an Empire-wide network of long-wave radio stations, with connections to all the principal members of the Empire at that time — Australia, Canada, South Africa and India — was under scrutiny by the government. One of the results of the share trading of the government members was to have the Empire network proposal substantially delayed so that the First World War intervened. The ultimate result of this delay was to ensure that this system of communication was never constructed. By the end of the war, technology had moved on, making the long waves obsolete.

However, in Germany, without the political problems to thwart an ambition to create a communication service to tie together its overseas possessions, a long-wave network was created that in many respects represented the next step in creating the prototype of the World Wide Web.

With the centre of the network at Nauen in northern Germany, this system had originally been proposed in 1906, and later had tentacles reaching to the extremities of the world. In the first instance, the network was a mixture of radio links and undersea cable connections, because the range of the radio transmission was limited. However, by 1914, due to progressive technical improvements, the range of the transmitter at Nauen had grown to 8000 kilometres (5000 miles), allowing signals to be received in South Africa, South America and China.

By 1914, the 24-kilohertz signal from Nauen was generated by a 200-kilowatt alternator using a system of frequency doubling produced with saturation transformers and all driving a quite gargantuan antenna system, roughly in the shape of a tent. By 1914, this antenna had grown in height to 260 metres (850 feet) at the centre, with an overall length of 1626 metres (about 1 mile). This length was later increased during the war years as dependence on the Nauen broadcast information increased in the face of progressive destruction of the cable elements of the network by the British and Empire forces.

In passing, the name of the company, Telefunken, could be translated as 'far sparks' and, by 1914, this was clearly an appropriate name even though, by that time, the naked spark had been abandoned in favour of more efficient methods.

The receiving position at VIS radio station at Pennant Hills, north of Sydney, Australia, in 1913 *(OTC)*

AUSTRALIA

Apart from the technical developments that were achieved in Germany, the rivalry of Telefunken and the Marconi Company were to have significant ramifications for the radio industry all around the world but particularly in Australia. In addition, the commercial rivalry that occurred in Australia after 1901 could be seen as a forecast of the furious clash that was to occur in 1914. In that respect, although it occurred 'before the storm', it certainly could not have been described as a period of 'calm' in the history of world communication.

Whatever the characteristics of the latter part of the decade under consideration, in the period immediately following 1901 Australia was gripped by a form of radio lethargy resulting from political indifference. Apart from efforts by the Marconi Company in 1905 to interest the Australian government in a telegraphic link to Tasmania and a number of individual experiments, there was little other significant radio progress.

A few years later, at the instigation of a businessman of German extraction, a new company was created to sell the apparatus of the German Telefunken radio organisation in Australia. This was the Australasian Wireless Company Limited and, following its establishment, an experimental radio station was built on top of the Hotel Australia in Sydney. This made use of the call sign AAA.

At last, in 1910, and following the loss of the SS *Waratah,* the Australian government decided to act and tenders were called for the installation of two radio stations, one to be located in Sydney, New

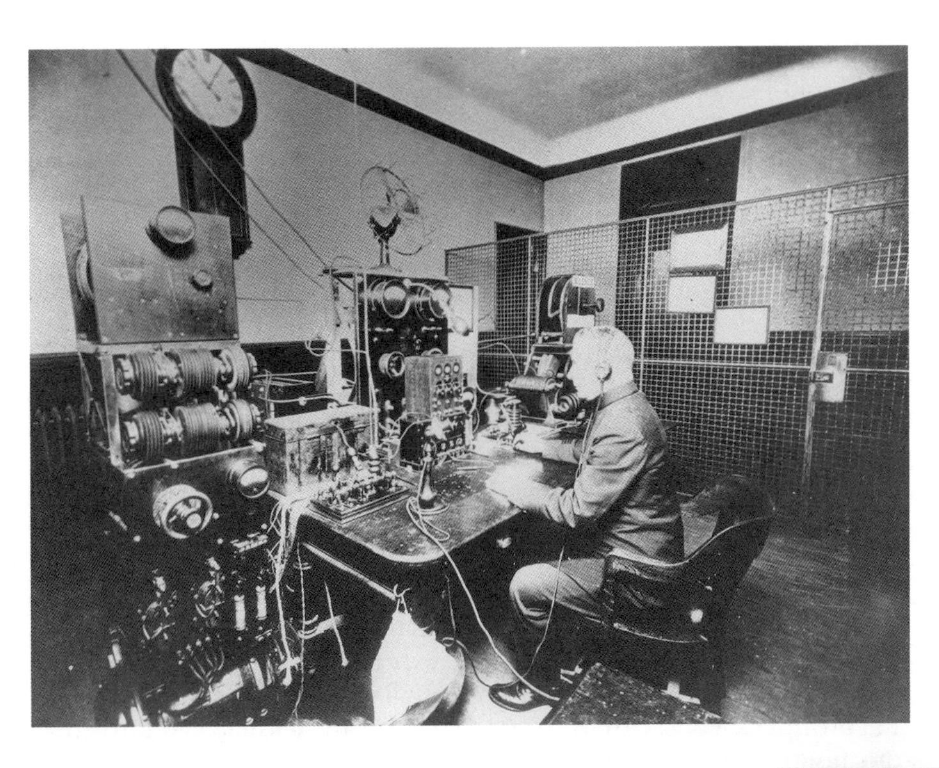

The original Telefunken quenched-spark installation at VIS, Pennant Hills, in 1913 *(OTC)*

South Wales, and the other in Western Australia, near Perth. When the tenders were received, the Marconi Company's bid of £19 020 was rejected in favour of that of the Australasian Wireless Company Limited, which had proposed a figure of £4150.

Such a significant difference suggests that the latter organisation was prepared to install the two radio stations at or below cost in order to secure a foothold in Australia. As the archival material at Chelmsford in England reveals, the Marconi Wireless Telegraph Company evidently took this rebuff by the Australian government as yet another example of what was called the 'Telefunken wall'. This was a way of describing what was perceived as an unfair advantage available to the Telefunken Company because it received special support from the German government and banks which was not available to the British company.

However, lurking in the tender of the Australasian Wireless Company was one significant difference from the tender of Marconi — where the Marconi Company was to provide the site and structures necessary to house the radio station, the tender of the rival company did not include this. This was to create major problems once the tender was accepted.

The VIS transmitter at a later stage using the quenched-spark blower units designed by Balsillie *(OTC)*

After some delay in agreeing where the stations should be located, sites were selected at Pennant Hills, north of Sydney, and at Applecross, south of Perth in West Australia. The navy had demanded that both wireless stations be located well away from the sea and out of range of naval guns. By comparison, the post office wanted them close to the sea, presumably where damper ground and better earthing characteristics could be guaranteed.

On the selected sites the huge masts of the long-wave stations were erected and below, in the site buildings, the wireless apparatus was installed. The Telefunken system used the quenched-spark gap based on the 'shock excitation' method of Wien. The transmitters were powered by 60-horsepower Gardiner diesel engines, producing alternating current at a frequency of 500 hertz. The spark was able to be operated at a power level of 25 kilowatts by the use of an air-blast system of cooling and was reported to produce excellent morse code at the same audio frequency as the generator, 500 hertz. The continuous wave energy was transmitted at the accepted international frequency of 500 kilohertz or 600 metres wavelength.

The station at Sydney was given the call sign POS initially, but later this was changed to VIS. The Perth station had the call sign POP, but later this was also changed and became VIP.

From the outset, both stations appear to have suffered from technical problems, so much so that, ultimately, the government decided to employ an independent expert. The expert who was appointed was a former Marconi employee, an Australian, John Graeme Balsillie. Following his work with the Marconi Company, Balsillie had formed an organisation in London called the British Radio Telegraph Company and had soon run into difficulties over patent violation with his former employer, Marconi. As part of the agreement with the Australian government, Balsillie undertook to design a new system of transmission which would not infringe the Marconi patents and, further, would pass it to the Australian government, once it was in operation.

Upon taking up his new post in Australia, Balsillie's first job was to discover and rectify the problems plaguing the Telefunken apparatus. It appears that this was found to involve defective generators on the one hand and inefficient earthing on the other. The first problem was relatively easily overcome. The second involved a good deal more ingenuity and finally required the introduction of a 'counterpoise' mat below the transmitting masts. Ultimately, the two Australian long-wave stations were able to operate over a range of more than 2000 kilometres (1250 miles) and at wavelengths between 300 and 3000 metres, although it was principally to the international distress frequency of 600 metres that operations were confined.

Encouraged by this success, and assisted by an ecclesiastical gentleman, the Reverend Father Archibald Shaw, who had established the Maritime Wireless Company in a radio factory at Waverley, a suburb of Sydney, Balsillie created a new wireless transmission system as he had promised the government. This involved a system of quenching the spark which he claimed to be quite novel, and in due course the patent for it was indeed given to the Australian government. Further, it was to be the basis for a number of radio stations to be built around Australia, commencing in Tasmania.

One of the significant results of Balsillie's work for the Australian government was to infuriate the major commercial enterprises, Marconi and Telefunken. Litigation commenced, again related to patent violation and, after much wrangling, a settlement was reached. In this it was agreed that a new company should be formed and in it the three commercial organisations would hold shares in the proportion Marconi 50 per cent, Australasian Wireless 44 per cent, and Telefunken 6 per cent. The resultant entity, Amalgamated Wireless (Australasia) Ltd, more commonly known as AWA, remains active to this day, although divested of virtually all of its radio interests in the face of economic realities of modern Australia and overseas competition. The small component of this company once owned by Telefunken was liquidated just after the start of the First World War.

UNITED STATES OF AMERICA

Following the invention of wireless communication in 1895 and its patenting by Marconi in 1896, intensive experimentation occurred in many places and no more intensively than in the United States. In the period 1896 to 1901, a number of people developed a variety of different techniques to make use of the Hertzian waves, including John Stone Stone, Shoemaker and General HHC Dunwoody. As was the case in Germany, much of the effort of the early pioneers was directed at discovering ways to send and receive Hertzian waves which would not infringe the extensive Marconi patents. However, this effort was to yield extremely important advances that ultimately would lead away from the generation of radio frequency energy by sparks. Soon it would be realised that the use of pure sine wave energy was the preferred basis for radio communication and that has remained true to the present day.

In passing, the revolution in digital communications that is presently occurring may in the longer term be just as revolutionary as the change from spark radio frequency wave generation to continuous wave carriers and telegraphic signals.

In a theoretical sense it is arguable that Reginald Fessenden was the most important of the rival inventors to Marconi who were working in the United States. He had, prior to 1900, developed a considerable distaste for spark-generated radio frequency (RF) energy, having correctly perceived that it was inherently broadband, noisy and wasteful of energy. Fessenden believed that the ideal form of RF energy would be sinusoidal in form and therefore be able to be tuned. In this context, his first important step away from the spark and coherer system of Marconi was to invent the heterodyne detector. He had realised that a pure sine wave would not be able to be detected by any known means in 1901 and, therefore, only by mixing it with another radio frequency signal so as to create a resultant audio frequency tone would it be possible to hear anything. In the longer term this was to prove an extraordinarily important concept, and would appear subsequently as the basis of a later generation of radio receivers. However, around 1900, technology was not yet ready to provide a reliable and coherent sine wave at radio frequency. At this stage, Fessenden used a carbon arc generator to produce the required signal but it was noisy and tended to wander around in frequency so as to be rather ineffective in the process of producing an audio signal by mixing.

The counterpoint to this revolutionary idea of heterodyne mixing was that, to produce pure and frequency-stable sine wave radio frequency energy, a source quite unlike the spark generator would be needed. At that time the only device that seemed to exhibit the correct characteristics was the alternating current generator, or alternator as it was more commonly called. However, what Fessenden soon had in mind was not something operating at 60 hertz as do the electric mains in most households but something generating alternating current at perhaps 100 000 hertz or even higher at 150 kilohertz.

In 1901 such a device was just a dream, but with the help of the General Electric Company and, in particular, a young Swedish engineer, Ernst Alexanderson, Fessenden had an experimental alternator constructed to test his theory. Completed in 1906, this machine operated at a frequency of 60 kilohertz and a power of 2 kilowatts and was used to transmit the first continuous wave carrier-borne telephony, much to the surprise of a number of naval wireless operators in the vicinity of the Brant Rock US naval radio station. At Christmas 1906, a further experimental program of music and sound was heard by a large audience in the United States.

The next person in the United States to have a profound impact on radio technology was an Australian-born engineer, Cyril F Elwell. He had been the driving force behind the creation of the Federal Telegraph Company of California and had been responsible for purchasing the rights to use the Poulsen system of arc transmission in 1909 for the substantial sum of US$250 000. Valdemar Poulsen, a Dane based in Copenhagen, had by 1908 created a method of generating radio frequency energy from an arc at relatively low powers and had also designed a method of detecting the otherwise inaudible continuous wave that the arc produced using heterodyne mixing.

Elwell, with a background in high-power electrical engineering and arc furnace design, believed that in the arc lay a valid and technically attractive alternative to the Marconi high-powered spark system. In addition, he had investigated Fessenden's alternator approach to producing high-power radio frequency energy and considered it to be impractical in the longer term. In many respects this conclusion was ultimately shown to be true. As it turned out, the arc, too, was to prove impractical in the longer term and, because of the invention of the thermionic valve, was to have a limited life, becoming obsolete after about 1918.

Much of the early development work associated with the arc had been carried out by William Duddell in the United Kingdom, although an American, Elihu Thompson, claimed to have patented a similar arc device in 1893. The research of Duddell had been limited to generating audio frequency energy up to about 10 000 hertz, and it was this limitation that Poulsen was able to overcome. By the use of a hydrocarbon or steam atmosphere to surround the arc and a magnetic field to 'blow it out', Poulsen elevated the working frequency to well over 100 kilohertz.

After the stunning cacophony of the static and rotary spark, the arc was an extraordinarily quiet device and in that respect anticipated the valve that would soon displace it. Initially, its power-handling capacity seemed to be severely limited, but in the period from 1912 to 1917, Elwell and his successor, the engineer Fuller, were able to improve the design of the arc so that ultimately it was able to produce a continuous wave with an input power of 1 megawatt. Operating at its optimum efficiency of 50 per cent, this allowed a power of 500 kilowatts into the antenna, a substantial value compared with the Marconi rotary spark system which was inherently inefficient and noisy.

Although less than a perfect generator of radio frequency sine waves, the continuous wave produced by the arc was sufficiently pure to allow the transmission of audio frequency energy riding along on its back. Modulation of the arc radio frequency carrier wave was initially achieved by placing multiple carbon microphones in the lead to the antenna although this was found to have an undesirable impact on the tuning of the arc. In this context, although seemingly a simple piece of hardware, the arc was found to have quite complex electrical characteristics and to be difficult to set up and operate at its optimum capacity.

Reception of the modulated arc carrier was possible with the solid-state detectors produced in the period 1901 to 1914, the carborundum detector of Dunwoody, the perikon detector and, perhaps most importantly and best known, the galena or cat's whisker detector.

By comparison, detection of the unmodulated arc energy used for telegraphy required another approach and for some time this was achieved by an electromechanical device invented by Pedersen and known as the 'ticker'. Ultimately, heterodyne techniques were used in all continuous wave receivers to create an audible frequency and this was made easier with the invention of the valve oscillator a short time later.

The next person in this chain of events has the distinction of being a highly controversial personality, quite apart from his inventiveness and the extraordinary technical revolution that this attribute was to spawn. This was Dr Lee de Forest, a young and extremely ambitious electrical engineer, who invented the three-electrode valve now known as the 'triode'.

As a graduate of Yale University, Lee de Forest received his doctorate in 1899 and worked initially with Western Electric in Chicago. While with that company, he got to know a telephone engineer, Edwin H Smythe, and together they explored new ways to exploit Hertzian waves and to achieve wireless communication. This collaboration did not last for very long and, by 1905, de Forest was working with some assistants to refine the diode valve of Fleming for use as a detector of radio waves.

In the archives of HW McCandles of New York, a maker of electric lamps, it is recorded that, in 1905, de Forest ordered the construction of a number of copies of the Fleming diode. It can be assumed that this purchase was for experimental purposes. Some time later, de Forest had the inspiration to include a third grid of wire between the heated filament and plate of Fleming's valve and thus created the start of a revolution in electronics. This new device, now referred to as a triode, had a low vacuum and was therefore extremely temperamental and difficult to get to work at optimum levels. However, before long it was recognised as capable of amplification of low-level signals and also of the production of radio frequency energy on a single frequency — this latter was the pure sine wave that Fessenden had looked for with the alternator.

Because of the obvious physical relationship to the Fleming diode, the de Forest triode, or 'audion' as he called it, was the basis of major litigation with the Marconi Company in later years. No doubt because of this, de Forest always denied foreknowledge of the Fleming valve prior to the invention of the triode. However, it is clear from the detective work of GF Tyne, reported in his book *Saga of the Vacuum Tube* (pp. 56 and 57), that this was either a major lapse of memory or simply untrue. Readers are left to come to their own conclusion, but the evidence seems compelling, particularly when coupled with the reading of a patent application made by de Forest just after the patenting of the Fleming valve. This application specifically refers to the Fleming valve, not only in the written statement but also in the circuit diagram forming part of the application.

Apart from this, it is apparent that de Forest had relatively little notion of how the triode operated and for a long time believed that the contamination of gas that remained after evacuation was fundamental to the proper operation of the audion. As was later discovered, on the contrary, it is the existence of a hard vacuum that makes the valve or tube perform correctly and gas destroys its most valuable capacity to transfer electrons from the heated cathode or filament to the plate. In this regard it is probably unreasonable to be too critical. As with many pioneers of this early radio age, de Forest was doing no more and knowing no less than others, Marconi included. This was very much an age of experimentation or 'suck it and see', with scientific theory and explanation of physical behaviour trailing along some way behind what was being found to work in practice through a process of trial and error or, in the case of the triode, a process of trial and success.

The last person to refer to from this pioneering age is, in many respects, one of the most important, not so much because of what he invented but in the way that he found that the newly created valve technology could be made to work. This was a young man, still studying for his degree at Columbia University who, in 1912, discovered the key to the first consumer age of radio. He was Edwin H Armstrong who, as an enthusiastic experimenter, noticed that if he coupled some of the energy from the output of a valve amplifier to the input, this could be adjusted to make the gain of the valve very much greater and hence increase its sensitivity for receiving faint radio signals. This was to be known as regenerative detection and, as Armstrong noted in describing this breakthrough at a later time, he was suddenly able to receive signals from not only the United States but from Europe with a simple single-valve radio which had been quite impossible until that time.

Ultimately, far more important than the discovery of regeneration was the realisation that if the regeneration, or feedback as it is now called, was increased beyond a certain point, the valve would go into a state of sustained feedback and would start to produce continuous radio frequency energy. This behaviour of a valve was to prove the vital element in the new method of producing radio waves and almost immediately was to make all the spark, arc and alternator-type radio frequency generators obsolete.

Apart from finding out about valve regeneration and the capacity of the valve to act as a generator of continuous waves, Armstrong was later to discover other profoundly important methods of using valves for radio frequency transmission and reception (these are described in a later chapter).

The process of regeneration as developed for use in early valve receivers was extensively used in simple radio receivers during the 1920s but finally fell out of favour because of interference that it caused to other listeners. However, this was to be a problem of the postwar years and the immediate impact of regeneration was that it almost immediately led to a new and highly effective method of generating the continuous sine wave radio frequency energy that had previously eluded other experimenters. This was a most timely discovery because it was to have almost immediate application to transmitting apparatus that was used during the war years, discussed in the next chapter.

Despite ultimate recognition by the American patent office that de Forest was the prior inventor of 'feedback' on the basis of some very sketchy experimental work undertaken at about the same time as Armstrong's work, the public and the radio experts have always believed that it was Armstrong who really invented regeneration. This dispute was conducted as a long and very bitter battle between these two giants of the first valve radio era, which was settled only with immense legal cost and animosity and left the two participants irreconcilable enemies.

END OF THE ERA

As the first decade of the developing global communications era came to a close, certain major trends could be observed. The days of the individual experimenter were all but over and the big companies and organisations were becoming involved with the application of major engineering and financial resources to exploit the new mode of communication. What was to galvanise this process into a state of frenzy was the world war that would very soon start. Suddenly the normal process of discovery, patent, development and commercial supply would be savagely interrupted and it would be a matter of extreme urgency that the best and most refined and useful radio and cable systems should be created. Unsurprisingly, perhaps, one of the first casualties of the conflict was to be cable telegraphy, and it was the vulnerability of both undersea and terrestrial cables that would demonstrate the advantages of wireless communications but also its vulnerability to interception. But before the disaster of the First World War, there was one further event in this early period of world communication which still has reverberations even now.

RMS *TITANIC*

Although unrest in the Balkans was in the air in 1912, at this time it was a shipwreck that was to capture the attention of the public in a quite unprecedented manner, and still does, spawning over the years more films and books about it than any other shipwreck. This, of course, was the sinking of the White Star liner *Titanic*.

This notorious sinking of a supposedly unsinkable ship on its maiden voyage demonstrated a number of serious deficiencies in merchant marine practice of the period. Perhaps what fascinates the reader or observer so many years after is that one can see this event as epitomising the high point in Edwardian arrogance and hubris in the face of the natural elements.

Top Typical naval radio station in the United States about 1913 *(MWT)*

Bottom Standard Marconi 1.5-kilowatt rotary spark transmitter and receiver as used on ships including the *Titanic (MAR)*

Until this event, it was very much a philosophy of Victorian and Edwardian life that mankind was capable of anything and could build infallible machines for any purpose. Suddenly, despite all the glamour and high expectations, here was a huge and beautiful vessel cast to the depths of the ocean, seemingly driven by what later was to be seen as either carelessness or stupidity. Nowadays, no one would go ploughing into an ice field at 22 knots and more, expecting to reach the other side with impunity. In 1912, Captain Smith thought this was not only possible but expected, because both he and the masters of other trans-Atlantic liners had done this routinely during countless voyages across the Atlantic.

The disaster of the *Titanic* was made the more poignant by two other factors. The first of these was the inadequate number of lifeboats provided for the passengers and crew but sufficient to achieve more than full compliance with British Board of Trade regulations. The second was the fact that a nearby ship, the *Californian*, did not maintain a 24-hour radio watch which, at that time, was not a legal requirement.

Not keeping a radio watch was perhaps excusable because there was only one radio operator who had finished his watch and had gone to bed, but the failure to respond to a fusillade of emergency rockets was not. Although later many excuses were made for the behaviour of Captain Lord of the *Californian*, the simple fact is that, when advised of the unusual sighting of rockets, he chose to ignore the disturbance and go back to sleep. If he had not and had ordered the magnetic detector to be wound up by the radio operator and set in motion, the story of the sinking of the *Titanic* would have had a far happier ending.

Fifteen hundred persons went down with the *Titanic* and the subsequent public inquiry made it quite plain that most of this loss of life could have been avoided. The *Californian* had lain at anchor only 20 or so kilometres (about 12 miles) away as the *Titanic* sank, and the radio SOS signals were unheard by the sole Marconi operator who was asleep.

This great marine tragedy still has the power to make even hardened contemporary audiences shed a tear, a common reaction to the dramatic ending of a $200 million Hollywood epic, but it is not the human tragedy that is of most interest to the communications historian.

The sinking of the *Titanic* had subtle and far-reaching consequences that would stretch forward over the next 40 years in the exploitation of the short waves, a previously unexplored region of the electromagnetic spectrum.

When finally the few survivors had been rescued by the *Carpathia*, including the second radio operator of the *Titanic*, Harold Bride, messages were despatched to Cape Race in Newfoundland by the *Carpathia*'s radio operator, Thomas Cottam. Although exhausted by his exposure on an upturned collapsible boat, Bride soon went to help Cottam in dealing with the avalanche of messages. However, their work was seriously hampered by the selfish behaviour of a number of radio experimenters. The interference that they caused to the handling of messages concerning the sinking was ultimately to see radio amateurs banished below 200 metres. Ironically, this was to be a beneficial banishment as it proved to be the path to long-distance radio communication at low power.

In the space of a few short years, all the colossal structures of the trans-Atlantic long-wave radio stations would be rendered obsolete and an entirely new system of beamed radio short waves would be introduced. In this turn of events it can now be seen that the sinking of the *Titanic* would lead almost directly to a revolution in world communication and broadcasting which would respond to the world demand for message handling and news for another four decades.

PRELUDE TO WAR

By the beginning of 1914, the storm clouds heralding the approaching war were clearly visible and the future combatants were busy making their preparations. The Kiel Canal in north Germany had been built, allowing German naval vessels to exit to the North Sea without negotiating the narrow channel between Denmark and Sweden at Helsingfors (Helsinki) and Helsingborg. The fuse was primed and all that was required was the fatal spark. It was not long coming.

Senior wireless operator on the *Titanic*, John (Jack) Phillips *(MAR)*

The *Titanic* under way *(MER)*

However, with the new science of wireless telegraphy and telephony, not a great deal of progress had been made by the armies of both sides since the time of the Boer War in 1901. Compared with this situation, the Royal Navy was well advanced and well prepared. To understand how this state of affairs came about, it is necessary to return to a time just after radio communication had first been demonstrated.

Shortly after presenting himself and his wireless system to William Preece, the Chief Engineer of the Post Office in London in 1896, Marconi travelled to Salisbury Plain to demonstrate the capabilities of what he had created. Among the observers at this demonstration were representatives of the army and the navy and it was to be the navy that would become an early client of the Marconi Company when that organisation was instituted in 1897.

Although the Royal Navy and marine interests would soon become enthusiastic in their acceptance of radio, given its obvious advantages to seaborne vessels, initially the British army was rather less than enraptured with what was available. Although Captain Kennedy of the army had been present at those initial trials of the new system on Salisbury Plain in 1896 and had been sufficiently impressed to write a favourable report, little else was to occur for some while. At that time a well-founded concern was expressed that messages would necessarily be 'broadcast' and the enemy might well be a recipient. Telephone lines were seen as intrinsically more secure. Despite representations from the newly created Marconi Company, the army did not immediately take out licences to operate Marconi apparatus or offer to purchase equipment as the Royal Navy was shortly to do, prompted by the support of Captain Jackson.

However, when the Boer War broke out in 1899, the Marconi Company once again offered its equipment to the army and by then successful trials of the new system had been conducted by the Royal Navy. As a result, it was decided that the proposal should be accepted and a number of sets were sent out to South Africa for the use of the army.

Contrary to the expectations that had developed from the naval trials, the wireless apparatus did not perform at all satisfactorily or as expected. Due to the general failure of wireless to provide successful communication for the army, the equipment was ultimately transferred to naval vessels involved in blockading the ports used by the Boers and, in this situation, the wireless performed admirably.

The strange difference between the behaviour of wireless on dry land and at sea was explained in material presented to the centennial conference of the Institution of Electrical Engineers in 1995. This research paper suggested that the reason for the poor performance of the Marconi apparatus in South Africa was associated with the difficulty of achieving adequate earth connections because of the low conductivity of the soil. In addition, because of the time of the year, the extent of lightning strikes in southern Africa was sufficient to desensitise the coherer detectors in use at that time. Also, the level of electrical noise in the ether would have effectively masked the sound of the spark-generated morse code in any case.

In contrast to this situation, when the wireless was installed in ships, the earth connection was immediately improved by many orders of magnitude. The conductivity of sea water was nearly 300 times better than the dry soil of the inland. In addition, by the time the installation on the ships was complete, the level of lightning activity had died away with the change of season.

To a modern eye and even assuming contemporary apparatus, this seems quite convincing and plausible. The need for a 'good' earth remains a priority to this day if one is using the form of antenna in use in South Africa — a 'Marconi' wire held up on a couple of poles. Equally, for radio amateurs operating in the tropics or in the subtropical parts of Australia, the impact of lightning produces noise levels and 'static' that render the lower frequencies almost unusable in summer time. This remains a potent source of interference to successful communications in the lower part of the high-frequency band.

Although it can be appropriately claimed that the first military use of radio occurred in South Africa in 1901, the reality is that it was not at all a success and not until another 12 years had elapsed was radio once again to appear on the battlefield in any serious fashion. During the intervening period, considerable efforts were put into improving and making lighter the cumbersome spark-based apparatus of 1900. By 1914, not only had a variety of portable equipment been developed but even apparatus suitable for transport into the air in the first 'string and wire' military aircraft was possible. Aerial spotting for the purpose of ranging the guns was an early use of radio even though initially the traffic was one-way.

As will be discussed in a later chapter, despite all the pre-war development, on the ground the combatants went to war remarkably ill-prepared with appropriate wireless equipment. At sea and in the Royal Navy, it was a completely different situation and the technological capabilities in communications terms were significantly better than were available to either the British or German armies.

SPARK TRANSMISSION METHODS

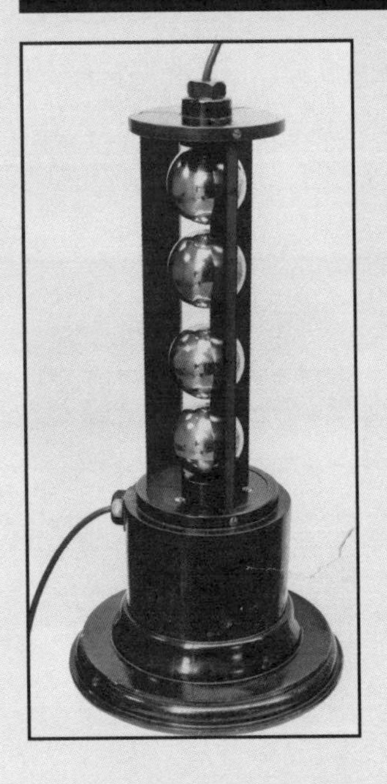

Spark gap as used by Captain Henry Jackson in 1896 *(FAR)*

A plain spark gap as used in the Marconi marine wireless system up to about 1910 *(PRJ)*

The development of radio communication using electric sparks as a means of generating a radio frequency signal is likely to appear a quaint if not obscure topic for the modern reader. However, as an element in the early history of the world communication system, its part is quite fundamental and, in a technological sense, intriguing. The burst of radio frequency energy that accompanies the discharge of high-potential electric energy across a gap must, at first sight, seem to be a rather unlikely basis for a system of signalling. However, it was this principle that allowed the first generation of radio communicators to operate and it is only the perspective of more than 100 years of radio development that makes the spark seem so peculiar as a source of radio energy.

The initiation of the radio age can be seen as springing from the work of the renowned experimenter Michael Faraday and later from the mind of a Scot, James Clerk Maxwell. In the middle of Queen Victoria's reign, Maxwell set out a theory of the propagation of electrical energy based on an analogy with the dynamics of fluid systems and vortices.

In the latter part of the Victorian era, while scientists were starting to accept that it might be possible for electrical energy to be able to move away from a source of electrical or magnetic disturbance, this theory was greeted with much scepticism. In 1886, Heinrich Hertz carried out the experimentation that not only demonstrated the existence of radio frequency propagation but also established its character. In a series of elegant and repeatable experiments, he showed that electromagnetic energy was identical to light energy and, perhaps equally important, travelled at the same speed, 300 000 000 metres per second approximately.

The basis of Hertz's experimental work was the use of a spark generated by an induction coil, at that time known as a Ruhmkorff coil. This work was developed by a number of experimenters including Oliver Lodge, Aleksandr Popov and Ferdinand Braun but it was the energy of a young Italian, Guglielmo Marconi, who translated the scientific experiments into a practical system of 'wireless' communication. All these experimenters used sparks to generate the radio frequency waves.

From 1895, the date of Marconi's earliest experiments, the spark transmitter was improved progressively until by 1918 it was an extremely sophisticated system that had reached the limit of its technical capability and was at that stage able to send a signal around the full circumference of the globe.

For the contemporary reader, it must immediately appear a conundrum that something as intrinsically random and impure (in a frequency sense) as a spark discharge could gradually be modified by any sort of technological means so as to be able to be used for signalling. Despite this, by 1918 Marconi, the principal user and promoter of spark-based signalling, was claiming that his system was able to produce radio frequency waves with equivalent characteristics to those produced by Alexanderson alternators or valves which had become available in recent years.

THE RADIO FREQUENCY (RF) SPARK AND QUENCHING

From the perspective of today and the sophistication to which radio communication has grown in over 100 years, the puzzle for a modern communicator is how this incoherent (in a frequency sense) discharge of electrical energy was harnessed to the production of radio frequency signals. Indeed, how did it work at all?

In the first instance, to generate the high-voltage energy required to strike a spark, a very simple device was used — a sort of primitive transformer, the induction coil. This consisted of a low-voltage primary circuit and a secondary circuit consisting of an enormous number of turns of wire which provided a series of very high voltage pulses when the primary circuit was activated with 'chopped' direct current, that is, rapidly switched on and off. Apart from the difficulties of constructing a device of this sort, which would not short over internally because of the high voltages in use, the transient nature of the pulses of high voltage produced meant that creating sustained energy in an antenna circuit was extremely difficult.

Replica of a Ruhmkorff coil (induction coil) *(PRJ)*

The diagrams (below right) showing spark and continuous wave energy illustrate the problem. The top diagram is the envelope of RF energy consisting of a sine wave oscillating at 1.8 megahertz, and this can be contrasted with the bursts or pulses of energy produced by the induction coil at roughly 60 hertz or 60 pulses per second. When one appreciates that in the very early days the 60-cycle pulsating noise was carried on the shoulders of the RF energy contained in those pulses of relatively short duration, again the question is: How did it work at all?

If one looks at the form of the pulse, where all the important RF energy was generated to carry the acoustic intelligence across free space, an appreciation of the problem of spark transmission is possible.

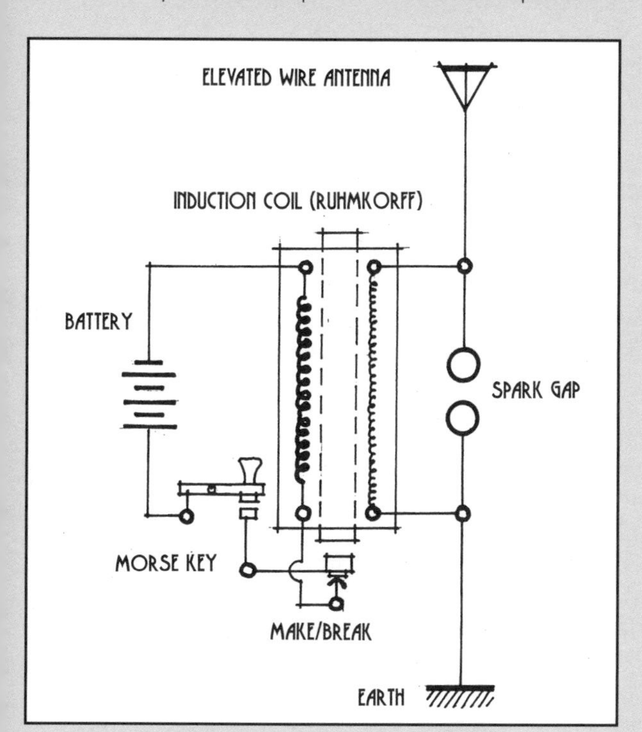

The first transmitter of Marconi in 1895 *(PRJ)*

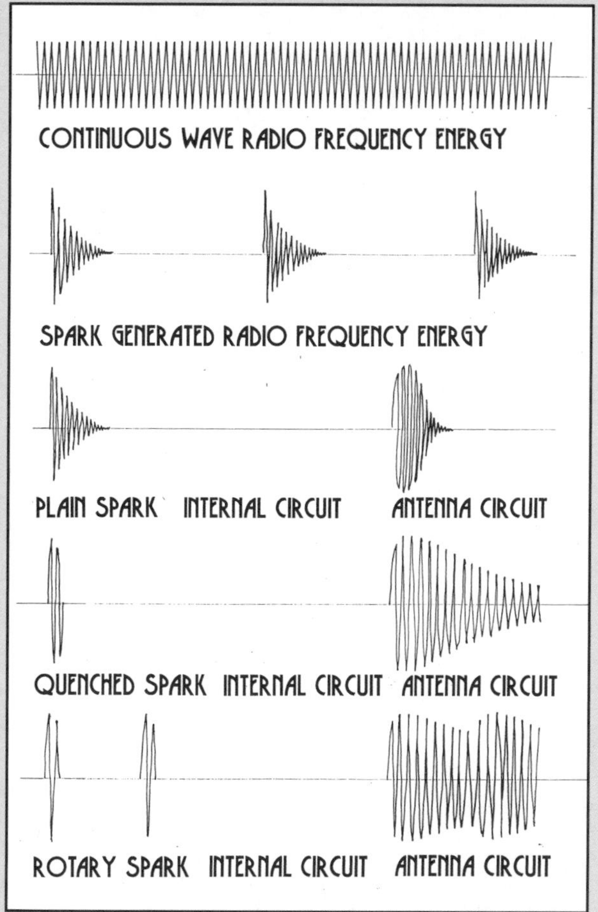

Diagrammatic representation of energy from plain spark, quenched spark and timed rotary spark transmitters *(PRJ)*

What is created in that instant of a spark discharge is a pulse of raw and largely untuned RF energy. If this is coupled to a resonant circuit containing inductance and capacitance, it will ring or resonate at the natural frequency of the circuit established by the components. By inductive coupling it is then possible to transfer some of the energy into a secondary circuit containing the antenna. The net result is a relatively inefficient but workable system for generating RF energy.

An emergency transmitter using a quenched-spark device *(PRJ)*

Over the first 15 years of development of the spark transmission technology, the reason for the inefficiency of the system came to be understood. During the striking of the spark, a virtual short circuit is created across the spark gap. This makes it possible for energy which has been transferred to the secondary circuit to leak back into the primary circuit and be dissipated there. The net result is a series of wave trains in both primary and secondary circuits which fall away in magnitude quite rapidly or, in the jargon of the time, had a steeply falling 'log decrement'.

In about 1902, the German scientist Max Wien discovered a partial solution to this problem in the 'quenched spark'. In this system, the burst of RF energy in the primary circuit is shut off by various technical means after only one or two swings of the RF signal. During this period, a substantial component of the RF energy can then be transferred to the secondary circuit including the antenna. Once the short circuit in the spark gap is removed, the energy in the secondary circuit continues to resonate without being able to leak away into the primary circuit.

Again the diagrams demonstrate this process. In particular, it can be seen that the quenched spark produces a wave envelope in the antenna or secondary circuit which fades away far more slowly than is the case for the plain spark gap. Indeed, as the frequency of operation of transmitters became progressively lower and lower during the first 25 years of radio, the interval between the striking of the spark and the RF signal became relatively much closer as well. Ultimately, it became possible to make the wave trains of RF energy generated by the spark join up into frequency-coherent packages. This then allowed the RF energy in the secondary circuit to be tuned up almost as in a conventional modern continuous wave (CW) transmitter.

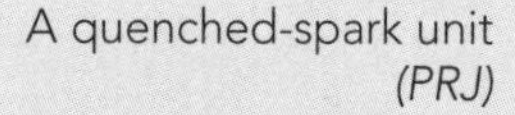

A quenched-spark unit *(PRJ)*

The 'quenched spark' device proved to be so robust and simple that it was used until well after the Second World War as a backup system for emergency communication. It appears that in some Australian coastal vessels the 'quenched spark' continued to be used as a primary means of communication during that period. Despite the extent to which this form of spark generation was used, it is unfortunately true of the 'quenched spark' that it was a potent source of interference to other operators. Because of its very nature — a sharp and transient burst of RF energy — it caused considerable problems with land-based stations including the early broadcast listeners and became very unpopular for that reason. In addition, it was well known to create sympathetic resonance in other antenna systems in the locality — they would resonate at their own natural frequency due to the introduction of this 'hand clap' of RF noise in the vicinity.

THE ROTARY SPARK

While in Germany and elsewhere the refinements associated with quenching the spark were being worked out so as to harness the spark and turn it into a more predictable and tunable source of RF energy, the Marconi Company tackled the problem in a somewhat different fashion. Fairly early on it was realised that there were considerable advantages in breaking the energy path between the spark faces at high speed. This appreciation was soon developed by Marconi into a rotating spark-gap system known as the 'rotary spark'.

By about 1910, both in land stations and at sea, the plain spark gap used in the early days had been almost entirely replaced by rotating spark systems of one sort and another. At sea, the alternators which had superseded the induction coil system were fitted with an extension plate and studs which spun synchronously with the alternator between spark poles. This produced something of an improvement in the quality of the RF signal but, inevitably, the audio frequency modulation produced by the studs was relatively low. At about 120 hertz or pulses per second, this was obviously quite close to the natural frequency level of interference from thunderstorms and the like. This could make the signal extremely difficult to hear and understand under poor operating conditions.

It was then realised that increasing the number of spark poles, which, incidentally, made the signal asynchronous, would result in an increase in the frequency of the acoustic modulation. This was found to be extremely beneficial for receiving purposes. Within a relatively short period, the frequency of the audio modulation was increased to about 800 hertz by adding spark poles to the rotating plate.

At sea this became more or less the standard Marconi system as installed from around about 1911 on. Such a system as this was installed on the *Titanic*, and is known to have been generally reliable. It was able to provide a working range of approximately 650–800 kilometres (400–500 miles) in daylight hours and significantly more than that at night time. In the period between about 1911 and 1918, ships crossing the Atlantic from east to west and west to east were able to keep in contact with one side or the other almost continuously, either by direct contact or by relaying via other vessels.

An Australian ship's radio room of about 1922 showing the quenched-spark device on the right and an AWA valved receiver on the left *(AWA)*

THE LAND STATIONS

By comparison with the marine alternating-current systems of rotary spark, the land stations developed firstly in Cornwall at Poldhu and later at Clifden on the west coast of Ireland and later again at Caernarvon in north Wales used direct current at about 15 000 volts. The electrical energy was generally stored in enormous batteries of primary cells or, in the case of Caernarvon, was achieved by direct conversion from the alternating current mains supply available from the National Grid.

At Poldhu and later at Clifden, the spark took the form more of an arc produced by direct current between the spark poles. However, for cooling purposes, the path of the discharge was between the surfaces of rotating steel plates. A large steel plate mounted on the shaft of a motor was spun at relatively high speed between the faces of two smaller round steel plates turning at a significantly slower rate. What this first system of rotary spark or arc discharge produced was something much closer to a continuous wave RF signal than was the case with later systems of rotary spark transmission. However, because of the characteristics of the discharge path, the RF sine wave, the wavelength of which was determined by the inductance and capacitance in the primary circuit, had mixed with it an enormous number of RF components. This made the signal difficult to tune and, because of the absence of pulsating acoustic information as available in, say, an induction coil, the signal proved to be extremely difficult to hear, particularly in the presence of static. The solution developed for this problem was no doubt borrowed from the rapidly developing technology of marine rotary spark transmission. Studs were added to the rotating plate and the more or less continuous wave energy became interrupted spark/arc once again with a pulse repetition rate of about 800 hertz.

The advantage of having audio modulation of around 800 or 900 hertz imposed on the spark-generated signal had significant operator advantages. Despite the poor frequency purity of the signal, which inevitably led to difficulties in tuning and in much of the RF energy being wasted in merely heating up the primary circuit and the antenna elements, audibility of the signal was good and a workable system was achieved.

ULTIMATE SPARK

Over the first 20 years of spark development, for various reasons, not least of which was the general success that was achieved, there was an inexorable move to longer wavelengths and lower frequencies. By about 1914, the frequency of transmission across the Atlantic was down to about 25 kilohertz, or just above the range of hearing. This had some important implications in terms of the ability to create a coherent series of wave trains in the spark primary circuit. When the repetition rate of the spark discharges was increased to about 800 hertz and the radio frequency of operation was at, say, 15 000 or 20 000 hertz, it was just possible to make the bursts of RF energy overlap or join up. When coupled to the secondary circuit, it was ultimately possible to produce more or less continuous RF waves in an antenna system. As seen on a modern oscilloscope, the wave form would be continuous but pulsating at a rate of 800 hertz.

The rotary plates in a 'timed spark' unit as used at Caernarvon, Wales *(MAR)*

During the development period of 1895 to 1920, there is no doubt that there was an extraordinary expansion of the scientific basis of radio and the technology of radio communication in an incredibly short space of time. By the same token, in the early days, it is quite clear that the electrical technicians and scientists were guided at times more by experimental results than by good solid theory. However, by 'cut and try' methods in many instances, the apparatus was made to work effectively. By 1920, only major changes of technology could displace the spark transmission system and one of the new developments was 'waiting in the wings' — it was, of course, the thermionic valve.

By about 1916, during the dark period of the First World War, the Marconi

Company had developed a remarkably sophisticated system of RF-energy generation by spark. Though the details of this system are somewhat clouded by the mists of time and the inherent security associated with those war years, it is possible to obtain some idea of what was developed. This last and greatest expression of the art of spark radio occurred at Caernarvon and was known as the 'timed spark' system.

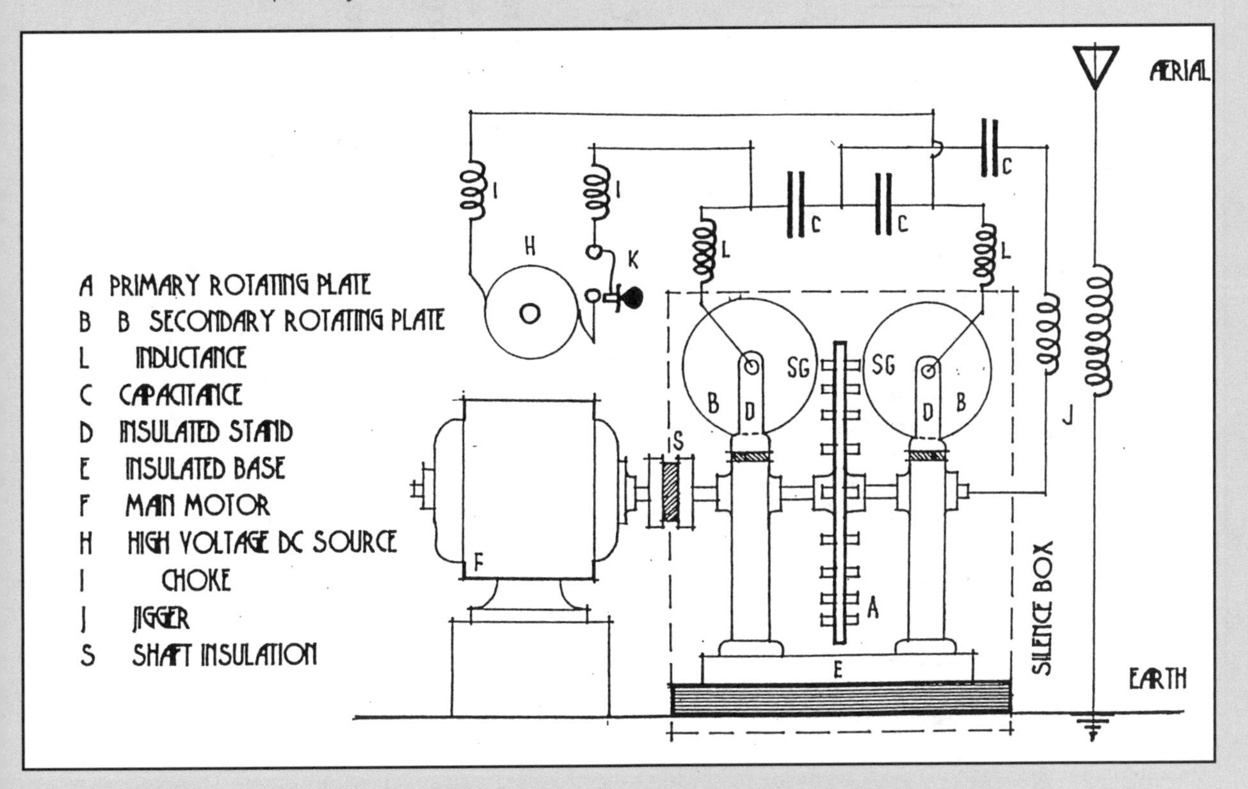

Diagram of the first rotary spark machine as installed at Clifden in Ireland in 1907 *(PRJ)*

The 'timed spark' system was developed in two stages, the first being described in considerable detail by Professor Fleming in 1916. The second more sophisticated stage is still the subject of a degree of speculation.

Grasping the potential advantages of producing an overlap of the RF signals developed in the individual pulses of spark-generated energy, the first 'timed rotary spark' system used a number of plates with rotating poles set in staggered array. The plates were mounted on a common shaft from the motor used to rotate them, but were electrically insulated from each other. As the plates spun at high speed, the discharges at each of the poles of the four plates occurred sequentially and added together in the primary circuit. When inductively coupled to the antenna secondary circuit, a pulsating but continuous wave was produced with all the advantages of tunability and audibility at the receiving end.

However, this first 'timed spark' system proved to have problems. In particular, the spark gap dimension was critical and the striking of the spark was very sensitive to atmospheric pressure differentials coupled with humidity variations. The net result was that the task of getting the spark pulses to join up to create a coherent continuous wave was subject to natural variables which were very hard to control. For this reason, a second-stage 'timed spark' system was developed and this was so successful that it was possible to send a signal halfway round the world. In 1918, a signal was sent from Caernarvon to Sydney, Australia.

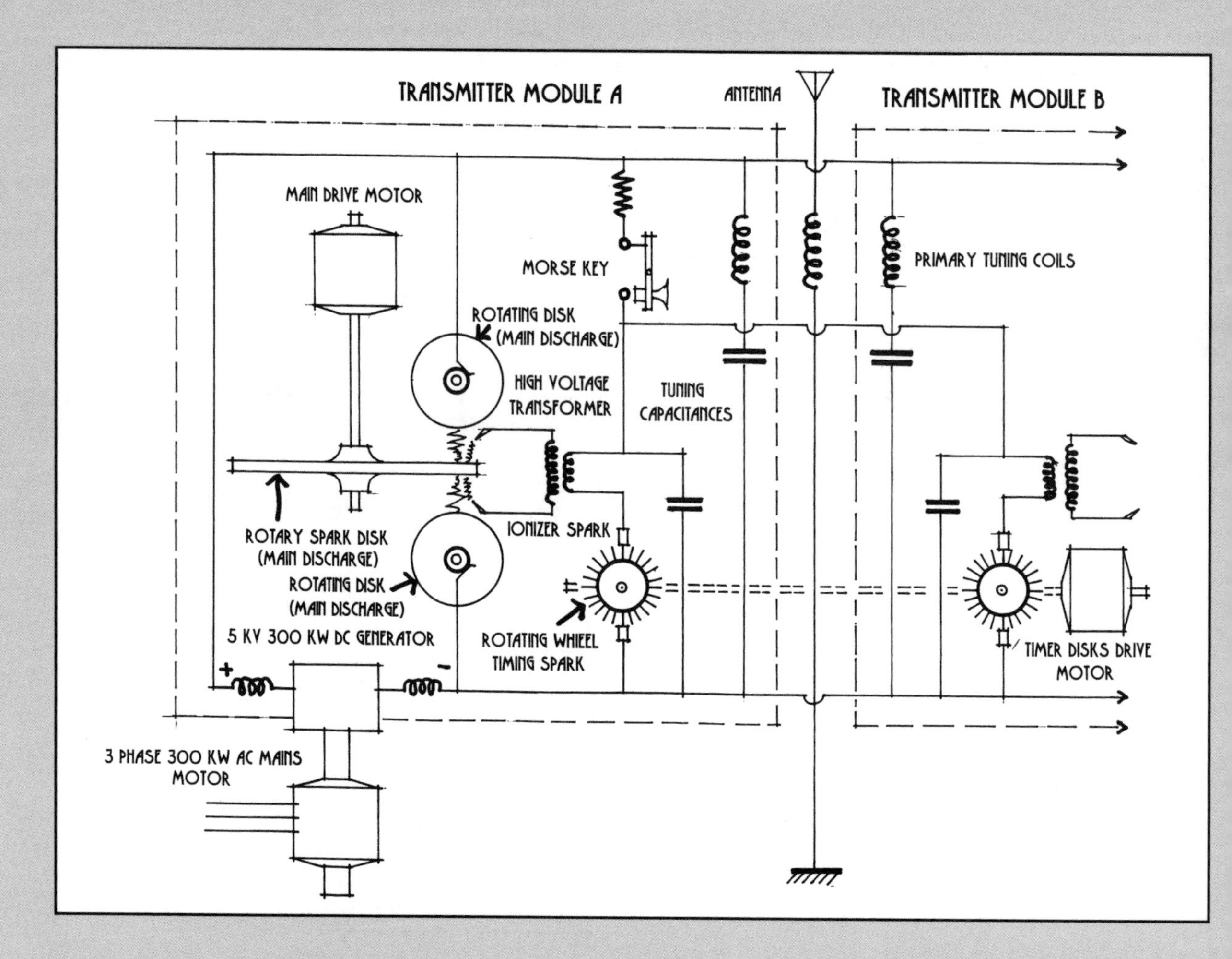

The ultimate timed rotary spark machine as installed at Caernarvon, Wales, in 1918 *(PRJ)*

In order to understand this last system of rotary spark, it is necessary to appreciate that the distance that a spark will jump is related most particularly to the characteristics of the path between the poles and the voltage imposed across the gap. At a high gas pressure, the voltage required to strike a spark is less than at low pressure but, more importantly, if the path through air is filled with ions then the spark will strike at a voltage far below the normal value for an air gap.

It appears that what Marconi developed in the last 'timed spark' system was based on the use of dual rotors. The main rotary gap, which did all the work and produced the high power required — 300 kilowatts — consisted of a simple plate rotating at high speed between rotating spark poles. However, these were set away from the rotating plate, which had no studs, to a distance where the operating voltage would not initiate a spark or arc. What initiated the spark was a second set of poles set close to the main spark poles and driven by the spark discharge produced by another rotating plate with multiple studs. This separate plate was, in effect, a timing wheel, rather like the timing mechanism in a conventional ignition system in a motor car. What it did was to create an ionised path close to the main spark discharge path. As soon as this pathway was established with a cloud of ions, the main spark would strike but would extinguish as soon as the ionisation was removed. The result was a series a of pulses at the main spark discharge faces, controlled in terms of repetition rate by spark discharges generated separately at very high voltage but low power by the rotating timing plate.

By using a very high voltage at low power and an initiation or ignition spark to ionise the air, the problems of natural conditions affecting the repetition rate of the high-power spark were overcome. More importantly, by achieving a very stable and accurate rate of repetition of the spark pulses, the wave trains of RF energy could be made to join together in a completely consistent fashion and the sought-for pulsating continuous waves at the antenna were able to be produced on a reliable basis.

THE VALVE

It is somewhat ironic that, while this great burst of ingenuity, directed at improving the characteristics of the spark, was occurring during the First World War, it was this same conflict that generated the successor to the timed spark, the valve (also known as the tube). Within only 4 years of the end of the war in 1918, the valve had been developed to a stage of reliability that made it possible to completely replace spark and electromechanical systems in the long-wave trans-oceanic stations such as Caernarvon.

At last the roar of the rotary spark was silent and the new age of 'quiet' RF-energy generation had arrived. Radio men from that earlier time described the rotary spark sound as 'awesome' — it is perhaps regrettable that, with the lack of awe associated with modern radio, some of the romance of radio communication has also vanished.

CHAPTER 5

1911 to 1912

The Mawson expedition

During the period before the start of the First World War, life in Europe continued much as it had in Victorian times. As 1914 approached, however, increasing tension and instability became more apparent, particularly in the area of the Balkans. In that same period, though in a relatively primitive state still, wireless, or radio as it was starting to be called, developed rapidly as a new and useful way to remain in contact where ground-based or undersea cables were not practical.

Unsurprisingly, radio was found to be particularly useful at sea as an aid to saving life, where vessels had been overcome by the savagery of the elements. Now when a cry for help went out, there was some chance that, via the radio waves, it would be heard by other ships and a rescue could be attempted.

MARINE EVENTS AND DISASTERS

A number of well-known and lesser known events involving ships at sea and shipwrecks were coloured by the presence or absence of radio during the first decade of the 1900s. In the category of disasters that would have been far worse except for the presence of radio was the sinking of the *Republic*, a White Star liner. This ship collided with the Italian liner *Florida* in January 1909. With a complement of 461 passengers and 300 crew, there was potential for great loss of life.

Radio amateur Paul F Godley meets ship's operator Jack Binns *(PRR)*

Due to the efforts of the Marconi operator on the *Republic*, Jack Binns, who distinguished himself by conspicuous bravery, this was averted. Although the *Republic* was so badly damaged that she ultimately sank, the wireless call for help ensured that all the passengers and crew were saved.

Based on radio signals that were passed between the masters of the two vessels that had just collided, a large number of passengers were able to be transferred to the *Florida* which was found to be in no danger of sinking. Later, with the arrival of the *Baltic*, it was possible to move the passengers and crew from both the *Florida* and the *Republic* to this ship which then sailed on to New York. Jack Binns and his use of the emergency call CQD received much publicity and ultimately he was presented with a gold watch to mark his gallant service in staying at the morse key until help was assured.

Equally notable in this period was the apprehension of the escaping Dr Crippen and his lover, Ethel Le Neve, who travelled badly disguised as the doctor's nephew. Dr Crippen had dispatched his disagreeable wife in small pieces to the basement of his house in London and, when questioned by the police about her disappearance, had fled to the continent with his lover. If it had not been for the wireless, he might very well have then made good his escape to America, but signals were sent to the *Montrose*, the ship on which he was travelling. As a result, a detective from Scotland Yard was able to overtake the fleeing couple in a fast ship and meet the *Montrose* at its destination.

When the *Montrose* finally reached port and met the Canadian pilot boat, the British police were able to climb aboard and arrest the startled doctor. No doubt Dr Crippen was exceedingly surprised by this turn of events — in July 1910, radio was a considerable novelty.

THE ANTARCTIC

In distant Antarctica, Dr (later Sir) Douglas Mawson, an Australian geologist of British extraction, undertook an expedition which was to be distinguished by great discovery, high adventure and, in the end, great personal tragedy for the leader of the expedition, Mawson himself.

Mawson's expedition was largely overshadowed by the tragedy involving the British explorer Robert Falcon Scott and the whole of his party. In addition, the start of the First World War was to have an adverse impact on what was, in other respects, a very successful venture. Mawson's expedition was also distinguished from other Antarctic exploration by its concentration on scientific discovery rather than on the establishment of international territorial rights.

Apart from that aspect, in the story of world communication, the use of a radio link from Australia via Macquarie Island to the Antarctic base is seen as particularly relevant to the development of the intercontinental system that now exists. Such a place as Antarctica was inherently remote and inaccessible and, before radio, could be reached only by a hugely expensive and unjustifiable cable connection laid across the seabed.

THE EXPEDITION IS PLANNED

Douglas Mawson, later Sir Douglas, rests on a sledge before beginning the disastrous journey; kneeling is Dr Mertz, who died from excessive vitamin ingestion *(MAW)*

In the early 1900s, the island continent of Australia was still a distant and largely unknown land for most Europeans. However, lying far to the south was an even more remote and unexplored place, Antarctica. First circumnavigated by Captain Cook during his second great voyage in the period 1772–75, by the latter part of the 19th century its shape and extent was still largely a matter of conjecture, unsupported by direct exploration. From the time of Cook, Antarctica had been visited by a small number of intrepid voyagers but it was not until the very end of the 19th century and the beginning of the 20th that concerted efforts were made to discover the extent of the continent and to explore and map it in detail.

In 1902, Robert Falcon Scott led the first of the modern expeditions into the interior of Antarctica, and this successful enterprise was followed by an expedition led by one of Scott's party, Ernest Shackleton, in 1908. One of the members of Shackleton's party was a young scientist, Douglas Mawson, who had been born in England in 1882 and educated in Australia. During this expedition, Mawson, together with Dr Forbes Mackay and under the leadership of Professor TW Edgeworth David, made a successful journey to reach the south magnetic pole.

Upon completion of the expedition, Mawson returned to Australia determined to return to the newly discovered continent as soon as he was able. As had many before him and, for that matter, many since, Mawson had become fascinated by the bleak and remote wilderness of ice and snow and had been enchanted by its unspoilt beauty. As the storm clouds gathered over Europe, heralding the approach of the First World War, both Mawson and his former leader, Scott, were making preparations to return to the southern continent. Whereas Scott's ambitions were fairly simple and territorial in nature, Mawson was preparing for an expedition which was, first and foremost, scientific in its complexion.

RADIO EQUIPMENT

Perhaps unsurprisingly, with such an interest, Mawson decided to include in his stores the newest of scientific discoveries, radio. Given its novelty, particularly in the Australian context, this was quite a bold step to take. At that time wireless telegraphy involved apparatus that, by the standards of today, was quite gargantuan and ponderous and inevitably would displace other stores in the restricted confines of a sailing ship's hold. Apart from the physical implications of carrying heavy and bulky radio apparatus, this form of communication had only just been introduced to the Australian mainland. As described in the previous chapter, radio was very much in its infancy in 1911, and at this time the Australian government had at last made a decision to have two radio stations constructed at Sydney and Perth. A little later the decision was made to construct a further radio station at Hobart in Tasmania.

It was at about this time that Mawson was preparing his expedition to Antarctica. Further, as described earlier, it was in the face of the technical problems and uncertainty associated with the newly installed Sydney and Perth stations that he decided to take Telefunken apparatus with him in his sailing vessel, the SY *Aurora*.

ANTARCTIC DISASTERS

Because of the tragedy that overtook Scott's expedition of 1911 and 1912, the name of Scott is far better known around the English-speaking world than that of Mawson. Captain Scott and his four companions all perished following their abortive attempt to be the first men to reach the South Pole. It was Amundsen who achieved this distinction, assisted by a pack of enthusiastic huskies towing the sledges and with the expeditioners following along on skis. In contrast, Scott's party had pulled their sledge themselves, and a slow and painful process it had been on restricted rations. This latter problem was to be the basis of their ultimate failure, when a blizzard prevented them from reaching a supply depot by a distance of just 20 kilometres (13 miles).

At a personal level, the expedition of Mawson was to be just as disastrous as that of Captain Scott and his party. In an attempt to explore the newly discovered coastline of Adelie Land, Mawson was to lose both his sledging companions, Xavier Mertz, a scientist, and Lieutenant BES Ninnis. The latter was lost when his sledge, fully laden with most of the food, together with the dog pack which was hauling it, broke through the roof of a crevasse and fell into the depths. Later, when all the remaining food was gone, Mawson and Mertz were forced to eat their own sledge dogs which progressively died from lack of food. After desperate efforts to sledge back to the main base at Commonwealth Bay, Mertz also died. Much more recently it has been suggested that he may well have been poisoned by an excess of vitamin B, ingested with the cooked livers of the animals.

However, apart from these dramatic and fatal events, the Mawson expedition was otherwise a considerable success. Much scientific work was conducted and, in particular, despite the primitive state of the technology, radio was to prove of some value in transmitting messages to the new wireless station at Hobart. One of the important longer term outcomes of the scientific work was that Macquarie Island was declared a protected sanctuary for wildlife where the seals were no longer butchered for their 'blubber'.

MACQUARIE ISLAND

Mawson had decided that, given the distances involved to the main Antarctic icecap, he would need to establish a radio relay station approximately halfway to Australia at Macquarie Island. This was, then, the expedition's first landfall after leaving Hobart aboard the *Aurora*, a substantial sailing vessel with an engine in addition to the sails carried.

One of the fascinating aspects of the Mawson expedition is that one of the members was the noted photographer Francis Hurley, usually known as Frank. One of the early exponents of cinephotography, Hurley took with him to Antarctica a 35 mm hand-cranked cinecamera and with it produced some extraordinary footage. Although his affection for the wildlife of Macquarie Island and Adelie Land (where the main camp was to be located) is evident in the abundant shots he took of penguins and seals disporting themselves on the lonely shoreline, for the radio historian, photographs of the erection of antennae on Macquarie Island have rather more charm.

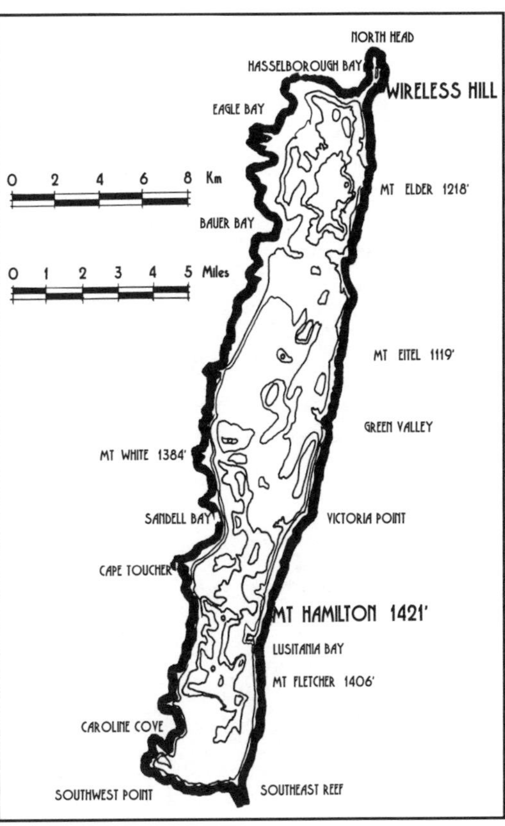

Map of Macquarie Island, showing Wireless Hill at the northern end *(PRJ)*

While the *Aurora* lay at anchor in Hasselborough Bay, a slightly less inhospitable inlet in the rocky shore at the northern end of Macquarie Island, the crew ferried a large quantity of material and food across the tumbling surf to the land. The radio apparatus was carried in barrels and the mast sections were simply floated across and, once landed, the crew set to work to erect the antenna system.

THE RADIO STATION

As the map of the island shows, the position chosen for the wireless station was on a rocky, flat-topped hillock at the extreme northern end of the island. This hillock is about 100 metres (330 feet) high with very steep sides and, in order to get the timber sections to the plateau above, a 'flying fox' was set up. With a block and tackle linked to a long rope running down to the beach below, it was possible with a large team of men to haul the huge 200-millimetre (8-inch) square sections of the masts to the top. With thirty or so men set on either side of the haul rope and pulling briskly, this group is revealed in Hurley's cinefilm looking like some giant centipede.

As it transpired, this was a thoroughly dangerous occupation and one of the crewmen, Frank Wild, was nearly badly injured when hit by

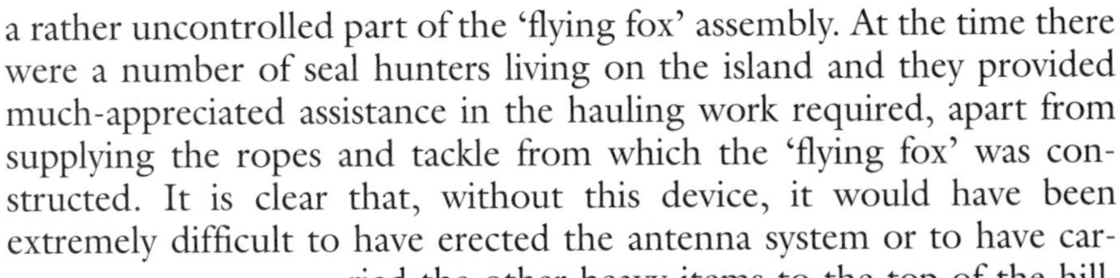

a rather uncontrolled part of the 'flying fox' assembly. At the time there were a number of seal hunters living on the island and they provided much-appreciated assistance in the hauling work required, apart from supplying the ropes and tackle from which the 'flying fox' was constructed. It is clear that, without this device, it would have been extremely difficult to have erected the antenna system or to have carried the other heavy items to the top of the hill. These included a petrol engine and dynamo, an alternating current generator and other apparatus associated with the wireless installation.

Erecting the wireless masts on the top of Wireless Hill, Macquarie Island, in 1912 *(MAW)*

The antenna masts were erected very much like flagpoles, in four sections made out of oregon and set one above another and held together with iron bands. Apart from setting the foot of each mast into the ground, holes were dug for 'deadmen', short sections of timber set horizontally and then covered with earth, to which were attached the guy wires. This method of fixing the guy wires was not very stable and the wireless operators had to make frequent adjustments because of the movement of the 'deadmen' in the sodden, peaty soil of the plateau.

About all that is left now of the antenna system on the lonely Macquarie Island hillock are the stubs of the old timber masts in an advanced state of decay and weathering and the original iron bands in a badly rusted state.

Close by on the hilltop, two sheds were erected, the smaller of the two housing the transmitting and receiving apparatus and the other the electricity generating machinery. By comparison, the main party on Macquarie Island was housed in a much larger hut which was built on the neck of land connecting the main island to the isthmus. A site was chosen in the lee of a large mass of rocks which projected from the grass-covered area of sand at the northern end of the spit and not far from the sloping side of Wireless Hill, as it was named. No doubt the daily climb to the top of the hill must have kept the radio operators very fit.

Centre Looking along the northern end of Macquarie Island towards Wireless Hill *(ANAR)*

Bottom The native inhabitants still in residence on Macquarie Island in modern times *(ANAR)*

Apart from the natural elevation of the antenna site and useful 'take-off' towards the mainland of Australia, the flat top of Wireless Hill was wet and boggy and an ideal 'earth' connection for the Marconi antenna system in operation. An early photograph shows the two masts on the hilltop, together with the guying system and the antenna wire hanging between them.

Five crew members were left to look after the station at Macquarie Island — GF Ainsworth, H Hamilton, LR Blake and the two radio operators, AJ Sawyer and CA Sandell. Sawyer, the chief wireless operator, was

26 and came from New Zealand. He had considerable experience and had joined the expedition from the Australasian Wireless Company. The other wireless operator was 25-year-old CA Sandell who came from Surrey in England and had most recently been working with the Commonwealth Branch of Telephony. Later, in August 1913, when Sawyer fell ill and had to return to Australia, Sandell took over complete responsibility for the operation of the wireless station.

A modern base at Macquarie Island looking down from Wireless Hill *(ANAR)*

ADELIE LAND

Having safely established the small party at Macquarie Island, the remaining members of the expedition set sail once more and headed for the unknown coastline of Adelie Land. At last, on 29 December 1911, the *Aurora* encountered the edge of the icefield and, after one abortive attempt to reach the mainland of Antarctica, the expedition sailed into what was soon to be named Commonwealth Bay.

Here the expeditioners stepped ashore onto a frozen land which lay pristine and untouched by the destructive human hand. Here was soon unloaded a small mountain of equipment and supplies, enough for the intended 2 years' duration of the expedition. Before very long, a building consisting of two connected pyramid-shaped huts had been erected to provide shelter from the almost ceaseless wind which came roaring in from the heart of the Antarctic icecap. So fierce and continuous was this assault of the wind that Mawson named this inhospitable place the 'home of the blizzard', which later became the title of his book telling the complete story of the expedition, including his own dreadful experiences.

The Mawson expedition hut at Commonwealth Bay in recent times, before restoration *(ANAR)*

Once the huts were constructed, it was possible to erect the antenna system and the supporting masts. Initially this consisted of an arrangement of three masts, with the centre mast made up from three sections and the two outer masts being a single section in length. The antenna was set up in what would now be called an 'inverted V' configuration, with the main conductor running from the high point in the 'V' to the transmitter and receiver. Later, when a particularly fierce blizzard had destroyed the upper section of the centre mast, the expedition was forced to erect an antenna on a similar pattern to the installation at Macquarie Island. Instead of the three masts, the old centre mast was reduced in height to two sections only and the other two masts were made into another single mast to produce a 'Marconi' style of antenna. In this the top section is horizontal and the conductor runs from one end to the transmitter and receiver.

Once the huts had been erected and sheathed with timber planking, a corner of the smaller of the two huts was set aside for the wireless installation which, in other respects, was identical to that previously established at Wireless Hill on Macquarie Island. The equipment was made by Telefunken and was a conventional 1.5-kilowatt quenched-spark system with crystal detector. As with all early wireless apparatus, the transmitter relied on brute force electrical energy to produce the signal from the spark gap. The receiver had no amplification and relied entirely on the field strength generated by the distant transmitter to produce an audible signal. When this was competing with a very high level of local static, the results were inevitably doomed to failure and, for the first year of the expedition, that is all that the radio operators were able to achieve.

The Mawson expedition hut with some repairs to the roof *(ANAR)*

RADIO IN ANTARCTICA

In Adelie Land the wireless operators were SN Jeffryes and W Hannam (Walter), and for most of the first year of the expedition they tried without success to establish two-way communication with Macquarie Island. What they were unaware of was that their signal was sometimes heard at Macquarie Island but any answering signal was lost in the roar of locally produced static interference. This resulted from the electrostatic charge developed by the movement of the wind-blown snow particles. When the wind was at its fiercest, sparks could be drawn from metal objects inside the huts, including the antenna elements. Under these circumstances, expecting a non-amplified system of signal detection to produce results was a rather forlorn hope.

The following year, when the conditions were more suitable and the wind speed had abated, at last contact was established with the Macquarie Island party. Sadly, the first information that was conveyed was the news of Captain Scott's party and their unfortunate demise. Then, just as sadly, the news of the deaths of Lieutenant Ninnis and Xavier Mertz, and Douglas Mawson's extraordinary journey back to the headquarters at Commonwealth Bay was able to be conveyed to Australia via the relay station at Macquarie Island.

Mawson recorded these events in his book *Home of the Blizzard* as follows:

Walter Hannam, radio operator, uses the Telefunken apparatus at Commonwealth Bay in 1912 *(AWA)*

> On the night of February 15, Jeffryes suddenly surprised us with the exciting intelligence that he had heard Macquarie Island send a coded weather report to Hobart. The engine was immediately set going, but though repeated attempts were made, no answer could be elicited. Each night darkness was more pronounced and signals became more distinct, until, on the 20th, our call reached Sawyer at Macquarie Island, who immediately responded by saying 'Good evening'. The insulation of a Leyden jar broke down at this point, and nothing more could be done until it was remedied.

Looking inland towards the Mawson expedition hut *(ANAR)*

At last on February 21, signals were exchanged, and by the 23rd a message had been dispatched to Lord Denman, Governor-General of the Commonwealth, acquainting him with our situation and the loss of our comrades and, through him, one to his Majesty the King requesting his royal permission to name a tract of newly discovered country to the east, 'King George V Land.' Special messages were also sent to the relatives of Lieutenant BES Ninnis and Dr X Mertz.

The first news received from the outside World was the bare statement that Captain Scott and four of his companions had perished on their journey to the South Pole. It was some time before we knew the tragic details which came home, direct and poignant, to us in Adelie Land.

Erecting the wireless masts at Commonwealth Bay, Antarctica, in 1912 *(MAW)*

The native residents of Commonwealth Bay on display *(MAW)*

Walter Hannam, who had, no doubt, been attracted to the expedition because of the unique opportunity to be involved in a new use of wireless, had already established himself as a radio expert. Not only had he assisted in the first use of radio by the Australian Army in 1910, surprising many sceptical officers, but he had also been involved in the establishment of the Wireless Institute of Australia, one of the earliest amateur radio associations in the world. This institution, founded in 1910 by a number of notable radio experimenters, remains extremely active today in an increasingly complex world where radio, television, cable and satellites provide the means of communication that once long ago was the sole domain of the electric spark.

VK2QI initially and, later, VK2YH and VK2AXH, as Walter Hannam was known to other radio amateurs, he remained interested and involved with the Wireless Institute until his death in 1964. In particular, it appears that he was one of the people instrumental in obtaining the Institute's former premises at Atchison Street, St Leonards, New South Wales.

Since his death, the whole field of communications including the activities of radio amateurs has seen a revolution. During this time, personal microcomputers have emerged as a potent new technology to be welded to the older systems of communication involving cable and wires. No doubt Walter Hannam would have found these developments as fascinating as those that drew him to Antarctica in 1912. Conversely, he may have concluded that the whistle of the quenched spark of so long ago represented part of the 'Golden Age' of wireless.

Looking down on the hut and antenna masts at Commonwealth Bay in 1912 *(MAW)*

CHAPTER
6
1914 to 1918

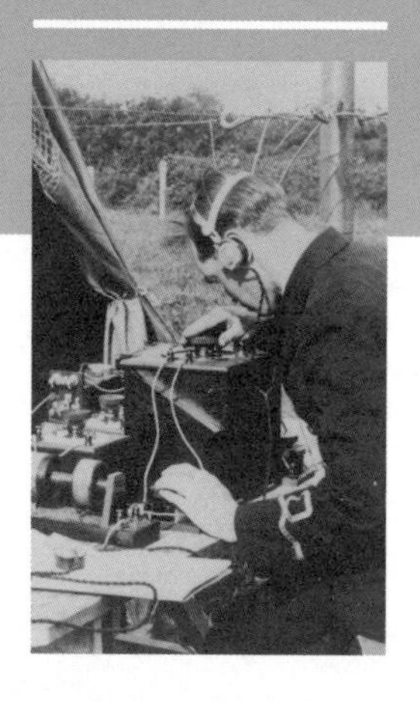

Wireless in the First World War

When the fateful shot rang out in Sarajevo at 11 a.m. on 28 June 1914 and Archduke Ferdinand lay dead, few could have envisaged the oceans of blood that would all too soon flow across the fields of Flanders in the north and in the heartland of France. As the armies of Europe expanded to meet the demands of the new 'trench' warfare, most of the young men who rushed to join the swelling ranks of soldiers may well have thought that they were going on a great and exciting adventure. There was a common delusion that the war would surely be over by Christmas.

Such delusions were totally shattered 4 years later. When at last the nations of Europe had disentangled themselves from the catastrophe of the Western and Eastern Fronts, the combined loss was some 11 million of the best and bravest left dead on the battlefield.

Those willing young volunteers who jostled excitedly to join a new crusade to crush their opponents could have had no conception of the unmitigated horror of warfare they would be exposed to. Their leaders had prepared for war, but of a very different, non-mechanised kind and one in which the notion of chivalry might have had some currency. Old-fashioned ideas of gallantry that the generals might have entertained did not last very long as the men grappled hand to hand and breast to bullet amid the mud and shellfire of Gallipoli and Flanders.

Soldiers in the trenches about 1916 with a Vickers machine gun and ready for a gas attack *(BOS)*

This was a war in which the generals seriously expected that horses would still be used in combat and, as in Napoleonic times, the enemy would be smashed after the first full charge. This was to be Omdurman revisited, when the Dervishes had been driven before the lances of the cavalry and scattered to the four winds. The reality was to be quite different.

Despite the obvious horrible lessons of the American Civil War which had so graphically demonstrated the impact of modern weapons on unprotected infantry, the delusions remained to blight the years after 1914. Here were no unskilled natives to be frightened by raw steel. On the contrary, here were opponents armed with similar weapons and trained with the same lethal objectives in mind. The reality of this crusade was the 'scythe' of the Maxim and the Vickers machine gun and high-powered bolt-action rifles. Exposed to the hail of bullets and shrapnel from high-explosive shells, the unprotected ranks of young men were mown down in their thousands in futile attempts to overcome other young men who stood in opposition.

WAR ON LAND

The general outline of the initial phases First World War in the west is quickly drawn and the basis of the disaster can be appreciated from the maps in this chapter. These show the position of the combatants after the initial thrust of the German army across the Netherlands and Belgium in August 1914.

An initial movement was ultimately arrested at the line of the river Marne. With this reverse, the German army was pushed back to a battle line that curved across the north of France from Verdun to near

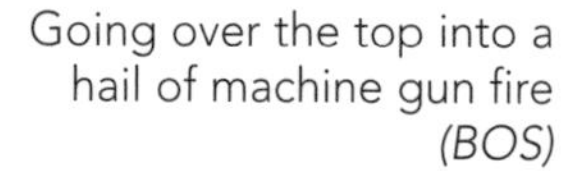

Going over the top into a hail of machine gun fire *(BOS)*

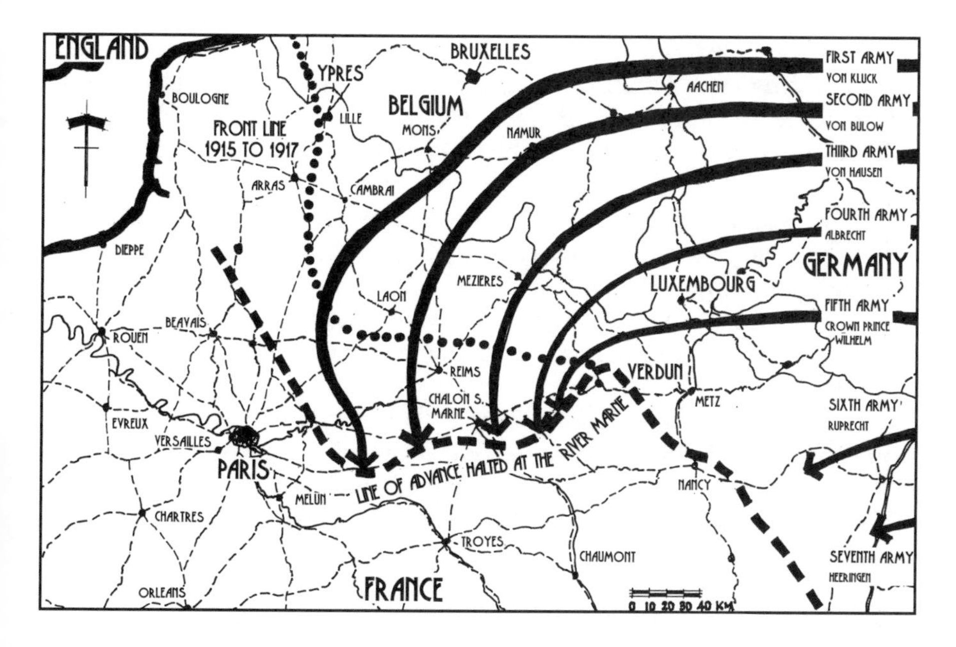

Map showing the German advance to the River Marne in 1914 and the retreat to the line of 'stalemate' in the trenches stretching across northern France to the sea *(PRJ)*

Dunquerque (Dunkirk); there the movement of the combatants effectively ceased for the next 4 years. The opposing forces found themselves bogged down in a sea of mud and despair where 'attrition' seemed the only way to achieve a victory. Small intrusions or salients would be forced in particular parts of the line but soon the other side would flatten out the bulge and perhaps create an intrusion in the other direction. In this awful process, thousands of the infantry were slaughtered. Finally, in a desperate attempt to break the stalemate on the Western Front, the German army launched a vast and furious assault against the Allied lines in 1918. This was finally arrested by an Allied army revitalised by the presence of forces from the United States. The exhausted German army then fell back as far as the Hindenberg line and capitulated. The Armistice was signed and the seeds of a later war were unknowingly sown.

The breakthrough — troops advance against the German front line with a tank carrying a trench-crossing device on its forward section *(25Y)*

It was modern armament that created the stalemate, but it was through technology in the form of a new and potent weapon, the tank, that the impasse was broken. Although all the potential of this new weapon was apparent in the first instance, initially its capabilities were squandered, mainly through ignorance and lack of vision by the commanders in the field.

The first use of the new weapon in 1916 clearly showed the way to its proper application. It was not until 1917 that tanks were at last used over appropriately stable ground and in sufficient numbers to make a real impact. In this instance, the bulge created by the Allies was about 65 kilometres (40 miles) in depth but lack of backup meant that the breakthrough was not capitalised on and the bulge was very quickly flattened out again.

Unfortunately, the lesson of this failed attack was learned by the German army rather than the British and then applied with devastating effect 20 years later as the Panzers blasted their way into Poland and demonstrated for the first time the full power of the 'Blitzkrieg' (lightning war).

However, not only in the field of weaponry was the First World War to be the catalyst for dramatic changes and advances. Other apparently benign technologies would also be pressed into service to help to win the war. From such pressure came the developments of aeronautics which later led to commercial passenger aviation. Another was the new science of wireless.

Despite the technical advances that had been made since the invention of wireless in 1895 and the extent to which it had been successfully used at sea, on land and in the army it was a very different story. When the British army went to war in 1914, it was assumed that wireless would be an adjunct of the cavalry. As a result, in the initial stages of the war in 1914, the British army had available a single wireless troop equipped with three horse-drawn Marconi spark transmitters and associated receivers.

With the loss of mobility that resulted from the descent into the trenches, the cavalry were taken off their horses and joined the infantry in the mud and gore of the front line. The wireless wagons remained at headquarters, and only when problems with conventional telegraph lines began to accumulate was the notion of wireless to receive much attention.

WAR AT SEA

At sea in the first few days of the war, an event occurred that was to have the most profound repercussions, not only in the First World War but much later in another desperate conflict. The cable network that had been laid since the pioneering work across the Atlantic in the 1860s had grown significantly in the intervening years, but in one important respect it remained very similar to the original installation.

The majority of cables that crossed the Atlantic still started the eastern end of their journey from the shores of the British Isles and this included cables that had their origin in Germany. Surprisingly, Germany had only five cables that did not follow this path via Britain and the relay station at Porthcurnow in Cornwall. As a consequence, German telegraphic traffic was extremely vulnerable to eavesdropping and interception in the period before the war and this situation continued during the war years. This was a fact that Germany apparently took some time to become aware of.

A Marconi wireless wagon in 1914 *(MAR)*

Within a few hours of the declaration of war by Britain, the cable-laying ship *Telconia* had been despatched to the German coast near Emden and the five

German cables independent of British relaying were dredged up and cut. This left Germany to rely on communications with the United States only via a Swedish cable and one that ran from Denmark. Remarkably, it appears that these cables also ran to the relay station at Porthcurnow, a fact that was not known to the German authorities. This pair of links remained and were routinely monitored by British personnel.

In the absence of the cable links, Germany was forced to rely on the long-wave wireless network that had been established in the years immediately before the war, with Nauen as its centre. However, in the weeks that followed this initial action of cutting the cables, further attacks were made on German wireless relay stations around the world. These were located at Dar es Salaam in Africa, Yap in the Caroline Islands, Nauru and Samoa, Kamina in Togoland, Rabaul, Augair in the Pacific region, and Duala in the Cameroons, also in Africa. These successful Allied actions had the effect of disconnecting Germany from communication with America and all other parts of the world, apart from the few cable links that had been allowed to remain under surveillance unknown to Germany. In addition, the few remaining wireless links were carefully monitored by the British.

The operating position on a wireless wagon with a drop-down work bench, morse key and headphones *(MAR)*

The consequences of this action were considerable. In the first instance, it forced Germany into using radio communication far more frequently for long-distance communication. At a time when the capacity of unintended listeners to intercept such transmissions was generally not as well known as in later times, this was very hazardous. It was to lead to much of the German traffic being intercepted and read by the British including a good deal of material that had been encrypted for distribution to overseas consulates and embassies.

Apart from this long-range wireless traffic, radio was used indiscriminately by the German navy in the early years of the war. Experts at the Marconi laboratories now discovered that signals emanating from the German fleet were able to be received in England. This was made possible by the use of the new valve radio receivers and amplifiers that were being developed by the staff of Marconi. As the first major impact of the amplifying thermionic valve that had been developed from the work of Fleming and de Forest, it was a reflection of the major development work that had occurred under the control of the Marconi Company experimenter, Captain HJ Round.

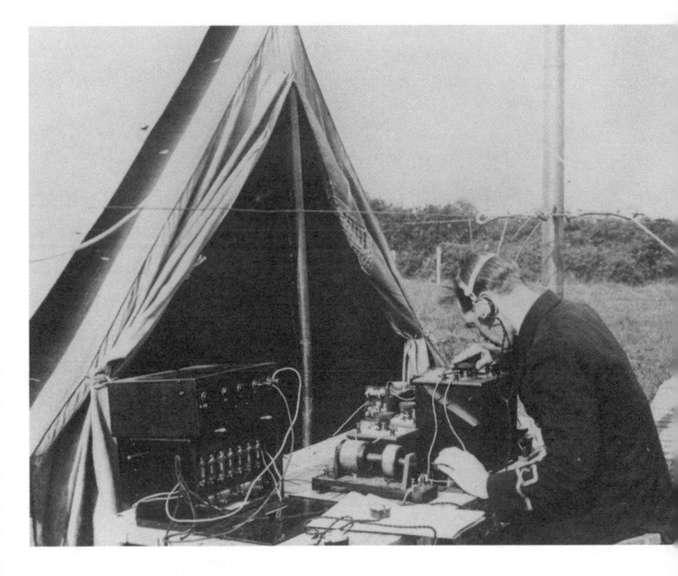

Operating a listening post in the field *(MAR)*

This new source of intelligence about the activities of the German fleet took some time to penetrate the consciousness of the British

Right Kaiser Wilhelm *(DKI)*

Far right Admiral Hall, who appointed Alfred Ewing to run the Naval Intelligence Branch known as 40 OB (Number 40, Old Building) *(DKI)*

Alfred, later Sir Alfred, Ewing, head of Naval Intelligence, 1914–18 *(40B)*

Admiralty. Because of this failure of perception and understanding, the potential for a massive victory at sea over the German fleet was largely wasted at Jutland. In fact, this rather untidy action was claimed by both sides as a victory, although to the latter-day reader it has the appearance of a major 'shambles' which did no one credit and certainly represented no advantageous application of the new radio technology.

NUMBER 40, OLD BUILDING (OB)

The other major implication of the action by the Royal Navy in cutting the German cables was to see a major increase in the use of coded communications. Later, ciphering techniques would become more common, both on land and in the field of battle. With this expansion would come the realisation that the means to break into enemy ciphers and codes was of critical importance and would provide intelligence of considerable value. The British response to this new development in intelligence gathering was to set up a specialist department in the Admiralty which was located in Room 40 of the old Admiralty building.

To run this new organisation, the Director of Naval Intelligence, Admiral Reginald Hall, known as 'Blinker' because of a facial tic, installed Sir Alfred Ewing who was, at the time, director of naval education based at Dartmouth. His new department, usually referred to as 40 OB, soon became an official subsection of naval intelligence and was known more formally as NID 25. This department was destined to become a most effective cipher-breaking organisation and later provided the nucleus of the Government Code and Cypher School in the years after the war. This organisation will be described at a later stage in the story of the World Wide Web.

Although 40 OB was responsible for many successful interceptions during the First World War, it is perhaps the Zimmermann telegram that stands out as a most significant piece of work. Deciphered by Nigel de Grey in January 1917, this material had the effect of precipitating America's entry into the war, thereby helping to end the stalemate on the Western Front.

THE ZIMMERMANN TELEGRAM

Reading a copy of this document as ultimately deciphered reveals why it received such an explosive and indignant response when its text was revealed to the citizens of the United States of America. What is even more remarkable to a modern reader is that when the text was revealed, Arthur Zimmermann, the German foreign minister, was to admit its authenticity. Previous suspicions that it had been part of a British deception to bring America into the war were immediately swept aside and the US President was able to take the crucial step that would ensure that American soldiers would go to the aid of the Allies on the Western Front.

The text of the telegram which Zimmermann sent to the German ambassador to Mexico, von Eckhardt, in January 1917 was as follows:

Arthur Zimmermann, who sent the infamous telegram that brought the United States into the war in Europe *(40B)*

> We intend to begin on the first of February unrestricted submarine warfare. We shall endeavour in spite of this to keep the United States of America neutral. In the event of this not succeeding, we make Mexico a proposal or alliance on the following basis: make war together, make peace together, generous financial support and an understanding on our part that Mexico is to reconquer the lost territory in Texas, New Mexico and Arizona. The settlement in detail is left to you. You will inform the President [of Mexico] of the above most secretly as soon as the outbreak of war with the United States of America is certain and add the suggestion that he should, on his own initiative, invite Japan to immediate adherence and at the same time mediate between Japan and ourselves. Please call the President's attention to the fact that the ruthless employment of our submarines now offers the prospect of compelling England in a few months to make peace.

With the arrival of new forces from across the Atlantic in 1917, the German army was finally beaten back and forced to capitulate. There followed the armistice in November 1918 that was later so bitterly resented as a betrayal of the German people.

It appears that there is a very unfortunate habit of British bureaucracies and government not to give proper recognition of effort by individuals. Despite his immense success, Ewing finished his war with effectively no acknowledgment of the role that he had played in defeating the German forces. In an incautious but entirely understandable fit of pique, Ewing delivered a lecture to a selected audience in 1927 on his 'war work', and suddenly the German navy and army realised what had been the basis of some of their most puzzling defeats during the years up to 1918.

The publication of this secret material was viewed with great dissatisfaction by the authorities and subsequently Ewing was forced to give an undertaking that he would publish no further revelations. He remained in England and contained his dissatisfaction until the end of his life.

A military field radio station including antennae and pedal-operated power source *(MAR)*

Whatever the justification, the revelations of Ewing were to have the most disastrous consequences for they immediately confirmed a developing German interest in mechanical cipher machines. At about this time, the German army had purchased a number of the early version of a commercial enciphering machine known as the 'Enigma' with a view to achieving communications that would be entirely incapable of decipherment. This will be discussed later as a prelude to the age of the computer.

WAR IN THE AIR

Somewhat surprisingly, it appears that the first effective use of wireless in the First World War was to occur from the air to the ground. The 'string and wire' aircraft of the Royal Flying Corps, at that time a branch of the army, were equipped with a few small spark transmitters manufactured by the Sterling Telephone Company, a subsidiary of the Marconi Wireless Telegraph Company, to be used for artillery-spotting purposes. These transmitters drew a power of about 30 watts from the aircraft power supply and transmitted via a wire that was allowed to trail behind the aircraft once it had left the ground.

The airborne transmitters were operated in conjunction with ground stations that initially consisted of carborundum crystal sets made by the firm ATM Ltd. Known as the Mark 3 Receiver, this device later had a valve amplifier added to it, which was designed to increase the range at which the aircraft signals could be received. Very few of these valve amplifiers were actually brought into service, no doubt because the basic crystal set was quickly superseded by valve receivers with significantly greater sensitivity without supplementary amplification.

Impressed by the success of these arrangements and in the face of problems with wired telegraph and telephones in the trenches, arrangements were made to try some of this equipment at ground level. During 1915, in conjunction with a 'lorry' or 'wagon' set, a number of

Below left The Sterling aircraft transmitter of 1914 used for aerial gunnery spotting *(PRJ)*

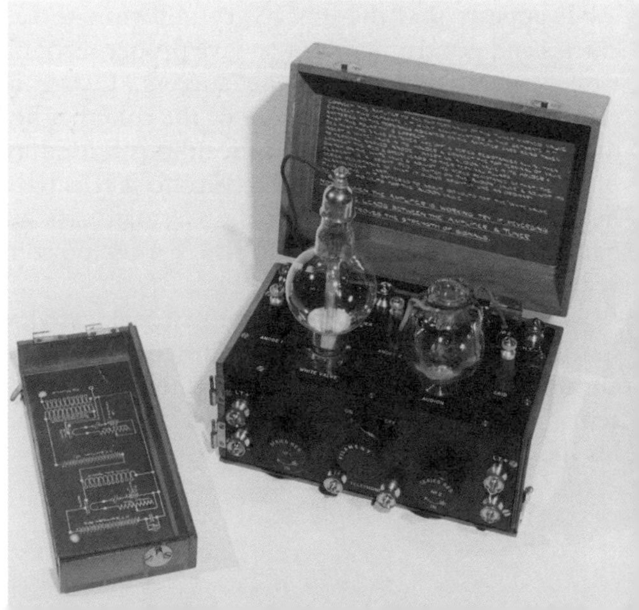

Below right Crystal set amplifier Mark 3 with a White valve and a de Forest audion *(PRJ)*

Sterling transmitters and ATM receivers were set up to provide communication between 5 Corps headquarters near Ypres and the front line. Due to interference with the artillery spotting apparatus in aircraft, the wavelengths to be used by the army and Royal Flying Corps were then assigned. The Royal Flying Corps was required to operate on 300 metres and the army was given 350 metres, 450 metres and 550 metres. At this time, this range of wavelengths was said to lie in the 'short' waves.

Royal Flying Corps aircraft lined up ready for battle *(25Y)*

Despite the success of these early airborne radio activities, they all suffered from a major problem, the use of morse code. For a pilot already busy concentrating on the skies and looking for the enemy, tapping out morse was not only slow but potentially hazardous and it did not take long for a demand for telephony to develop. Based on a request by the commanding officer of the Royal Flying Corps (RFC), Major-General Trenchard, experiments began and apparatus with a range of about 2 kilometres (1 mile) were flown in aircraft to demonstrate what was possible. The development work was placed under the control of Major Prince of the RFC who liaised with the Marconi Company which had been commissioned to produce the radio hardware.

At this time, the Royal Engineers were responsible for radio installations, even in aircraft, and it seems that Major Prince had some difficulties obtaining a satisfactory system until he enlisted the support of another senior RFC officer, Major Dowding. This was a fortunate relationship in the longer term, for although radio telephony in aircraft made rather little progress in the First World War, at a later time it was to prove one of the vital elements in avoiding defeat in the Battle of Britain. The supportive major was later better known as Air Chief Marshall Hugh Dowding, who was in overall control of the fighter squadrons during this later conflict and who relied on radio telephony to direct the movements over the English Channel and inland.

Far left Crystal set Mark 1 as used with the Sterling transmitter for aerial reconnaissance and artillery spotting *(PRJ)*

Left Military crystal set as used in the trenches and now located at the Marconi archives, Great Badow, near Chelmsford *(PRJ)*

Top Valved aircraft transmitter *(PRJ)*

Bottom Typical early trench transmitter and receiver *(PRJ)*

RADIO DEVELOPMENTS

By the start of the conflict in 1914 there had already been profound changes in the technology of radio compared with the simple apparatus sent to South Africa for the Boer War. Perhaps more importantly, some of the fundamentals of radio propagation were much better understood so the problems encountered in that earlier exercise no longer remained unsolved and communications could be achieved even under the most difficult conditions. Among other things, earth mats of copper wire were provided to the signallers to compensate for inadequate or excessively dry and non-conductive ground.

When, in the trenches of Gallipoli and later the Somme, the freshly laid telephone lines were torn up by concentrated shellfire, some other method of communication had to be found. Unsurprisingly, a 'wireless' solution was intrinsically appealing even if the antenna masts did tend to become targets of enemy fire. Beyond that, it was also discovered that the enemy was able to listen to conversations on the telephone lines without a direct connection being made, using newly developed valve amplifiers. The supposed security value of wired transmissions was therefore an illusion, and 'broadcasting' of messages by wireless just as insecure. Encoding of messages then became the accepted means of securing communications for both wired and wireless-distributed messages.

Pressure for improvements to radio to make it simpler and more reliable saw a series of significant changes. In particular, the spark was rapidly displaced by the newly available valves based on French designs, with hard vacuums as compared with the earlier audions of de Forest. By 1916 such valves were being used to provide continuous wave (CW) signals, with the attendant advantages of much greater power efficiency, improved selectivity and reduced interference and, as a result, much greater range of operation. Despite the advantages of the new valve sets, it appears that they were extremely prone to faults of many different kinds and army signallers had to become much more than simply morse code senders and receivers.

Early valved trench transmitter and receiver *(PRJ)*

During the 4 long years of the First World War, the old spark telegraphy became obsolete and was replaced by valve or tube-based radio. In almost every area of radio, the advances were profound, and by 1920 radio was able to be turned to the provision of popular entertainment because of the advances that had been made in that war period. Many of the improvements that came from the adoption of the valve for transmission and reception and the refinement of radio were to have profound repercussions both during the war and shortly after. Broadcasting for the public was only one of these later developments.

Lord Balfour with Winston Churchill in 1916 *(40B)*

GALLIPOLI

Despite such technical advances and in spite of the best efforts of the army, by the end of 1915 the opposing forces had ground to a complete halt and the terrible 'stalemate' of the Western Front was established. In the face of this impossible situation, with no clear way out of the impasse, the First Lord of the Admiralty, Winston Churchill, cast around for a means to break the 'deadlock'. As a student of history and no doubt being aware of the part that the Gallipoli Peninsula had played in the Crimean War, it was to this location on the boundary between Europe and Asia that his gaze turned. Here perhaps was a place to attack that would prove to be the 'soft underbelly' of the German army and an easier target than the Somme.

For all Australians, Gallipoli has come to represent what is frequently referred to as an 'icon', however incorrect such an expression is to convey the notion of a landmark episode in history. Perusal of the many documents and books that describe this epic and disastrous campaign makes it clear that it was an operation in which the British, French, Australians and New Zealanders were all to play a part and in which the Anzac Cove landings were but a component of a larger action undertaken on a number of fronts. However, given its place in the short history of Australia and New Zealand, most people in these countries have a fair idea of where Gallipoli is and what it was like. For the descendants of other participants in the campaign, time and distance may well have made this place less than immediately familiar.

The merest glance at this long finger of land running down in a southwesterly direction evokes the reaction 'Surely an attack on such a place would have to be a military impossibility'. One would imagine that the difficulties of defeating a well-prepared and entrenched enemy on a narrow front would have been self-evident. Something so fundamental seems to have been ignored and perhaps it was the use of the navy to run the waters of the Dardanelles initially that was somehow to commit the land troops to the peninsula. The naval attack was a failure and no doubt this was to doom the subsequent land attack to failure as well. For a fine word picture to supplement the photographs in this chapter, one can turn to the words of John Masefield, the author of the book *Gallipoli*:

> The Peninsula of Gallipoli, or Thracian Chersonese, from its beginnings in the Gulf of Xeros to its extremity at Cape Helles, is a tongue of hilly land, about fifty-three miles long, between the Aegean Sea and the Straits of the Dardanelles. At its north-eastern, Gulf of Xeros, or European end, it is four or five miles broad; then, a little to the south of the town of Bulair, it narrows to three miles, in a contraction or neck which was fortified during the Crimean War by French and English soldiers. This fortification is known as the Lines of

The First Sea Lord, Lord ('Jacky') Fisher *(40B)*

Below left Anzac Cove as seen from the heights above *(GAL)*

Below right Map of the Gallipoli Peninsula *(PRJ)*

Bulair. Beyond these lines, to the south west, the Peninsula broadens in a westward direction and attains its maximum breadth, of about twelve miles, some twenty-four miles from Bulair, between the two points of Cape Suvla, on the sea, and Cape Uzun, within the Straits. Beyond this broad part is a second contraction or neck, less than five miles across, and beyond this, pointing roughly west-south-westerly, is the final tongue or finger of the Peninsula, an isosceles triangle of land with a base of some seven miles, and two sides of thirteen miles each, converging in the blunt tip (perhaps a mile and a half across) between Cape Helles and Cape Tekke. There is no railway within the Peninsula, but bad roads, possible for wheeled traffic, wind in the valleys, skirting the hills and linking up the principal villages. Most of the travelling and commerce of the Peninsula is done by boat along the Straits, between the little port of Maidos, near the Narrows, and the town of Gallipoli (the chief town) near the Sea of Marmora… The seashore, like the Straits shore, is mainly steep, with abrupt sandy cliffs rising from the sea to height of from one hundred to three hundred feet. At irregular and rare intervals these cliffs are broken by the ravines and gullies down which the autumnal and winter rains escape; at the sea mouths of these gullies are sometimes narrow strips of stony or sandy beach… Viewed from the sea, the Peninsula is singularly beautiful. It rises and falls in gentle and stately hills between four hundred and eleven hundred feet high, the highest being at the centre.

In the Cape Helles district it is mainly poor land growing heather and thyme; farther north there is abundant scrub, low shrubs and brushwood, from two to four feet high, frequently very thick. The trees are mostly stunted firs, not very numerous in the south, where the fighting was, but more frequent north of Suvla … Viewed from the sea the Peninsula looks waterless and sun-smitten; the few watercourses are deep ravines showing no water… The soil is something between a sand and a marl, loose and apt to blow about in dry weather when not bound down by the roots of brushwood, but sticky when wet.

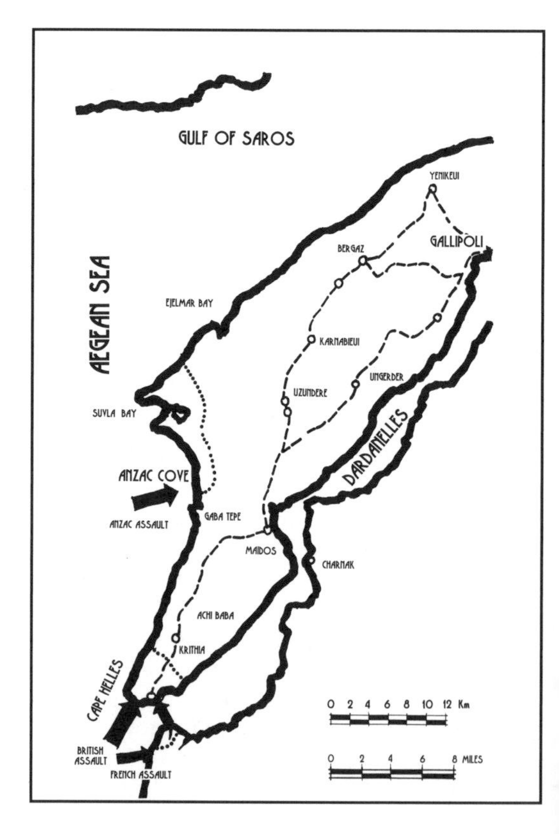

Notwithstanding the visually appealing attributes of this peninsula, in a military sense it was inevitably a most difficult proposition to wrest it from its Turkish defenders. For months before the landing, activities of the Royal Navy in the Gulf coupled with an almost unbelievable lack of secrecy by the army had allowed the Turkish army to be about as well prepared for invasion as could be imagined. As a result, a formidable system of trenches and fortifications awaited the invaders as they leapt ashore from their boats.

The troops on the beach at Anzac Cove in 1915 *(UNK)*

Eight months later, when it was brutally apparent that a breakout from the beach heads and the immediate hinterland was never likely to succeed, the decision was made to abandon the attack. With commendable skill, the Allies then removed themselves from the unconquered peninsula and sailed away to other battlefields even more appalling than Gallipoli.

If nothing else, the lessons of Gallipoli were to be applied over 30 years later in the preparations for the landings in Normandy in the Second World War. Normandy was a huge success where Gallipoli had been a dreadful failure.

Anzac Cove as seen from the sea in 1916 *(GAL)*

SIGNALS AT GALLIPOLI

The distribution of the invading forces coupled with the physical form of the peninsula made communications difficult from the very outset. Communicating across the rocky and inhospitable foreshores and hinterland of the Gallipoli Peninsula was initially undertaken with conventional means. Telephone lines were strung around the battlefield and, where this was not practical, despatches were distributed by hand. However, with the loss of cables to shellfire, the possibilities of wireless

Australian troops rush into Turkish machine gun fire at Gallipoli *(GAL)*

A view of Anzac Cove from above in 1996 *(WIN)*

were considered and soon the newly developed trench radios were brought up. With short antennae, short-range communications were possible.

Among important wireless innovations on the battlefield was the introduction of a system of electricity generation using a tandem bicycle frame. Given the inefficiency of spark transmitters, producing the energy to operate them must have been exceedingly hard work even for two infantry men. Among the heat and the dust and flying shrapnel, pedalling hard enough must have been a particularly thankless task. However, it may well have sowed the seed of an invention that was to hatch in the fertile mind of a young Australian radio amateur at a later time, Alfred Traeger.

As photographs of the period reveal, here was a particularly unforgiving terrain even as compared with the Somme and the Western Front. Modern photographs of the area demonstrate even more clearly the ground over which the Allies had to operate — a dry and extremely harsh environment, with scrubby plants and grass providing virtually no 'cover'. Trenches were an inevitable response to what the invaders encountered and in the heat and dust they were dug as the best response to flying lead and bursting shells.

Months of futile activity were to end in a tactical withdrawal which, in its skilful handling, ensured that the remaining troops were not exposed to fire during their departure. In that respect, what was achieved in this manoeuvre made an extraordinary contrast with the 'shambles' of the initial assault. Perhaps if the same careful and deliberately deceptive techniques had been used at the outset, the outcome of the Gallipoli campaign could have been quite different.

The view north towards Suvla Bay from above Anzac Cove in 1996 *(WIN)*

THE WESTERN FRONT

Despite the months spent in the disaster of Gallipoli, the troops were soon deployed back to one of the other crucibles of the First World War, the Somme and the area around Ypres. Forming a small part of the full extent of the Western Front, the area around Ypres seemed to attract the most extraordinary expenditure of German and Allied effort. At this place, the front made an elongated salient towards the north, and for the best part of 4 years the German army fought to straighten out the bulge known as the Ypres Salient. Every metre that was achieved in advancing south by the Germans was soon reclaimed by the most extreme expenditure of effort and lives by the Allies. Perhaps only at Verdun, located at the eastern end of the Western Front, was such concentrated savagery to be seen in attempts by both sides to achieve realignment of the front line.

Above The camp at Suvla Bay and Anzac Cove during the campaign *(GAL)*

Below The cliffs behind the beach at Anzac Cove in 1996 *(WIN)*

Above left On the Somme waiting in the trenches to go 'over the top' *(BOS)*

Left In the trenches after going 'over the top' *(BOS)*

As photographs show, the result of this activity was to reduce the thriving and historic town of Ypres to rubble. Unbelievably, the industry of its inhabitants has allowed Ypres to be rebuilt so that the devastation visible in 1918 is now completely hidden. Without the evidence of contemporary photographs it would be very hard to imagine that once, many years ago, this small Belgium town, now known as Ieper, had been destroyed about as effectively as was Hiroshima at a later time.

PORTABLE COMMUNICATION

At the start of the war, the wireless equipment suitable for the trenches was generally very heavy and clumsy and required a minimum of four strong soldiers to transport it. The various elements of the 'K' type knapsack set had been developed by Marconi as a supposedly portable radio station, but in practice this was a somewhat optimistic description. Four strong men were required to carry, respectively, the antenna mast and battery, copper earth nets, the transmitter and receiver as separate units and, lastly, the aerial gear.

This apparatus is described in the *Marconigraph* of August 1912 as follows:

> The transmitter, which consists of an ordinary ignition coil requiring a pressure of only six volts, is contained in a square wooden box weighing 11 pounds. The receiver consists of the ordinary carborundum receiving circuit with jigger, tuning condenser, and four dry cells, which may be switched on and off as required. This is also contained in a box which weighs six pounds.

Top The Belgium town of Ypres (Ieper) in 1913 *(YM)*

Bottom Ypres in 1918 with the remains of the Town Hall in the distance *(YM)*

Ypres Town Hall in 1997 *(PRJ)*

British soldiers in the trenches near Ypres in 1917 *(YM)*

By the time of Gallipoli, a radio transmitter that drew some 50 watts of power had been developed and was known as the 'BF' set, suggested to be related to the level of 'foolishness' required for the acceptable operation of the apparatus. Working over a range of frequencies centred on 450 metres, it was officially known as the British Field Set and was generally successful in operation. In tribute to its unofficial name, the set proved simple to operate in the field.

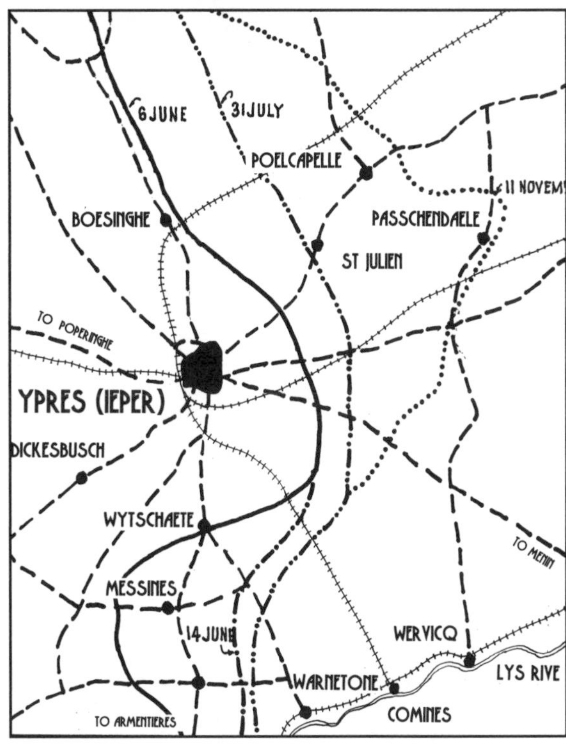

The moving front line at Ypres in 1917 *(PRJ)*

Built into a wooden box, the transmitter and receiver weighed a healthy 14 kilograms (31 pounds) and were powered from a 10-volt battery which was also a substantial load for the long-suffering signaller. With 5-metre (15-foot) masts for the antenna, the set was able to communicate over a distance of about 4 kilometres (2½ miles), depending on the level of spark-generated interference received from other radios in the vicinity. Sparks were intrinsically broad-spectrum devices and electrically very noisy so that interference was almost inevitable.

This early cumbersome spark set was to be replaced within 2 years by apparatus which used the newly developed valves in the receiving side, although sparks were still at the heart of the transmitter. Such sets operated on a much shorter wavelength at around 80 metres or a frequency of 3.5 megahertz. For that reason, much shorter antennae could be used than for the earlier 'BF' set operating on 450 metres. The complete system consisted of 'forward' and 'rear' stations operating on different frequencies of 65 and 80 metres. By this means, mutual interference was to some extent removed.

Given the paucity of wireless apparatus available to the British army at the start of hostilities, it is surprising to discover how well equipped the Australian Imperial Force was by comparison. Shortly after war had been declared, the Australian government purchased six 500-watt pack sets from the Marconi Company to be distributed to the States on the basis of one set to Queensland, two sets each to New South Wales and Victoria and one set to South Australia.

The prototype of the Marconi military pack set as used at Gallipoli; located at Great Badow *(PRJ)*

A formidably cumbersome array of equipment came with the basic transmitter and receiver components, weighing in total 250 kilograms (550 pounds) and designed to be carried by four packhorses. In addition, it was anticipated that six signallers would be required to set up the wireless station.

The transmitter consisted of a rotary spark machine coupled by a drive shaft to a Douglas 2½-horsepower water-cooled motor, set on either side of a metal pack frame. Needless to say, the drive shaft could only be installed once the pack was off the horse. The transmitter was designed to be set up some distance from the receiver cabinet which was intended to be set on top of the alternating current transformer. This was fed with low-tension alternating current from the drive motor. This low-voltage current was keyed at the receiver cabinet and the high-tension leads taken back to the transmitter.

The rotary spark plate was equipped with 24 studs and fed with 50–60 hertz alternating current at high voltage. This allowed six sparks of varying amplitudes to occur every half cycle and, as a result, the station produced a 720-hertz 'singing' note. This was sent on a wavelength of 700 metres, although the apparatus was designed to be operated anywhere between 300 and 1000 metres.

This station used two 10-metre (33-foot) steel sectionalised masts set 100 metres (330 feet) apart with a simple Marconi antenna taken down to the transmitter and receiver. The receiver used the conventional and reliable if somewhat insensitive carborundum crystal detector which required a local battery to set the operating bias for the device. The voltage delivered to the crystal was in turn controlled with a potentiometer.

The two sections of the Marconi pack set as used at Anzac Cove and Cape Helles; operating console on the left and power supply and rotary spark unit on frame for packhorse mounting on the right *(MAR)*

Installation of the apparatus so as to commence work apparently could be achieved in 5 minutes in an emergency, although it seems that 7 minutes was the usual time to begin signalling. Messages were always sent using 5-letter groups and were enciphered. Plain text was strictly forbidden.

Following the declaration of war, the First Signal Troop, Australian Engineers, attached to the First Light Horse Brigade, First Australian Imperial Force, was formed on 19 August 1914. This force departed for the Middle East on 20 October 1914, and after a 6-week journey arrived in Egypt where desert training was undertaken.

With the decision to undertake the Gallipoli campaign, two wireless sections from the First Light Horse and two sections from the Second Light Horse were attached to British Headquarters to attack the peninsula at Cape Helles on 25 April 1915. During the landings, this station was quickly established and provided contact and ranging advice to the naval gunners supporting the landings.

The 500-watt pack set was designed to have a daylight range of 55 kilometres (35 miles) but, in practice, the signallers found that they could frequently achieve three times that range. At night time it was found that ranges of between 300 and 500 kilometres (200 and 300 miles) were not uncommon.

Following the Gallipoli campaign and the transfer of the Australian and New Zealand troops to the Western Front, a number of wagon sets were purchased by the Australian government. These sets incorporated 1.5-kilowatt rotary spark transmitters and required the double-limbered wagons to be drawn by six horses. By comparison with the pack set, the wagon set was powered by a 7-horsepower Douglas engine.

Major Armstrong in France in 1918 at the time that he invented the superheterodyne system of radio reception *(PRR)*

AMERICA AT WAR

The United States entered the war in 1917 and there can be no doubt that this was the vital factor in persuading the German generals that they must seek an armistice. This occurred in November 1918 and saw the end of one of the bloodiest episodes in history. However, in that last 2-year period, there was one technological advance that was to have a most profound impact. As a basis for the reception of radio signals, this development is still in evidence to the present day. This was the discovery of the 'supersonic heterodyne' principle of radio detection and amplification. Its discoverer was Edwin H Armstrong (whom we met in an earlier chapter), who had found out about regeneration (also called feedback) and continuous oscillation.

In the early part of 1917, before going to France with the US forces, Armstrong had delivered a paper on the rather arcane subject of heterodyne mixing in valves. Later that year, he joined the army as a captain and in France had the need to find more sensitive ways to detect enemy aircraft. What came out of this need was the 'supersonic heterodyne', or 'superhet' as it is now more commonly called.

Later there was to be some controversy over who had been the primary inventor. A French army soldier, Lieutenant Levy, had also been experimenting with something similar to the superheterodyne principle, but ultimately Armstrong was accepted by the French government as the being the prior holder of the patent rights.

In many respects, apart from the technical development of the valve, the most important radio development of the years 1914–18 is seen as being the 'superhet'. As will be discussed at length later, this was to be a device that would ultimately see application up to the present day and in all radio receiving devices at virtually every part of the radio frequency spectrum. Truly were 'swords' of invention of the First World War beaten into the 'ploughshares' of radio broadcasting in later times.

RECEPTION BEFORE THE VALVE

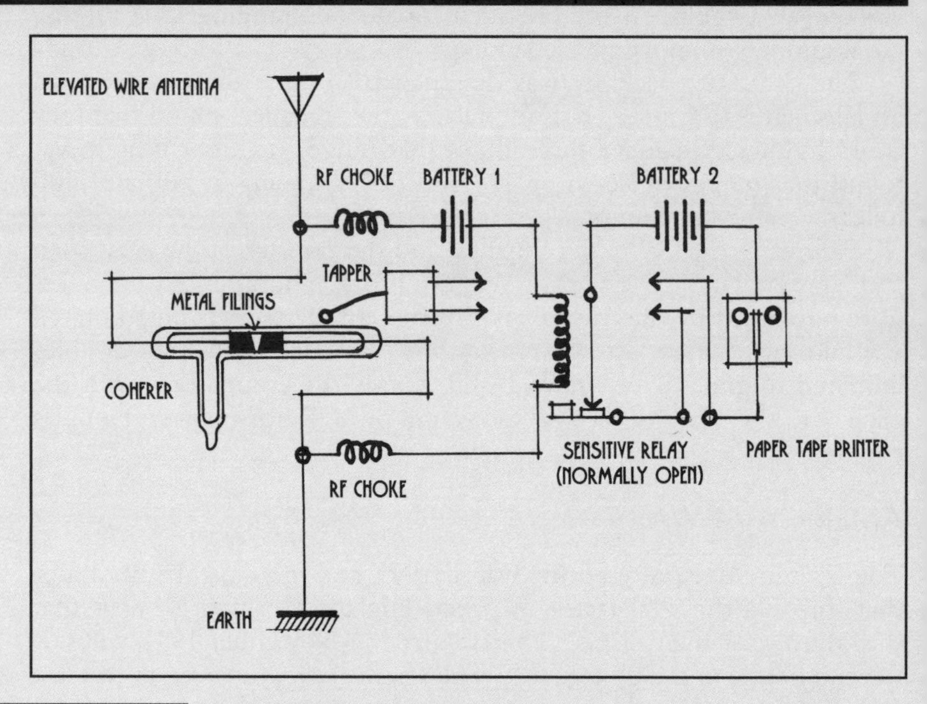

Schematic diagram of the first coherer receiver of Marconi, 1895 *(PRJ)*

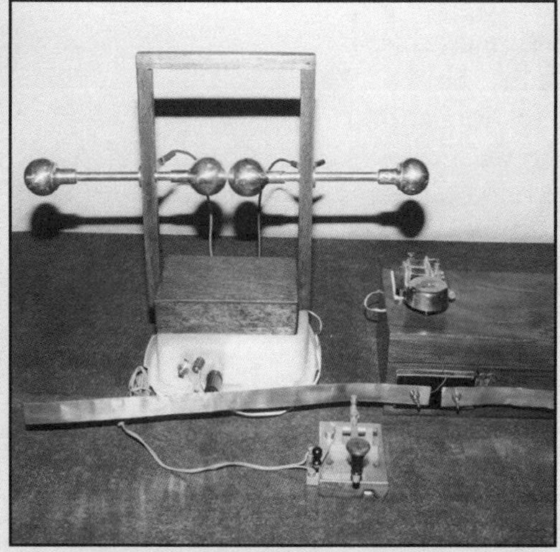

Replica of the first Marconi transmitter and receiver of 1896 *(PRJ)*

To the present generation of electrical experimenters and communicators, it may be hard to imagine a time before the use of solid-state devices, but for the best part of 60 years the vacuum in a bottle — the thermionic valve or tube — provided the basis of telegraphic and telephonic communication both by radio and through wires.

It may therefore be a surprise to learn that, in that remote time at the beginning of the radio era and before the invention of the valve, there was another significant period when radio waves were detected by a variety of, to modern eyes, quite peculiar devices. Of these, the coherer, magnetic detector and crystal detector are perhaps the most important, although a surprisingly large array of different methods was assembled following the experiments of Heinrich Hertz in 1886 and before the exclusive use of the valve for detection commenced in about 1916.

THE COHERER

In his vital experiments of 1886, Hertz made use of simple loops of wire with spark gaps to detect the radio frequency (RF) energy from his induction coil transmitter. A minute spark gap in the ring, which was adjustable with a screw-threaded contact and viewed with a magnifying lens, allowed him to see the tiny spark when resonance was achieved. This evidently was not a very easy or attractive device to use for experimental purposes, let alone for signalling, and other experimenters looked for better means to detect the RF energy.

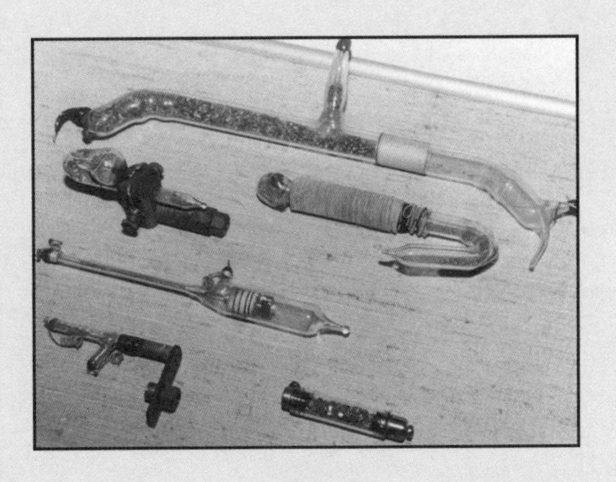

A collection of coherers as used by Oliver Lodge in 1894 *(PRJ)*

In even earlier times and as a means of protecting telegraph lines from lightning strikes, the scientist SA Varley had invented a device that was based on a peculiar property of granular conducting material in the presence of RF energy. When placed between the poles of a circuit including a battery, under normal circumstances granular metallic filings exhibit a very high resistance which appears to a low-impedance relay much like an open circuit switch. However, if a high-voltage pulse is created in the vicinity of the filings, with great suddenness they will stick together or cohere and the resistance through them drops to a very low value of ohms so as to appear almost a short circuit.

This effect was investigated in the late 1880s by a number of experimenters, such as the French scientist Professor Branly, the British scientist Oliver Lodge and the Russian scientist Aleksandr Popov. From this work came a primitive RF-activated switch which was known as a 'coherer', the name given to it by Oliver Lodge.

Although a pulse of RF energy would turn this switch on, once the particles of metal had 'cohered', the switch was locked on. The way to unlock it and open the circuit was to physically 'de-cohere' the metallic particles and this was done by lightly tapping the glass tube in which the particles were contained. Soon this was done by the trembler of a bell mechanism activated by the relay in circuit with the battery and coherer — an early form of 'feedback' control.

The coherer receiver as made by Captain Henry Jackson in 1896 *(PRJ)*

The action of the early receiver was therefore to turn on a relay in the presence of a radio signal which in turn activated a bell mechanism which immediately turned the coherer switch off again. A burst of RF energy, whether short or long, would create a series of very short pulses as the coherer was switched on and off by the relay and bell mechanism in the circuit. These short pulses would join together to synthesise the longer burst of RF energy as audible noise in headphones or as a continuous streak on a papertape morse code recorder as used in terrestrial telegraphy. However, this arrangement was intrinsically limited in terms of the speed at which morse code could be received. It was rarely possible to exceed about ten words a minute using a coherer and, even compared with terrestrial telegraphy of 1895, this was very slow.

The coherer was an incredibly crude and insensitive device by modern standards and one has to wonder how such a system worked at all. For the first 10 years of radio, the coherer was the basis of most commercial systems until displaced by other more sensitive devices after about 1905.

THE MAGNETIC DETECTOR

The RF switch provided by the coherer proved quite incapable of detecting the minute energy of the signal sent from Poldhu in 1901 and, in fact, the signal was received on a primitive 'solid-state' device, the Italian naval coherer. This used a globule of mercury and a plug of iron as the poles of what at that time was called a self-restoring coherer. In other words, one did not have to tap it to make it detect after the receipt of each telegraphic symbol. It is now known that this detector was actually operating as a solid-state diode.

Replica of the magnetic detector designed by Marconi in 1902 *(PRJ)*

It is clear that Marconi came back from Newfoundland dissatisfied with the ordinary metal-filings coherer that he had also used and immediately set to work to find something a lot more sensitive and robust. Basing his experiments on the work of the scientist Ernest Rutherford in the field of moving magnetic fields, Marconi produced a device that he called the magnetic detector. To radio operators over the next 15 years, it was to be known more familiarly as the 'maggie'.

The magnetic detector looked rather like an old fashioned reel-to-reel tape recorder, a device that would also appear quite unfamiliar to many people brought up on the technology of digital CDs. A continuous wire rope travels around two reels passing through a double coil of fine wire and the poles of two horseshoe-shaped magnets. The reels are driven by a handwound clockwork mechanism and if the operator forgets to wind up the device it stops detecting.

One coil goes to a set of earphones and the other connects the antenna system to earth. When an RF signal passes through one coil of wire, a separate burst of energy at audio frequency is induced in the coil attached to the headphones and the signal becomes audible.

Interestingly, despite modern investigations of the mechanism of this device, the exact physical process remains the basis of argument among experts. In this regard, a number of useful discussions are referred to in the list of references. Of these, the most recent by O'Dell is probably the most useful, although it appears that his explanation is not accepted by everyone.

Although the magnetic detector worked best with the signal produced by the discharge of a spark, it was not entirely insensitive to other forms of RF energy which were more like the continuous wave energy that is now used for radio transmission. For this reason its use persisted until spark was finally abandoned in the period around 1920 and full continuous wave transmitters using valves became the norm. In passing, it was the primary method of reception on the *Titanic* although, by 1912, there was available a receiver that used a Fleming diode valve. It appears that even at this stage the 'maggie' remained the preferred device for normal operations.

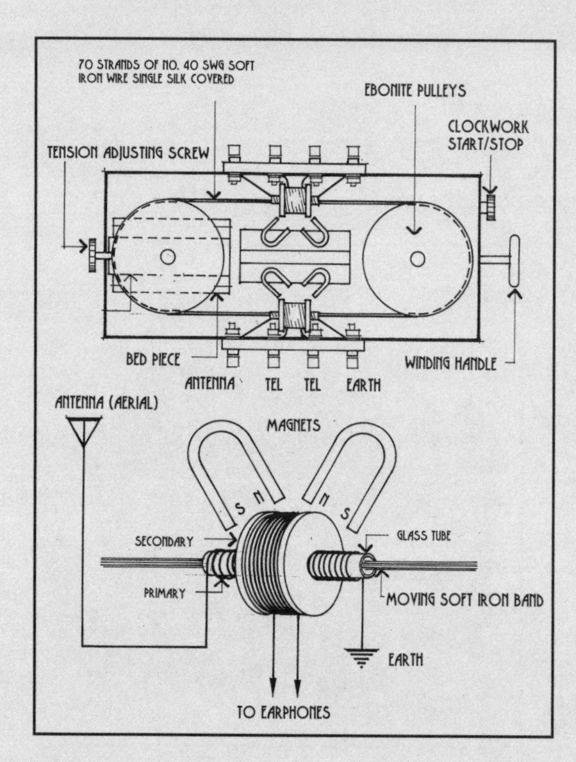

Schematic diagram of the magnetic detector of 1902 *(PRJ)*

CRYSTAL DETECTORS

In many respects the solid-state detectors that appeared after 1905 represent the most interesting of the pre-valve detecting devices. Although largely replaced by valves after about 1924, the crystal detector was to have a long and meritorious life, not only as the detector in much commercial radio telegraphy but also in the first generation of radio receivers as used after about 1920. From then until 1940, the crystal detector was generally set to one side as an unused relic of a former age.

With the development of radar, it was found that the solid-state diodes were extremely suitable for detecting the very high frequency energy reflected from distant aircraft and other targets. Many thousands of solid-state diodes

were made for this purpose and a great many were used by young experimenters after the Second World War for crystal sets to receive radio programs. For this experimenter, 'Dick Barton, Special Agent' via a radar diode and headphones was the gateway to a lifetime of communications activity.

The first successful crystal detector was based on a synthetic material, carborundum, and was discovered by the American experimenter General Dunwoody. Although stable and reasonably sensitive, the carborundum crystal detector required a battery to set the operating point of the device. For this reason, although it was used in a number of commercial radio systems, including those of Marconi and Telefunken, other methods were pursued and led to the realisation that a large number of naturally occurring minerals would operate to detect RF energy. Of these, undoubtedly the best known is the so-called cat's whisker which used a coiled silver wire in contact with a crystal of galena or silver lead sulphide.

Apart from the galena crystal, other combinations included the use of iron pyrites (iron sulphide), bornite (copper iron sulphide), molybdenite (molybdenum sulphide) and silicon. In addition, a dual crystal arrangement with a crystal of zincite (zinc oxide) in contact with a crystal of chalcopyrite and known as the perikon detector was very stable and easy to use as well as being quite sensitive. Stability was an attractive feature of this combination because, in other point-contact detectors, the slightest 'nudge' would lose the sensitive spot on the crystal and reception would cease.

Top row (left to right):

Replica of the multiple tuner designed by CS Franklin in 1904 *(PRJ)*

A multiple tuner manufactured by AWA in Australia *(PRJ)*

A cat's whisker detector unit from the 1920s *(PRJ)*

Bottom row (left to right):

A cat's whisker crystal set replica *(PRJ)*

A coherer receiver made by the Marconi Company for use in the trans-Tasman Sea experiments in Australia in 1905 *(PRJ)*

A Siemens coherer receiver, captured from the Boers in 1901 by British forces in South Africa *(PRJ)*

CHAPTER 7

1918 to 1922

Radio dinosaurs

By the end of the First World War, technical advances had created all the new devices that would be needed for the next generation of radio communication. However, it was apparent that the big organisations such as Marconi and General Electric in the United States had invested enormous capital sums in developing and refining the technology of 1912. Given this and despite the impending obsolescence of the alternator and the rotary spark gap as generators of radio frequency energy, many transmitters had been built based on such principles and the urge to press them into service was inevitable.

At the Marconi Company, the technology of spark had been progressively developed with the implicit intention of producing purer and less energy-wasteful radio frequency waves. This would enable the spark energy produced to be more like the 'continuous wave' defined by researchers in the radio field as desirable. By comparison, based on the theoretical and practical work of Fessenden, General Electric, with the technical guidance of engineer Ernst Alexanderson, had developed the alternator to a highly reliable producer of pure sine wave radio frequency energy. In the United States also, the use of the Poulsen arc transmitter had been improved and developed by Elwell to produce high power and relatively pure radio waves. This had made it possible to create long-range communications links that had been used by the American navy during the last few years of the war.

Looming over all this development was the new wonder device, the valve (or tube). Silent in operation and having about the same efficiency as the Telefunken quenched spark when operating at its optimum, the valve did not have the highly undesirable attributes of the 'shock excitation' employed in spark-quenching methods. In the face of this impending technological revolution, in a sort of Jurassic period of the first world communications age, the doomed radio dinosaurs battled to justify their existence and the capital that had been sunk into their creation.

In order to appreciate the basis of this impending battle, it is necessary to go back to the period just before the start of the First World War to review the technology that was being developed for radio communication at the time. Of the various systems that had been developed, it is not unfair to suggest that the most commercially and internationally significant was that based on the work of Marconi and his team of experts. This Marconi system relied on the generation of radio frequency energy that accompanied the striking of a spark, and it was the refinements of this approach that were the workhorse of long-distance communications for the first 15 years of radio communications.

THE MARCONI SYSTEM

Despite the use of 'spark' in so many of the early transmitters, it did not take long for the new breed of radio engineers and scientists to realise that 'spark' systems were intrinsically very noisy and wasteful as a means of producing radio frequency (RF) energy. More seriously, because the spark created a burst of RF energy at no particular frequency, the result was a broad band of noise that spread all over the spectrum as pollution or interference to other users.

Initially, the experts who worked for Marconi, such as Dr Fleming, were of the opinion that the spark was the only way to create the disturbance in the ether which would spread out as an expanding wave of energy, able to be used for signalling purposes. As we now know, this assumption was erroneous.

Over a 15-year period from 1906, more and more experts came to realise that any high-frequency alternating electric current was capable

Left Residence of the Chairman of AWA, ET Fisk, at Wahroongah, north of Sydney (*PRJ*)

Right Memorial commemorating the first direct radio signal to Australia from Wales in 1918, located next to the former Fisk residence (*PRJ*)

The FIRST DIRECT WIRELESS MESSAGES from ENGLAND to AUSTRALIA.

The Right Hon. W.M. Hughes. P.C, K.C. LL.D.

The Right Hon. Sir Joseph Cook G.C.M.G., P.C.

AMALGAMATED WIRELESS (Australasia) LIMITED
WIRELESS HOUSE, 97 CLARENCE STREET, SYDNEY, N.S.W.

No. 1. Wahroonga OFFICE Sept 22nd 1918
Handed in at London — Carnarvon (direct) 1.15 pm E T Fisk

To Sydney for Publication

I have just returned from a visit to the battlefields where the glorious valour and dash of the Australian troops saved Amiens and forced back the legions of the enemy, filled with greater admiration than ever for these glorious men and more convinced than ever that it is the duty of their fellow-citizens to keep these magnificent battalions up to their full strength
W. M. Hughes. Prime Minister

AMALGAMATED WIRELESS (Australasia) LIMITED
WIRELESS HOUSE, 97 CLARENCE STREET, SYDNEY, N.S.W.

No. 2. Wahroonga OFFICE Sept 22nd 1918
Handed in at London — Carnarvon (direct) 1.25 pm E T Fisk

To Sydney for Publication

Royal Australian Navy is magnificently bearing its part in the great struggle. Spirit of sailors and soldiers alike is beyond praise. Recent hard fighting brilliantly successful but makes reinforcements imperative. Australia hardly realises the wonderful reputation which our men have won. Every effort being constantly made here to dispose of Australia's surplus products
Joseph Cook. Minister for Navy.

These messages were transmitted by arrangement with Senatore G. Marconi, G.C.V.O., D.Sc. & Godfrey C. Isaacs Esq. Managing Director, Marconi's Wireless Telegraph Company, Limited, from the Marconi Transatlantic Station. at Carnarvon, Wales, at 3.15 a.m. & 3.25 a.m. (Greenwich mean time), September 22nd, 1918.

Received instantaneously at 1.15 p.m. & 1.25 p.m. (Sydney time) by Mr. E.T. Fisk, Member Institute of Radio Engineers & Managing Director, Amalgamated Wireless (Australasia) Limited, — at his Experimental Wireless Station, Wahroonga, New South Wales, with apparatus designed and manufactured in Sydney by Mr Fisk and the Staff of Amalgamated Wireless (Australasia) Limited.

Trans-Ocean Wireless Transmitting Station

Experimental Station Wahroonga, Sydney

Senatore G. Marconi G.C.V.O. D.Sc

E.T. Fisk Esq. Receiving the messages shown above.

Copy of the first radio message sent from the Marconi radio station at Caernarvon in Wales to Australia in 1918 (*MAR*)

of being propagated as RF energy. More important was the related conclusion that a sine wave form of continuous wave was required if tuning or syntony was to be achieved — for this reason, spark transmission was a technological 'dead end' that could never be entirely successful.

Despite its intrinsic limitations, major efforts were made during this period to improve the characteristics of the spark. By the end of the First World War in 1918, Marconi was claiming that the apparatus of the company could produce a signal with a rotary spark generator that was equivalent to a continuous wave. In reality this was demonstrably not true and the actions of the Marconi Company to respond to the threat of new technology made it clear that, despite the rhetoric, the days of the spark were numbered.

In the last years of the First World War, negotiations led to the purchase and installation at Marconi-owned radio stations of a number of Alexanderson alternators built by General Electric. This clearly revealed acceptance by the Marconi Company that spark could never produce a totally acceptable result in the longer term. In a commercial sense and given the huge investment of time, effort and capital in the technology of spark, this was no doubt an unpalatable and expensive conclusion to have been forced to reach. In terms of the battle of the dinosaurs, evidently Fessenden's creation had triumphed but, in the end, it was to be a short-lived victory.

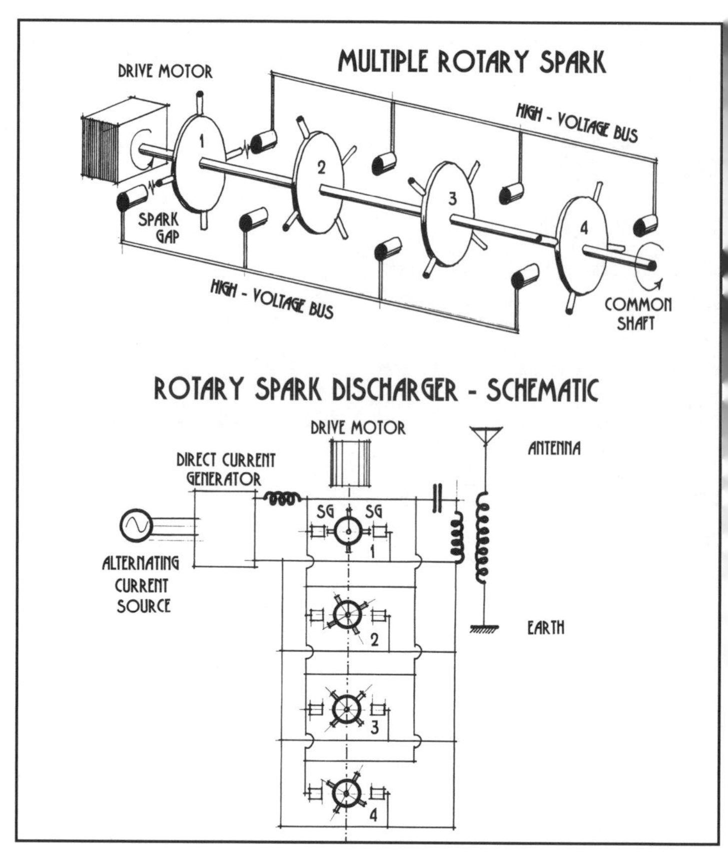

The main coupling helix for the timed rotary spark transmitter at Caernarvon (*MAR*)

Diagram showing the operation of the first timed rotary spark transmitter at Caernarvon (*PRJ*)

TIMED ROTARY SPARK

Marconi's most clearly documented method of creating something approaching a continuous wave from spark energy depended still on mechanical rather than electrical means. Although perhaps not his last system, Marconi now had a rotary spark machine created in which multiple spark wheels were coupled together so that each burst of spark energy added to that which had gone before. In addition, because of the rapid rotation of the spark wheel, each burst of spark energy was made very short — 'quenched' in fact. The succession of short bursts of energy had the effect of shocking the antenna system into resonance at its natural frequency and what was produced was something akin to a continuous wave, with some residual modulation at the frequency of the spark excitation pulses.

In order to make the pulses of RF energy add together to produce a continuous sine wave, the frequency of operation needed to be very low. Ultimately, long-wave, timed rotary spark stations operated in a band from just above the audio frequency spectrum starting at 12 kilohertz and going up to about 30 kilohertz as a maximum. This was the basis of the system that was ultimately installed at Caernarvon in Wales, the last great trans-Atlantic long-wave station to use a spark to create its radio signal. This station is also noteworthy because it was the first European station to send a signal direct to Australia in September 1918 using the timed spark transmitter.

THE ALTERNATOR TRANSMITTER

As discussed earlier, one of the experts who saw that the way ahead lay with other methods of RF energy production was a Canadian, Professor Reginald Fessenden. As early as 1899, he had stated with great clarity that the ultimate method of generating RF energy would involve waves of continuous alternating current. This perception built on the foundation established by Sir Oliver Lodge who had very early realised the importance of tuning and resonance in the context of radio frequency energy.

While the Marconi Company clung tenaciously to spark technology, Fessenden sought for ways to produce continuous wave energy and finally concluded that the best method would involve the development of a radio frequency alternator. During 1905, Fessendon approached General Electric with a proposal to have an experimental alternator transmitter developed. This work was put into the hands of a young Swedish engineer, Ernst Alexanderson, and towards the latter part of 1906 he had successfully created such a machine. With a power of about half a kilowatt at a frequency of up to 75 kilohertz, it generally fitted Fessendon's requirements.

Despite its low power output, this first alternator transmitter proved to be a potent device compared with the rotary spark transmitter that Fessenden had been working with until then. When in was installed in the US naval radio station at Brant Rock, Massachusetts, experiments were conducted using continuous wave radio frequency energy and communications with a station in Scotland at Machrihanish took place. Earlier in 1906, Fessenden had established two-way communications with this same station using a synchronous rotary spark transmitter so that a direct comparison with the alternator could be made.

Later, a carbon microphone was inserted into the antenna lead of the alternator transmitter and, with it, the transmission of sound as amplitude-modulated radio frequency energy was possible. On 24 December 1906, this arrangement was the basis of what appears to have been the first radio sound broadcast. Material that was transmitted during this remarkable event included a selection of various musical pieces and a talk by Fessenden himself.

Much to the surprise of a large number of radio operators who had never heard anything on the ether before but morse code, the signal reached up and down the Atlantic coast and was heard as far away as the West Indies.

The reason it was possible for these operators to receive the amplitude-modulated signal was that, by 1906, the advantages of crystals as RF detectors had become known. Many were incorporated into receivers at radio stations with capabilities that few appreciated. Apart from demodulating the rough energy of a spark station, a galena or carborundum detector was also capable of converting the modulated RF signal to audio that could then be heard in the headphones.

With this dramatic event, it was quite clear that Fessenden had created a source of pure sine wave RF energy and radio communications would never be the same again. More than that, the observations of Lodge made in 1894 were now confirmed.

Following the Brant Rock demonstration, development of the alternator continued at General Electric under the supervision of Alexanderson. By 1918, this high-frequency alternator, which is now coupled indissolubly with the name Alexanderson, was progressively developed into a machine with an input power of 400 kilowatts and an operating frequency at around 16 kilohertz. A number of these machines were ultimately sold with perhaps the most appropriate sale being to Sweden, the homeland of its developer. As finally installed at Grimeton, its Swedish home, there were two machines in use with one on 'standby'. These machines were able to create a power level in the antenna system of around 200 kilowatts at a frequency of 17.2 kilohertz.

Via e-mail in February 1996, it was possible to obtain the following technical specifications of the Grimeton alternator from Bengt Willander, SM7BKH, of Swedish Telecom:

• Station call sign	SAQ
• Power input	400 kilowatts
• Power output	200 kilowatts
• Rotor diameter	1600 millimetres
• Number of teeth	976
• Rotation speed	2115 rpm
• Frequency of CW	17.2 kilohertz
• Antenna system (TX), 6 towers	127 metres (416 feet) high
• Antenna system (RX), 'Beverage'	12 kilometres (8 miles) long
• Operation commenced	1925

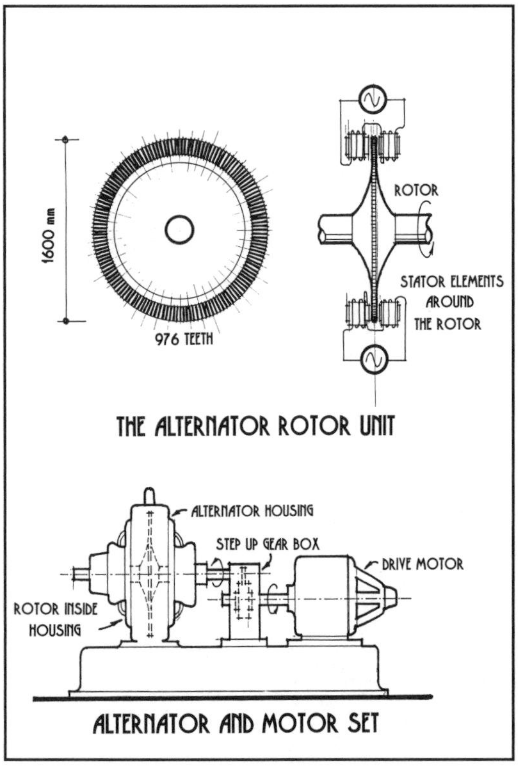

Diagram showing the main elements of the Alexanderson alternator (*PRJ*)

THE VOICE OF GRIMETON

In September 1995, the British Institution of Electrical Engineers held a centenary conference in London to mark the first wireless transmission of Marconi in Bologna in 1895. The conference covered a substantial array of topics relating to the history of radio.

During the course of the proceedings, it was announced that there would be a segment involving a presentation of slides about an alternator transmitter at Grimeton. This segment turned out to be a video film made in 1985 and was a good deal livelier than a slide show could have hoped to have been. Towards the end of the presentation, it was announced that shortly the delegates would be able to hear the voice of Grimeton. By courtesy of a BBC monitoring station in the United Kingdom, a message in impeccable morse code was soon heard in the auditorium, extending greetings to the participants in the conference from the oldest and only remaining Alexanderson alternator transmitter in the world.

To hear such a beautiful continuous wave signal

transmitted by an electromechanical machine was quite intriguing. At that stage, the devices that provided the transition between rotary spark and valve transmitters were very much a mystery to this author, so Grimeton became a must to visit.

Above left The radio station building at Grimeton, Sweden (*PRJ*)

Above right The T-shaped antenna towers at Grimeton, Sweden (*PRJ*)

The radio station at Grimeton is located 100 or so kilometres (60 miles) north of Helsingborg in Sweden, close to the town of Varburg. On the skyline the unmistakable T-shaped antenna masts of the Grimeton station can be seen, with the radio station building situated close by and the huge external antenna coupling helix on its concrete base. The antenna cable stretches away into the distance, looping over the huge lattice masts.

Grimeton radio station supervisor, Bengt Dagas, standing next to the antenna loading helix (*PRJ*)

The first impression of the Alexanderson alternator is that it is huge. Photographs do not provide a sense of scale, but given the 400-kilowatt power input of the machine, one should not be surprised. When Bengt Dagas, the station supervisor, started up the machine, its size was very much mirrored by the incredible noise that was generated — a high-pitched and powerful scream of the rotating electric motor and alternator coupled with the roar of the cooling pumps and overlain with the clatter of the keying relays.

As may be obvious from the name 'alternator', the means of producing continuous waves from an electromechanical device involves techniques which derive directly from the generation of alternating current as found in the household mains supply. However, in the case of the Alexanderson alternator, this is no trivial 60-hertz alternator but one of 17.2 kilohertz involving a power of rather more than 200 kilowatts in the antenna system as derived from the input power of 400 kilowatts.

Left The Alexanderson alternator radio frequency transmitter (*PRJ*)

Right A close-up inspection of the Alexanderson alternator (*PRJ*)

THE LAST DINOSAUR

In many respects, the story of the Grimeton alternator verges on the tragic. Almost at the moment that an alternator transmitter was able to successfully demonstrate that Fessenden was correct in his predictions, other technology in the form of the power valve had made mechanical transmitters obsolete. That continuous wave radio frequency energy was the way of the future there could be no doubt. Now the high-power transmitting valve was to make all the earlier devices — the plain, rotary, quenched and multiple spark with timing, the arc and, lastly, the alternator — completely obsolete and very soon shuffled off to extinction.

With the advent of the power valve, why the Grimeton transmitter was not consigned to the 'scrap metal' merchants in the late 1920s, as happened elsewhere, is not entirely clear. Although valves were able to produce high-power radio frequency energy a good deal more efficiently than an Alexanderson alternator, a substantial investment had been made at Grimeton. With all the publicity that had attended the installation of the two machines there and commencement of the trans-Atlantic telegraph service, there was a good deal of national pride at stake. No doubt this led to considerable reluctance to accept that the world of radio had moved on, let alone a decision to invest in further expensive foreign-developed technology. Whatever the reason, for a considerable number of years, an effective if energy-inefficient telegraph service was provided to North America from the installation at Grimeton.

The load-shedding water-cooled resistors (*PRJ*)

In later years, one of the inherent advantages of the long-wave energy available from the Alexanderson transmitter was put to a new use. This was to provide communications with the Swedish submarine fleet which at limited depths was able to receive the signal from Grimeton without coming to the surface. During this later period, the alternators came under the control of the Swedish navy. The care that was able to be lavished on these fine machines by the naval ratings has undoubtedly allowed Alexanderson's work to be seen so many years later.

Apart from its use at Grimeton, the alternator transmitter was to displace the timed-spark system developed by Marconi at Caernarvon,

The main switchboard, starting controls and relay controls (*PRJ*)

Below The main output coupling coil from the transmitter to the antenna system (*PRJ*)

and two alternators were installed at that trans-Atlantic station in 1919, one kept as a 'standby'. However, despite their successful operation in Wales, within only a couple of years they, too, were shut down as an array of valves was brought into action to produce the station's 150-kilowatt signal, and the roar of the alternator died away forever.

Only at Grimeton is it still possible to see and hear the voice of this last remaining dinosaur from the early radio age. And what a splendid beast, too! It is a worthy inhabitant of a latter-day Jurassic radio park and worth a visit by any radio historian or amateur with an interest in the history of world communications.

Left Close-up of the end of the alternator (*PRJ*)

Below Coolant water pumps for the load-shedding resistors (*PRJ*)

CHAPTER

8

1918 to 1938

Communications developments

As one who is deeply attached to the sea I am proud to have been able to render this service to the sea-going community as part of my life's work.

G Marconi

Marconi's 40th anniversary message to the world of telegraphists and mariners in 1935 *(MAR)*

To a contemporary student of communications and computing history, the early years of the 20th century have a strangely familiar appearance. Developments in radio communications during the two decades between 1918 and 1939 were extraordinarily rapid and, more importantly, this was a period in which the technology of radio was at last to become available to the general public.

First there would be a vanguard of radio amateurs and experimenters, but hard on their heels would come the great mass of the public, eager to experience the wonders of broadcast music and speech as it became available after about 1920.

In this sudden conversion of a rather arcane technology into consumable goods available to the public at large, the similarities to a much later era were pronounced. This time of galloping change after 1920 can be seen to have a great similarity to the microcomputer era that started some time before 1980.

From 1918 on, new radio techniques and hardware stimulated by the First World War were developed by major commercial enterprise and became available to the wider public — a public ready for the wonders of mass entertainment offered by radio as an expansion beyond its simple message-handling capacities up until that time.

In the 20-year period after 1918, radio broadcasting was developed and almost immediately became as much a part of people's lives as had electric light, telephone, the water closet and piped water. Nothing was ever quite the same again and the expectations of the public in relation to news and entertainment assumed a complexion completely alien to the immediately preceding generation of Edwardian and Victorian ladies and gentlemen.

First home of the BBC at Marconi House in London *(BBC)*

After a rather slow start and one that had involved the blind alley of mechanical scanning, television also would be developed to a system of image display that would be instantly recognisable to a modern viewer. Lacking only colour and the enhanced detail of today's television broadcasting, by 1936 a full service was available to the British public, although at that time it involved expensive and quite complex receivers compared with radios then available.

Another immensely important development was the discovery of short waves and the exploitation of this new part of the radio frequency spectrum, not only for broadcasting but also for long-distance telephony. During this period, attempts to emulate the connection of the continents by telegraph cable with a system of wired telephony were begun but with little success. The ability to span the gulf of the Atlantic and Pacific oceans would have to wait for another era of technical development, when voices would be able to travel below the waters over a distance of several thousand kilometres. For the time being, it would be radio alone that could pass telephone messages from continent to continent.

Apart from some notable exceptions, this was an era of big business interests and the involvement of the large engineering corporations. The lone inventor, putting together the latest bright idea on the kitchen table, was largely displaced by the research resources and engineering talent of organisations such as Marconi, Radio Corporation of America (RCA), Telefunken, and Amalgamated Wireless (Australasia) (AWA). This new, heavily engineered and resource-rich approach to the development of radio broadcasting, television and telephony was to have profound repercussions, particularly in the rate with which new ideas were developed and applied.

One of the notable exceptions to this changing environment was Edwin H Armstrong, who had invented 'feedback' only to have his success snatched away from him by the American courts in favour of de Forest. His work on the superhet during the war years was of even greater importance in the longer term, being a method of achieving selectivity and sensitivity in radio receivers still in use to the present day.

During the next era of radio development, Armstrong went on to invent perhaps the most important new system of radio transmission and reception since the work of Marconi. This was frequency modulation, now referred to as FM and used for broadcasting where high-fidelity music is required, with freedom from interference by static, which remains the bane of conventional amplitude-modulated (AM) radio broadcasting.

First transmitter for the BBC at Marconi House in 1922 *(BBC)*

NATIONAL BROADCASTING

There can be little doubt that the most significant development after 1920 was the advent of broadcasting. Some experimental work had been undertaken many years before in 1906 by Fessenden, using his newly developed alternator as the source of pure continuous radio waves. With the application of an audio signal to modulate the radio frequency energy, he had been able to send a signal to naval operators listening on crystal sets in the vicinity. However, this had been a short-term experiment and difficult to develop to anything more significant on the basis of technology available at that time. Until the valve was converted into a reliable commercial component during the years of the First World War, audio transmission was a possibility rather than something that could be realised in any practical or effective manner.

Early BBC microphone *(PRJ)*

With the cessation of hostilities, many people who had served in the armed forces had caught the bug of radio transmission and reception and were now able to obtain war-surplus equipment and become experimenters, or radio amateurs as they were later to be known. Not content merely to listen, some of the more enterprising of these new radio users started to present programs of music and other material and very soon found that there was an enthusiastic audience just waiting for such a development. Apart from anything else, in the early days, making up a crystal receiver required only a handful of components, some of which the average handyman found easy to build and, suddenly, the wonders of the airwaves were free for the taking.

In the United Kingdom in 1920, early experiments by enthusiasts working for the Marconi Company were very quickly subsumed into the new British Broadcasting Company, later to become the British Broadcasting Corporation. This initial exercise of broadcasting, undertaken via a special temporary licence from the British Post Office, had allowed programs of music to be produced. In particular, in a concert broadcast on 20 June 1920, the voice of the famous Australian soprano Dame Nellie Melba had been heard across the British Isles and part of Europe, too. The reaction was immediate and enthusiastic.

Despite the initial attempts by the Post Office to stifle this new upstart activity, public demand could not be silenced and in January of 1922 the Marconi Company was issued with a further licence and established the experimental station 2MT at Writtle, close to

BBC House in London in 1997 *(PRJ)*

Chelmsford, the home town of the company. This was followed by the establishment of the much more famous radio station 2LO in Marconi House, London, in November of the same year. The British Broadcasting Company had been formed shortly before that on 18 October 1922.

With this public recognition of the demand for broadcasting, in Britain the private enterprise of individual broadcasters was also brought under government control and radio amateurs were rigorously regulated in the way that they were allowed to use the airwaves.

In America, a spirit of free enterprise motivated the field of radio broadcasting from the outset and, in the main, programs were distributed by commercial organisations for economic gain. The famous station KDKA in the city of Pittsburgh began operations in November 1920 at a power of 100 watts which was soon increased as the demand escalated. In that initial period, there appear to have been between 5000 and 10 000 receivers capable of receiving KDKA, but by May 1922 this had grown to 750 000 receivers in the United States. By 1924, the number of broadcasters had grown from one to 1400 — an extraordinary rate of growth.

In contrast to these two models, Australia had an 'each-way bet'. Although initially the main Australian broadcasters were private commercial entities, a national broadcasting organisation was also set up, the Australian Broadcasting Commission.

METHODS OF RECEPTION

Lord Reith, first General Manager of the BBC (1922) and later first Director General (1927–38) *(BBC)*

Initially, reception of the new broadcasting stations was, for the most part, on very simple apparatus of the crystal-set variety, but as triode valves became less expensive and more freely available, the regenerative circuit of Armstrong was set to work. With a nearby radio station, this form of receiver was able to drive a loudspeaker to an audible level, allowing more than one person to listen to the program. Previously it had been a matter of getting access to the headphones or doing without.

Early short-wave transmitter using the newly developed power output valves, built by S Newman, located at the Pennant Hills radio station, north of Sydney *(MAR)*

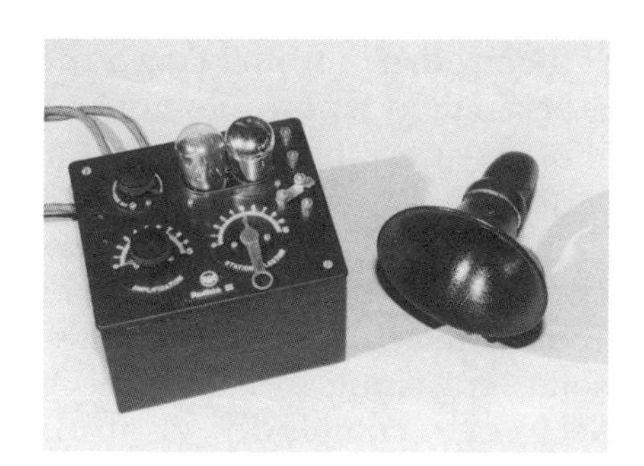

Above left to right:
A bakelite-encased crystal set made by Brownie, the Brownie Number 2, from 1925 *(PRJ)*

Early commercial two-valve/tube radio, the Radiola III *(PRJ)*

Early valved radio set for receiving the BBC broadcasting service *(PRJ)*

Unfortunately, the regenerative detector, for all its sensitivity, was capable of causing much interference to other listeners if incautiously used. The difference between optimum levels of feedback and a valve briskly oscillating at the frequency of reception was very slight and whistles and howls would be broadcast by offending operators to annoy their neighbours.

For a while in the mid-1920s, a more sophisticated form of tuned radio frequency (TRF) circuit was introduced which did not use positive feedback. On the contrary it used negative feedback, or neutralisation as it was known, to stabilise the otherwise difficult-to-use triodes of that era where full amplification was to be achieved. The Hazeltine 'neutrodyne' circuit was to provide the solution to the need for radio frequency amplification without regenerative feedback and the associated tendency to put annoying signals on the air where the receiver was in the hands of an inept operator. However, although this managed to overcome one problem, there was another major problem with this type of receiver — lack of selectivity.

As anyone who has operated a TRF receiver of the period after 1924 knows full well, in the absence of regeneration to sharpen up the selectivity, such radios had selectivity characteristics that were as 'broad as a barn door'. In the earliest days of broadcasting, when stations were few in number and widely spaced in the frequency band between 550 kilohertz and 1500 kilohertz, later expanded to the present allocation, broad selectivity did not matter very much. However, with the spate of new radio stations and increasing levels of broadcast power, adjacent channel selectivity became vital. In the short space of 5 years, the

Below left to right:
The Marconi V2 radio receiver, for receiving the BBC and frequencies between 185 metres and 3200 metres *(PRJ)*

Broadcast radio kit set made in Australia, using a battery supply and a horn loudspeaker *(PRJ)*

Atwater Kent 'bread-board' radio set from the United States *(PRJ)*

Cathedral radio set of 1931 made by Airzone of Australia *(PRJ)*

Stromberg Carlson 'autodyne' — local oscillator and mixer produced the intermediate frequency with a single tube *(PRJ)*

Typical broadcast radio from the mid-1930s with short-wave bands covered *(PRJ)*

Hazeltine 'neutrodyne', from being a useful solution to adequate radio reception became, in its own fashion, a small dinosaur of the first age of radio broadcasting. It simply could not separate the closely spaced radio signals that now filled the broadcast band from end to end, and produced an acoustic version of 'the Tower of Babel'.

Something much better was needed and it did not take the broadcast receiver manufacturers long to realise that there was an elegant and feasible solution waiting to hand. This was the Armstrong 'superhet'. By now its level of complexity was manageable with the availability of new, cheaper multi-electrode valves but, more to the point, it was not only far more selective than the neutrodyne of Hazeltine but also much more sensitive. Beyond that, its earlier difficulty of operation was overcome by the introduction of ganged tuning controls. As ultimately sold to the public in massive numbers, with the exception of short-wave receivers, the ordinary broadcast radio set had two controls, a tuning control and a volume control.

These sets did not have the on/off switch of modern sets because of supply authority regulations. Because a two-wire connection to the lighting supply was frequently used together with an unpolarised bayonet plug, it was quite possible to have the radio on the active side of the mains and so put into a lethal state even when apparently switched off. For this reason, radio sets had to be turned off by switching at the power plug or light switch rather than at the radio as is now common.

The superhet of the late 1920s, which now appeared in millions and was sold around the world to an enthusiastic public, was a radio that was simple enough for a child to operate and, with a loudspeaker built in, could provide entertainment to the whole family. A social revolution in family activities occurred in the space of 10 years and the upright piano, as the previous centre of family entertainment, was immediately relegated to being a sort of elaborate ornament rather than an indispensable piece of furniture in a fashionable home.

THE SHORT WAVES

In his earliest experiments with radio communication in 1895, Marconi had been operating at frequencies that would now be called VHF (very high frequency) at around a wavelength of 2 metres or a frequency of 150 megahertz. For the next 20 years, much of his effort was expended in the pursuit of long-distance communication on the assumption that this could be best achieved with longer and longer wavelengths and ever-increasing levels of power. By 1918, power levels in the trans-Atlantic radio stations were at about 300 kilowatts and frequency was down to about 25 kilohertz.

In the latter part of the First World War, Captain Round of the Marconi Company began experimenting with wavelengths at about 100 metres. So successful was this work that the construction of a short-wave station at the old Poldhu site was commissioned under the capable hands of engineer CS Franklin. In 1919, Marconi purchased a 700-tonne steam yacht which was renamed *Elettra* and, after it had been fully equipped as a sort of marine laboratory, set out on a voyage of experimentation. At Poldhu a parabolic reflector had been built to operate on a wavelength of

Far left The Ekco radio of EK Cole Ltd designed by the architect Wells Coates about 1935 *(PRJ)*

Left McMichael Duplex 4, Type 8, portable radio with battery power, 1928 *(PRJ)*

97 metres and signals were passed between it and *Elettra* as the yacht steamed further and further away. At a distance of 2230 nautical miles from Poldhu, a 1000-watt signal was far more powerful than the long-wave signal sent out from Caernarvon with a power of 300 kilowatts.

On 30 May 1924, this experiment was repeated with an increase in transmitter power to about 17 kilowatts, with the result that the signal was heard with great clarity in Sydney, Australia, and a good-quality voice message was able to be sent by Marconi.

Further experiments soon demonstrated that an optimum wavelength for long-distance communications was reached at around 32 metres. At this wavelength and with a power of only 12 kilowatts, daylight communications were possible to Sydney and to a number of other cities closer to Poldhu. In the space of just 4 years, the long-wave services of all nations had been made obsolete.

In the 3 years immediately prior to these later results of Marconi, the repercussions of the *Titanic* sinking began to be felt among radio amateurs. In particular, the enforced confinement of the radio amateur community below a wavelength of 200 metres produced a new experimental drive among the more adventurous, with very surprising results. Radio amateurs started to experiment with the higher frequencies and shorter wavelengths, and soon found that they involved remarkable advantages. Contrary to the best scientific information of that period, it was found that lower power and higher frequency did not equate with shorter ranges for communication. Instead it was found that the higher the frequency in use, down to a limit of about 10 megahertz, the greater the range at which successful contacts could be made, even with the low power that was able to be used by radio amateurs.

By now, valves were the basis of a number of amateur radio stations and though they could produce continuous power only at about the 100-watt level, at 10 megahertz this did not seem to matter all that much and long distances were able to be covered. Short-wave communication had arrived and long distance, or DX as the amateur abbreviation has it, had been achieved.

Bakelite-encased Philco 'People's set', Model 444, 1936 *(PRJ)*

Pre-war technology in 1950 — the bakelite-encased Bush radio, type DAC 90A *(PRJ)*

Top Parabolic wire beam antenna at Inchkeith, Firth of Forth, in the early 1930s *(PRR)*

Bottom Marconi's floating radio laboratory, the SY *Elettra (PRR)*

Initially, much of the effort involved in opening up this new part of the radio frequency spectrum came from the United States, and in 1921 a noted American amateur, Paul F Godley, with the call sign 2XE, went to England to attempt to receive a signal sent across the Atlantic from amateurs in America. To maximise his chances of achieving success, he took his apparatus to Ardrossan near Glasgow on the coast of Scotland and on 7 December 1921 received the first signal from the United States. Appropriately, this was a 60-cycle synchronous rotary spark on 210 metres, given the work of Marconi just 20 years before in receiving a spark signal at St John's in Newfoundland from Poldhu in the United Kingdom.

During the next 10 days, Godley was to hear some thirty US stations using a sensitive superhet receiver. One of the more interesting aspects of this exercise was to demonstrate the clear superiority of low-power continuous wave signals produced by valves compared with the much higher powered spark signals that were being sent from America. Of the thirty signals received, about two-thirds were valve-generated whereas only one-third came from spark systems. The days of spark were now clearly over and this was well noted by radio amateurs on both sides of the Atlantic and also by the Marconi Company as confirmation of its own experimental work. The one-way traffic of Godley's experiment was soon turned into two-way radio amateur traffic on 8 December 1923 and that connection between the United States and the United Kingdom has continued ever since.

These various strands were all brought together by Marconi and one of his most prolific assistants, CS Franklin, in a system of short-wave transmission and reception using the newly invented beam antenna system. The much-vaunted long-wave, high-power system that had been offered by Marconi to the British government in the years immediately before the First World War was clearly obsolete. Despite the well-advanced negotiations to install the long-wave Empire wireless system, abandoned in 1914, Marconi, in a quite remarkable act of commercial responsibility, now offered to create a world communication system for the Empire at a fraction of the cost of the old system that had been proposed.

Despite rigorous government requirements for data transfer rates or 'throughput' as it would now be called, a new system was put into operation and provided the telegraphic and later telephonic links to Australia, Canada, India and South Africa. This was initiated with a short-wave link between a station at Ongar in the United Kingdom and Berne in Switzerland in 1922, with the Empire links being initiated by a connection to Canada in October 1924.

Marconi transmits a message to Australia in 1935 *(MAR)*

Although this was an immense technical triumph for the Marconi Company, its commercial repercussions were in many ways quite tragic. The very success of this new method of long-distance communications was so great that the cable companies immediately began to see a major loss of revenue. From the perspective of a government intent on retaining the strategic value of this resource and mindful of the adverse impact that the loss of the telegraphic cables had produced in Germany during the war, pressure was brought to bear to achieve a rationalisation. Despite its seeming to be in an unassailable position of strength, the Marconi Company allowed itself to be pressed into an amalgamation with the cable companies. From this union came the well-known entity Cable and Wireless Limited.

In spite of all the incredible effort that Marconi had put into achieving a worldwide communications network, at the instant that it was finally established it was effectively snatched from his grasp. For the next few years until his death in 1937, his attention moved to other aspects of the use of radio. In particular, during the short period that remained to him, his interest moved back to the shortest wavelengths and to the radio frequency pastures above 200 megahertz that he had roamed as a youth in 1895.

In 1923 there was a final 'twitch' of the long-wave monster when the British Post Office decided that it should build a long-wave station with a worldwide capability. The result of this decision was to see the erection of a huge radio station on 370 hectares (920 acres) of land at Hillmorton near Rugby in the Midlands. With an extraordinary array of 12 masts at the remarkable height of 250 metres (820 feet) and spaced at approximately 400-metre (1300 feet) intervals, this station was to become known throughout the world to almost everyone involved in radio communications. Its maximum 550-kilowatt signal was broadcast at a frequency of 16 kilohertz or a wavelength of 18 740 metres with the call sign GBR, providing timing signals that were immensely valuable to many users on land and at sea.

Following the opening of the long-wave part of the Rugby station in 1926, a short-wave radio telephone service to America was opened in 1927, allowing two-way conversations to occur. This service was located at 60 kilohertz and had the call sign GBT. Later, in 1928, this service was significantly expanded with the creation of short-wave radio telephone communications. Although transmission was concentrated at the Rugby site, reception was initially at Wroughton in Wiltshire but later at Cupar in Fifeshire, a much 'quieter' site.

Perhaps the most pervasive use of the short waves was to occur towards the latter part of the 1930s as propaganda began to be a significant part of the output of both Axis and Allied stations. Soon the grating voice of Lord Haw-Haw would be heard for the first time, to be countered by the mellifluous tones of the British Broadcasting Corporation announcers.

As the 1930s moved to their conclusion, preparations for the next war were well in hand on all sides. In particular, in a broadcast speech of February 1938, no one could have had much doubt as to the real intentions of Germany's Chancellor as Adolf Hitler harangued his subjects via the airwaves for all of 3 hours. 'The writing was on the wall' as, in reality, it had been ever since the pages of *Mein Kampf* had first become available. Europe was about to reap a very bitter harvest born of 20 years of complacency and inertia with regard to matters military.

FREQUENCY MODULATION

In this same period, yet again that indefatigable inventor Edwin H Armstrong was to create something quite new and with far-reaching consequences. His invention was an entirely novel method of transmission and reception using frequency modulation (FM). Initially this was a response to the need to combat the problem of atmospheric interference in radio broadcasting but later it was realised that FM had other outstanding advantages.

For broadcasting it offered the possibility of high-fidelity audio transmissions because the very high frequencies (VHF) used to carry the signals allowed much more generous bandwidth than the conventional broadcast band. This was a fundamental requirement for achieving such high-fidelity results.

For portable operations, VHF transmission also offered the advantage of physically much smaller antennas and the possibility of high efficiency and gain which at longer wavelengths were simply not feasible. This latter capability was to be incorporated into new forms of portable transceivers used by the American army in particular, with significant advantages that would become clearly apparent on the battlefield. By comparison, on the ground, Britain stuck to the use of high-frequency (HF) army radio equipment which was all too soon to prove extremely troublesome for short-range communications in thickly vegetated country.

For Armstrong, his work on frequency modulation was to bring him nothing but sorrow. As happened to others in the 1930s, he became embroiled in a fight with a big corporation. In this instance it was with the Radio Corporation of America (RCA) and its formidable boss, David Sarnoff. Inevitably, RCA brought to bear its enormous resources and did its best to crush small players such as Armstrong. In particular, it resisted his efforts to establish the primacy of his patents for this new form of transmission. This produced so much distress with the ongoing litigation that, ultimately, he committed suicide just after the Second World War. His widow, incensed by this, continued the fight with RCA for another 15 years, finally being vindicated by the courts. This resulted in her being awarded the disputed royalties and licensing fee for the use of the Armstrong system of frequency modulation for broadcasting.

The mechanical scan television set devised by John Logie Baird, the 'Televisor' *(PRJ)*

TELEVISION

Since the latter part of the 19th century, a number of people had tried to harness electrical methods to convey visual images over a distance. Of these, probably the most significant was the work of the German experimenter Paul Nipkow, who invented a mechanical system of scanning that would appear 40 years later in the work of John Logie Baird, a Scot.

Baird was to have the misfortune of expending enormous financial and personal resources on an electromechanical and radio frequency system of television only to have the prize of success taken from his grasp right at the end of the race to open a public television service. Unfortunately, his method of mechanical scanning, based on the Nipkow disk, was obsolete before he began work on television in the early 1920s because its successor had been described as early as 1911. To that extent, one could say, perhaps uncharitably, that Baird was largely the architect of his own misfortune. If he had pursued a system more likely to achieve success as would have occurred with an electronic method, the result of all his hard labour might have been quite different. Unfortunately he did not and, however hard he might try, a mechanical system, whether using 30 lines or 240 lines as it had become by 1936, simply could not compete in flexibility or performance with an all-electronic competitor.

Despite these inherent problems, Baird's energy resulted in the BBC sending out television signals generated by his apparatus, commencing in July 1926. In September 1929 this was converted into a regular service by the BBC, involving 30-minute programs on 4 days of the week. The signal, referred to rather optimistically as 'low definition', involved a 30-line scanning system generated by Nipkow disk, with a frame repetition rate of $12\frac{1}{2}$ frames per second. To a modern viewer, the resultant images certainly involved movement and a degree of light and shade but, compared with the modern version of television, flickered quite badly and lacked detail to any degree.

John Logie Baird, television pioneer *(25Y)*

Baird with the first mechanical scanning television transmitter housed at the Science Museum in London *(WIT)*

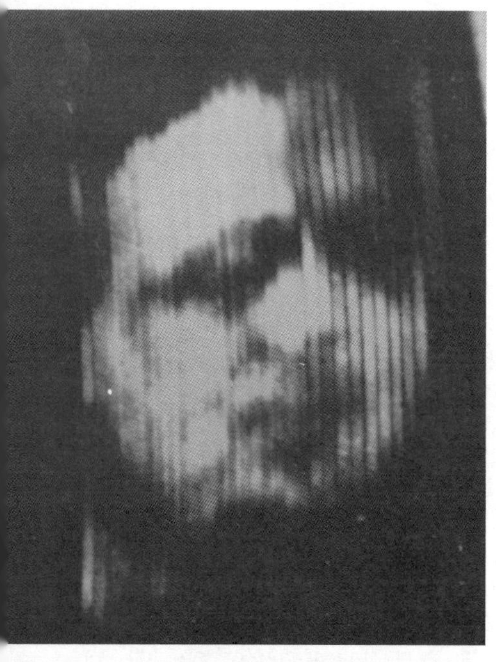

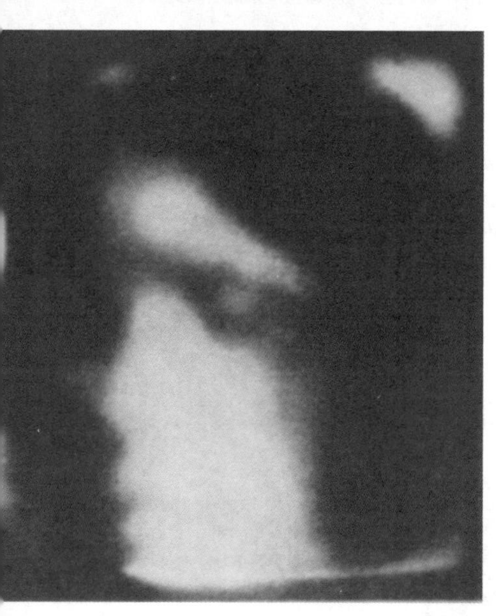

Top The image of a face produced by the Baird mechanical scanning television *(WIT)*

Bottom Typical image of a face produced by the Jenkins mechanical scanned television system of 1932 *(USM)*

By 1935, Baird had improved this electromechanical system to one involving a scan rate of 240 lines and a frame repetition rate of 25 frames per second. This was probably about the limit that such a system was capable of achieving. The result of this was the presentation of moving images that were somewhat closer to what would now be considered acceptable.

During the early 1930s, parallel with work that was being undertaken by RCA in the United States, the newly created British company Electric and Musical Industries (EMI) set out to create an all-electronic system of television. Under the leadership of Isaac Shoenberg, who had joined EMI from the Marconi Company, a team of engineers was assembled including such notables as JD McGee and AD Blumlein, who was later killed in the Second World War. Earlier, in December 1931, Blumlein had patented a system of audio reproduction he had described as binaural. This was to become the basis of a later revolution in sound recording and reproduction that is now known as stereo hi-fi, or high-fidelity sound.

By 1934 this group had developed an electromechanical system along similar lines to Baird but at a significantly higher resolution than the 30-line system. However, it became apparent to EMI during this period that such an approach would never achieve an entirely satisfactory result. For that reason, a parallel project was undertaken to explore the all-electronic approach being developed by RCA and suggested much earlier by AA Campbell Swinton. This method incorporated a camera tube that was called the 'emitron', which in principle was extremely similar to the 'iconoscope' of Vladimir Zworykin who at this time was working for RCA.

The decision by EMI to pursue an all-electronic system of television appears to have coincided with a developing attitude of the Marconi Company regarding the possibilities of television. This was to lead to direct technical collaboration, bringing together the technical radio broadcasting skills of Marconi engineers with the developing skills of the television group at EMI. In March 1934 this relationship was formalised in the creation of the Marconi–EMI Television Company Limited. It proved a very fruitful integration of common interests, for by 1936 it was apparent that two rival television systems were available, the one proposed by Baird and the other by Marconi–EMI. Faced with this situation, the BBC decided that in order to overcome the seeming impasse, it should institute a long-term test.

As presented to the British Broadcasting Corporation for these tests, the new, all-electronic system designed by Marconi–EMI produced pictures with a horizontal scan of 405 lines at a repetition rate of 50 frames per second. Although by this date Baird had available his electromechanical television system with a scanned display involving 240 lines, the Marconi–EMI system produced very much better images. In particular, the use of interlaced scanning in the electronic televison had overcome the problem of flickering images

Far left A Jenkins mechanical scanning television as developed in the United States, using a horizontally rotating perforated band *(USM)*

Left Vladimir Zworykin in 1932 with a cathode ray tube used for television reception, the 'kinescope' *(USM)*

which was hard to eradicate in the slower scanning rate of the Baird system.

Apart from this advantage, the Marconi–EMI system involved a relatively compact camera and this was a distinct advantage when the practical problem of producing studio and outside broadcast material had to be accommodated. By comparison, the Baird system was essentially static and could not be moved around. With a sad inevitability, the Marconi–EMI system was judged to be outstanding and became the basis of the 'high-definition' service introduced by the BBC in 1936.

As a relatively late starter, Australia had the benefit of all the technological effort that had gone into making a reliable all-electronic television system in the period before the Second World War. For this reason, from the introduction of black-and-white television by TCN 9 in Sydney in 1956, a new higher definition system was used involving a scan rate of 625 lines. This was to be superseded by colour some years later, with a compatible system also using 625 lines on the phase alternate line (PAL) mode, and this remains the current method to the present day. In the United Kingdom, the 405-line display would remain as a standard that many would remember from the early days of television broadcasting in the 1950s. The 405-line service was finally shut down in 1985, a remarkable testimony to the high quality of definition that it had embodied at the outset.

What this saga of the early television age demonstrated was that, in the world of electronics, the time of the determined amateur inventor was effectively over. Expecting that an enthusiastic but untrained person with no expertise in science or electronics would be able to succeed in competition with an organisation such as EMI, let alone RCA, was simply ridiculous with the benefit of hindsight. Baird was in competition with the technical experts of the Marconi–EMI group and inevitably it was not a race that could be won by a 'tortoise', however determined.

A little-known element of this saga was that in the final years of Baird's desperate fight to control the developing path of television, British Gaumont took a controlling share of his company. In 1935 when it became apparent that Baird would be compelled to compete with EMI and an all-electronic system, Gaumont insisted that he should convert his 240-line-based electromechanical system to an all-electronic system. Evidently it was justifiably feared that, if this was not done, Baird would not have any hope of competing with EMI. As a result, the Farnsworth-developed 'image dissector' camera tube, discussed later, together with cathode ray display tubes from General Electric Company, was incorporated into the Baird television system. This exercise involved a licensing fee of $100 000 but, in the end, was all to no avail. High definition was the way ahead, and a 240-line system simply could not compete — EMI became the preferred supplier in the new BBC 'high-definition' service.

In an extraordinary effort of prediction, Campbell Swinton, the engineer who had helped Marconi meet William Preece (the chief engineer of the Post Office) in 1896, set out the essentials of a purely electronic system of television in a lecture to the Roentgen Society in 1911. The salient points of his proposal are well conveyed in the diagram that accompanied his lecture. The main features of a modern electronic television system are all included in that diagram:

- electron beam scanning of image
- photoelectric pick-up camera tube
- cathode ray tube image display
- magnetic coil electron beam deflection
- synchronisation signals between camera and display.

However, in technological terms and also in terms of the frequencies that were able to be used, Campbell Swinton was well ahead of his time and the development of the valve and the higher frequency portions of the radio spectrum would be required before practical television could become a reality. By about 1925, with the advance to ever-higher frequencies, the time was ripe and the experimental work began in earnest in both England and America.

Despite the intellectual leap of Campbell Swinton, the further development of all-electronic television continued far away in the United States. This was based on the work of two 'parents', who in many respects were to be made rivals through the efforts of Radio Corporation of America (RCA) and its formidable leader, David Sarnoff.

Primacy in the development of an idea is always a very vexed question, as the ongoing dispute over the discovery of radio communications demonstrates. Despite the best evidence which favours Marconi, Russians still fiercely promote the erroneous notion that it was Aleksandr Popov who was first to demonstrate wireless telegraphy and had the glory taken from him by the political machinations of western capitalism. In passing, Popov himself acknowledged the work of Marconi as having been first and one wonders what better tribute could be expected.

Philo Farnsworth with his all-electronic scanning television camera, the 'image dissector', 1932 *(USM)*

In the development of electronic television, it is certain that a Russian emigre to the United States, Vladimir Zworykin, was to play a very considerable part, and in late 1923 he patented a design for such a system while working at the Westinghouse Corporation. Later, the primacy of this patent was to be overturned by the work of a native American, Philo Farnsworth, who in 1926 patented his own system of electronic television together with a demonstration model to back up his claim. (See the section 'Farnsworth and Zworykin', pp. 134–37.)

This latter system was announced in the press in September 1928 and Zworykin, who by now was working at RCA, was sent off to investigate by the Managing Director of the company, David Sarnoff. After several days of careful inspection, which to a latter-day reader has all the hallmarks of industrial espionage or 'snooping', Zworykin went back to Sarnoff to report. Shortly after this, in a quite unprecedented action, Sarnoff invited himself to visit Farnsworth's laboratory to inspect what had been developed. This must have been very impressive because, during the course of the visit, Sarnoff offered to buy the whole installation for $100 000, a huge amount given the date of the offer.

This amount was refused and apparently Sarnoff's reaction was to depart with the words 'There is nothing here that we need'. In retrospect he was quite wrong, as future events were to demonstrate. The patent battle that broke out between Farnsworth and RCA a little later was ultimately to go in favour of Farnsworth in July 1935. As a result of this, on 1 October 1939, RCA was forced to sign an agreement to pay Farnsworth royalties and a licensing fee for the use of his patents. It appears that this was the first time RCA had been forced into such a position, as usually it was the organisation collecting the fees from other organisations. Some would say with a degree of pleasure, having seen the 'bully boy' tactics of the corporation employed in relation both to Farnsworth and to Armstrong, 'What poetic justice!'

Unfortunately, the end of the story of Farnsworth was to be rather less satisfying. It seems that the battle with RCA had been sufficiently taxing to have strengthened his dependence on the bottle. Although he was to see a television apparatus construction and broadcasting company formed which carried his name during the years of the Second World War, he retired to his farm and later died in relative obscurity. His company was ultimately sold to the International Telegraph and Telephone Company in 1949. However, before Farnsworth departed the television scene, his work and his company were to have an important impact.

During the late 1930s and in the early years of the Second World War, Farnsworth's laboratory and television studio were located at Green Street in Philadelphia. Only 2 kilometres (1 mile) away from this establishment, a young graduate of the Moore School at the University of Pennsylvania, J Presper Eckert, was starting to apply the electronics

The home of the BBC television service in 1937, Alexandra Palace *(BBC)*

he had learnt to practical problems. The young enthusiast bought an experimental television set from Farnsworth's commercial rival, RCA, and got it operating. This set was used to watch and listen to the broadcasts from Farnsworth's station. When the studio ran out of material to broadcast, Eckert was one of the enthusiasts who would visit to provide additional material while friends watched the show from the television set at his home. The youthful Eckert was soon demonstrating electronics at the Moore School and there made a significant friendship with a new academic member of staff, John Mauchly. Both their names will recur at a later stage in the story of the Web.

Whatever the result of the patent battle in 1935, the television camera that was developed by Zworykin was to prove the way ahead for television. It seems that the reason for this was mainly to do with its sensitivity compared with the camera developed by Farnsworth. This was related to the method of storing electric charge on the photosensitive and emissive mosaic which was scanned with the electron beam.

By comparison with the Zworykin tube, which was called the 'iconoscope', the Farnsworth 'image dissector' seems to have operated in the manner of an electronic analog of the Nipkow disk. Despite certain advantages over the 'iconoscope', the low level of sensitivity of the 'image dissector' was ultimately to shunt it into second place in the race with RCA to achieve a viable television system. It appears that the level of sensitivity of the Farnsworth tube became progressively worse as the level of definition increased, and no doubt this was part of the problem that Baird had to deal with in the competition which was conducted with EMI in 1936 using 240 lines at 25 frames per second. As previously noted, the 'emitron', as developed by McGee and others for EMI, was a very similar device to the 'iconoscope', although it appears that it was created quite independently of the work at RCA and, as used in the 1936 BBC tests, was operated at a line frequency of 405 lines at 50 frames per second.

Continuing research work on the iconoscope and the emitron in the latter part of the 1930s led to camera tubes with increased sensitivity and optical dynamic range. These later devices were known as the 'image iconoscope' (an interesting tautology given that 'icon' means 'image') and the 'super emitron'.

By 1936, following the test of the competing systems of Baird and EMI, a workable all-electronic television service was initiated in the United Kingdom, transmitting a signal on 41.5 megahertz for the sound signal and with the video signal centred at 45 megahertz. This BBC television service was based at Alexandra Palace in north London and made use of the new 'emitron' camera to produce the 405 lines of this first electronic system.

The BBC 405-line television service in operation in 1935 *(25Y)*

In the United States, based on the work of Vladimir Zworykin and his 'iconoscope' camera tube of 1924 and his 'kinescope' receiver tube of 1925, RCA produced a fully electronic television system which was to be the basis of NBC television broadcasting, commencing on 1 July 1940.

The Second World War was to effectively postpone the acceptance of television for nearly 10 years. It was not until the late 1940s in America and the early 1950s in Britain that once again television appeared to captivate the public. Late though it was, this time television quickly came to dominate the broadcasting scene, in the process displacing radio as the main broadcasting medium.

Meanwhile, all the experimental work had not been wasted. The development of scanning techniques applied to cathode ray tubes and the use of very high frequencies for communication purposes were all to bear fruit in new and significant applications of communications technology. Together with the newly invented method of modulation, frequency modulation (FM), there would be long-term impacts on methods of radio and television broadcasting and communications. More importantly, in the short term, these developments would provide a basis for dealing with a new and critical military problem — the need for a method of detecting the approach of hostile aircraft and the provision of early warning. This was radar, an acronym of 'radio detection and ranging'.

DEVELOPMENT OF RADAR

Referred to in the early days by the British scientific and military establishment as radio direction finding or RDF, what was later to be far better known as radar was not by any means a new concept. It is apparent that the German scientist Christian Hulsmeyer had created a system of detection of reflected radio waves as early as 1904. This remarkably

Left A concrete acoustic reflector for detecting the approach of enemy aircraft in 1932; located at Hythe in the United Kingdom, photographed in 1996 *(PRJ)*

Right A long wall acoustic reflector for early aircraft detection from 1934, seen in 1996 *(PRJ)*

prescient system using a spark-signal generator anticipated, in the form of its antenna horns and metal tube antenna feed system, methods that were to be rediscovered and developed the best part of 30 years later in the valve-based transmission systems.

Unfortunately for Hulsmeyer, it would take the speed and power of the aircraft to generate a demand for a new system of long-range rapid detection of distant moving objects. For this reason, his very advanced system was effectively forgotten, only to be found lying in the German patent office many years later, to be admired by radio historians.

Towards the end of the First World War, a significant if restricted indication of what was likely to occur in the future had been provided. This was the intrusion into British skies of German Air Force Gotha bombers in September 1917. This new form of aerial bombardment was to culminate in a substantial raid on 19 and 20 May 1918 when 22 of these aircraft flew from Ghent in Belgium to attack London. By this stage, the British tactics of response had been considerably developed and 7 of the invading aircraft were shot down.

The message of this attack, with high-altitude bombing and the difficulty of preventing aircraft from overflying defended territory, was not lost on either the British or German authorities. The assertion by the British Prime Minister, Stanley Baldwin, in 1932 that 'the bomber will always get through' summed up the general attitude to the probable future of aerial invasion at that time. Further, it made the subsequent drive for peace at whatever cost and the appeasement attitude of the 1930s somewhat easier to understand. Few but a manic, small group of political opportunists who gathered around the new German demagogue, Adolf Hitler, could have had much taste for a new 'blood bath' of the kind that had been seen on the Western Front in the First World War. Aerial bombardment would make the prospect of war even more unpalatable.

Despite the apparent impossibility of preventing the incursion of enemy bombers, attempts were made to find some means of defence. High-velocity anti-aircraft guns could provide some degree of protection, coupled with the newer high-altitude aircraft that were developed after the conclusion of hostilities in 1918. However, it was soon realised

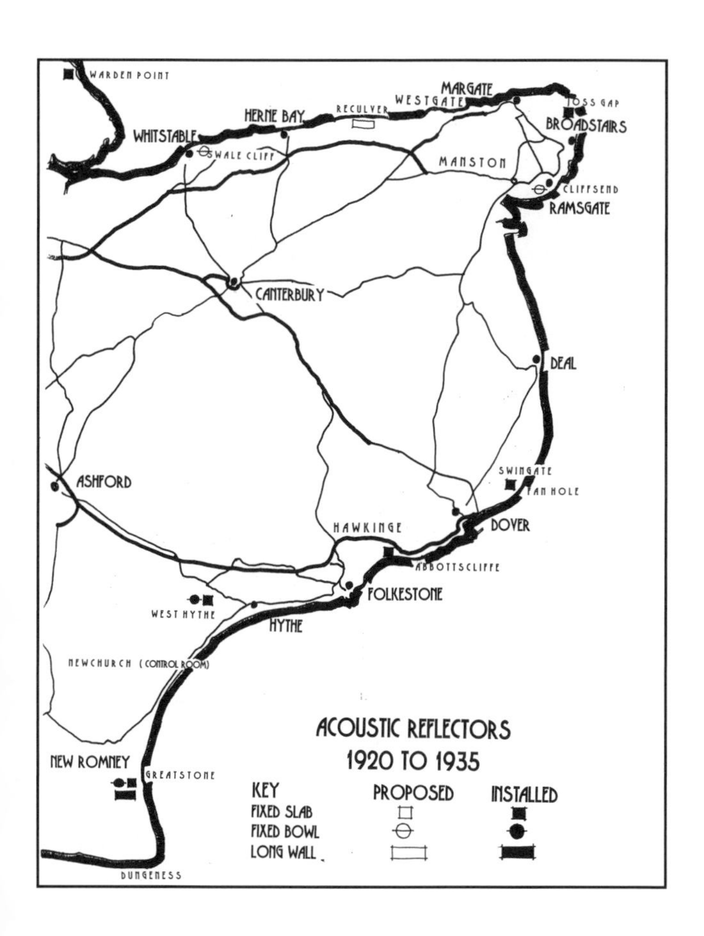

Left Map of the location of the remaining acoustic reflectors in the south-east of England *(PRJ)*

Right King George V is shown a military portable acoustic aircraft detector station *(BAW)*

that aircraft would not be very useful if the only way that approaching enemy could be detected was by standing patrols of airborne fighters — very wasteful of resources and not likely to be very successful in any case. What was needed was some system of early warning.

During the First World War, some attempts had been made to provide early advice of the approach of enemy aircraft using a system based on acoustic means, that is, by listening for the sound of approaching engines. In this period, a number of acoustic devices were tried, some portable and some static. In the postwar period, although money and enthusiasm for things military were significantly lacking, some further work on acoustic aircraft detection was continued on the south coast of England and in particular at the research station at Hythe.

In 1932, HE Wimperis, Director of Scientific Research at the Air Ministry, was a visitor to this station at Hythe and later, in 1934, Dr Robert Watson Watt, superintendent of the radio department of the National Physical Laboratory, also came to examine the acoustic reflectors that had been built at Hythe near the coast. AP Rowe, personal assistant to Wimperis, was also present and his reaction to what he saw may have been particularly significant and influential in the immediate future.

After this visit to see the acoustic reflectors in action, Rowe is known to have advised his superiors in the Air Ministry that, unless some new system of air defence was developed, Britain was likely to lose a war that might commence within the next 10 years. It seems highly

A Heyford biplane bomber as used in the radio direction finding (RDF) experiments of 1935 *(AOB)*

The first production model of the magnetron as used for centimetric radar (radio direction and ranging) designed by Dr JT Randall and HAH Boot at Birmingham University *(SAW)*

probable that this expression of concern led directly to the development of radar in Britain over the next 5 years up to the outbreak of war in 1939. (As an aside, AP Rowe ultimately came to Australia as the Vice-Chancellor of Adelaide University.)

In coming to the conclusion that was expressed in his warning to his department, Rowe had noted the limited range of the acoustic reflectors and more seriously their susceptibility to jamming by extraneous noise sources. In particular, wind noise was a major problem as was the noise of passing ships' propellers. Also serious was the inevitable masking of the approach of enemy aircraft when standing patrols of friendly fighters were flying in the vicinity.

With the benefit of hindsight, the concerns as expressed by Rowe following his visit to the Hythe acoustic experimental station were very well justified. With modern aircraft flying towards the coast at about 500 kilometres (300 miles) per hour, as was possible during the beginning of the 1939–45 conflict, the 30-kilometre (20-mile) range of the wall reflectors would allow a bare 4 minutes warning to get aircraft into the air to respond. Clearly something quite new was required and fortunately the solution lay just around the corner.

Soon after the 1934 inspection of Hythe, it is known that a comprehensive review of aerial early warning systems was undertaken by Wimperis and Rowe. In October 1934, Wimperis proposed the establishment of a committee of research on air defence to which HT Tizzard, later Sir Henry, was appointed as chairman. Rowe was made the secretary of this committee.

Meanwhile, Wimperis had consulted with Watson Watt as to the possibility of developing some form of 'death ray' to attack invading aircraft. During the early 1930s, 'death rays' had become a matter of popular concern and therefore the question had some rational basis. Wimperis in turn consulted with AF Wilkins of his department who carried out some simple calculations over the space of about half an hour. His conclusion was that, on the basis of contemporary apparatus, the concept of a 'death ray' was simply unfeasible. However, in reporting back to Wimperis, Wilkins made the suggestion that though 'death rays' were a fantasy in 1935, the reflection of radio waves from aircraft might provide a possible basis for a new method of detection. This notion was later presented to the first meeting of the Tizzard Committee by Watson Watt, who was asked to pursue the suggestion as a matter of urgency. This was to occur and soon he had produced the outlines of a system to detect the distance, bearing and height of aircraft to a range of about 160 kilometres (100 miles).

In February 1935, tests involving a Heyford bomber flying at 3000 metres (10 000 feet) and the detection of reflected short-wave radio energy were undertaken. One of the transmitters at the short-wave broadcasting station at Daventry operating in the 49-metre band was used. With the success of these tests, money was then set aside by the government for a research and development program. From this, in a remarkably short time, came the British system of radio direction finding (RDF), later known as radar (radio direction and ranging).

It was a system that was to be of vital importance, particularly in the early months of the Second World War when the capacity of Britain to resist aerial attack was extremely limited compared with the might of the Luftwaffe. There can be little doubt that, compared with acoustic methods of detection, radar was to make the difference between victory and defeat that had so much concerned Rowe in 1934.

As developed initially in the British air defensive warning system, known as 'chain home', the 160-kilometre (100-mile) range anticipated by Watson Watt was attained. However, a complementary system of reporting and plotting was essential. At an earlier time, associated with the acoustic method of aircraft detection, a sophisticated and easily comprehensible manner of reporting and presentation had been developed. This was now revived and refined to respond to the impact of the new chain home radio detection system. In seems quite conceivable that without this element of the new radar-based early warning system, the Battle of Britain could not have been won.

The acoustic reflectors of Hythe and Dungeness may have been rendered obsolete by new technology but the system of reporting and coordination that they led to was not. The light tables and gridded maps with coloured markers were to be a vital element in the first great test of the capacity of the Royal Air Force to protect Britain from attack — the Battle of Britain. Fortunately, radar made Stanley Baldwin's prediction that the bombers were unstoppable no longer valid. This ensured that, in 1940, Britain was not doomed to become simply a German colony.

As is now known, Germany, in the work of Hulsmeyer and of its scientists in the 1930s, was very well advanced in the development of radar. In the main, this work was related to systems that used centimetric wavelengths, whereas the British system was initially centred

The antenna towers of the chain home and chain home low RDF (later known as radar) system of 1942 *(SAW)*

Civilian use of chain home towers at Dover, Kent, in the south-east of England in 1992 *(PRJ)*

around the pre-war television broadcast frequency of 45 megahertz. For a defensive system this was to prove very effective when coupled with an integrated system of detection and plotting as was developed for Fighter Command.

A Royal Air Force plotting room for controlling the position and movement of fighter squadrons during the Battle of Britain *(BAF)*

By comparison, the German system was designed initially with an offensive capability in mind and for this the shorter wavelengths were more useful. However, the German centimetric system was to be hampered by lack of power output from the valves then available. As things turned out, it was the British who were to overcome this problem with the invention of the cavity magnetron.

This extraordinary new thermionic device, with its many kilowatts of pulsed centimetric wavelength radio frequency energy, was to provide an immense improvement in what radar was able to do both on land and in the air. Unfortunately for the Allies, an aircraft carrying the new device was shot down very shortly after the magnetron was put into service. As a result, German scientists were able to make copies for use in their own radar systems which, even without the high power available from the new device, had been very effective.

The output of the magnetron when projected through a suitable high-gain antenna was physically like an invisible beam of light. As a result, it was possible to use it to create maplike images on a cathode ray tube, known as a plan position indicator (PPI). This allowed night flying to become the basis of a new, deadly form of bombing. In addition, mounted in fighter aircraft it made possible effective mid-air night interception. As the war went on, this would become more and more essential as night bombing became more frequent.

One other important device that was developed for use in radar might appear to be of relatively little significance at first sight, being based on a principle well known to radio engineers at that time. This was an adaptation of the crystal detector of 20 years earlier, which was found to be particularly effective in rectifying and detecting the centimetric radar reflections that were produced by the magnetron. In an endeavour to make this form of rectifying detector more stable and predictable in its operation, experimentation led to the production of what would now be called a 'junction diode'. In this device, neatly contained in a small ceramic case with gold-plated end contacts, the junction between the wire and the crystal involved a connection between the two surfaces stabilised with hard wax. Unlike the famous 'cat's whisker', jolting or gunfire would not dislodge the connection.

However, in the development of the wartime junction diode was the seed of an incomparably more important device — the 'transistor'. In only 2 short years after the cessation of hostilities, the physics of solid-state devices would at last be discovered. Just on 40 years after the discovery of rectification in crystaline substances, the scientific principles would be determined and lead to a new class of revolutionary amplifying devices which would herald the start of a new electronic age.

In terms of the story of the Web, radar may at first sight seem rather far removed from the mainstream of development. This is, in reality, an illusion, for what radar did was to open up the ultra high frequency (UHF) part of the radio frequency spectrum and create familiarity with its peculiarities. Soon this knowledge was put to work in communications networks as microwave links were developed for use on land and as a substitute for expensive terrestrial cables. UHF would also be used in the satellite communications systems that were later developed, having a greater bandwidth and a far greater capacity to carry information than had lower frequency systems.

Centimetric radar installed in an aircraft with the plan position indicator (PPI) screen visible *(SAW)*

However, in the creation of a new version of an apparently simple but well-known detecting device as applied to the detection of the ultra high frequency energy used in radar, one can now see the next crucial step in the chain of events leading to the Internet. The junction diode would pave the way for all the transistors and microchips used in modern computers and communications systems. In this respect, radar can be seen to have made an enormous contribution to the technical development of communications as used in the Internet.

CALM BEFORE THE STORM

In the early 1930s, a new and baleful presence on the international scene was to erode resolutions to avoid further warfare. In Britain, America and France, the terrible impact of the First World War had produced a high degree of public revulsion for warfare coupled with a rise of overt pacifism. More seriously, in America a public attitude of isolationism and a desire to have nothing to do with the problems in Europe prevailed. However, in Germany, the continuing repercussions of the First World War were to produce extreme social instability.

In the 1920s resentment at the terms of the enforced surrender by Germany and a hunt for 'scapegoats' to explain the debacle of the treaty signed at Versailles led to the emergence of a new and thoroughly unpleasant force in German politics, the National Socialist Party. Under the leadership of a highly motivated and resentful former frontline soldier, Adolf Hitler, this party were to be far better known as the Nazi Party, a convenient contraction of the German name for this group.

By the middle of the 1930s it was apparent to those who were not blinded by pacifist sentiments that a new war was inevitable. Once again the implications of radio communications in time of war, including ciphering and the need for portability and light weight, were to have a significant impact on the work of the big corporations and the form of equipment and modes of operation. In many respects, the preparations to meet a new cipher war were in the longer term to be the most crucial of the steps taken during this period of comparative calm before the new storm about to blow across the world.

As the storm clouds of war gathered ever more thickly over Europe, the British government, realising that the battle of the ciphers was likely to be crucial, set about finding a safer base for the codebreakers than had been available in that earlier conflict. Between 1914 and 1918, as earlier described, British Naval Intelligence had located its codebreaking group in Room 40 of the old Admiralty building (40 OB) in central London.

With the development of air power and the heavy bomber in the period before 1939, it was anticipated that aerial bombing was likely to be a very serious problem. As the Blitz was to demonstrate, this was a well-founded concern. With this anticipated threat in mind, it was considered that central London was inevitably a potential target for the bombers. Accordingly, the government searched for alternative locations for sensitive and vulnerable activities such as code breaking, well away from London.

As a new base for the cipher personnel, an ideal place was found, strategically located on a line joining the university towns of Oxford and Cambridge on the main rail line north from London — a mansion close to the small town of Bletchley.

The large, late-Victorian mansion known as Bletchley Park that was located at the edge of the small town had been built by the wealthy philanthropist Sir Herbert Leon. Following his death in 1926, his wife continued to live on the estate until her own death in 1937. The property had then been sold to a group of businessmen led by Captain H Faulkner. These developers had purchased the property with the intention of converting the house and grounds into a new housing estate.

The intervention of the Foreign Office through what appears to have been a compulsory purchase of the premises brought an abrupt halt to these plans but not to the building activity on the estate. On the contrary, over the next 5 years, the site was a hive of activity as the many structures that were required to house the ultimate population of the grounds were progressively erected. This influx of personnel who were to be stationed at Bletchley Park and employed in the work of breaking the German codes was ultimately to number in the thousands. The story of their work is taken up a little later

FARNSWORTH AND ZWORYKIN

The demonstration of electromechanical television by John Logie Baird in January 1926 was noted in many places and, over the next 5 years, many people and organisations imitated this initial work. In the majority of instances this merely involved refinements of the original Baird apparatus. Despite the very limited results that a mechanical system could achieve inherently, the observations of AA Campbell Swinton after 1911 concerning the need for an all-electronic system did not seem to be widely known or to produce an appropriate response.

In January 1931, in a very perceptive article in the American periodical *Radio News*, A Dinsdale presented a blistering attack on the electromechanical systems then being actively promoted in the United States. Dinsdale called for the introduction of some new principle of television and noted:

> Television as we know it today is capable of giving only a head and shoulder view of the person seated before the transmitter and that more or less imperfectly.

Noting the earlier writings of Campbell Swinton, Dinsdale suggested that it was imperative that television needed to be improved, firstly by increasing the size of the image as presented to the viewer and secondly by increasing the amount of detail available. He also correctly observed that increasing the number of scanned elements as suggested would also inevitably lead to a marked increase in the bandwidth needed to broadcast a television signal.

Using as a basis for his calculation a displayed image 304 millimetres by 304 millimetres (1 foot square), Dinsdale observed that scanning this image so as to produce a picture with acceptable detail would involve 100 lines to every 25.4 millimetres (1 inch). This, he observed, would produce 1.4 million picture elements, or what would now be described as pixels. If this array was

Back view of a Baird televisor showing the Nipkow scanning disk, 1928 *(WIT)*

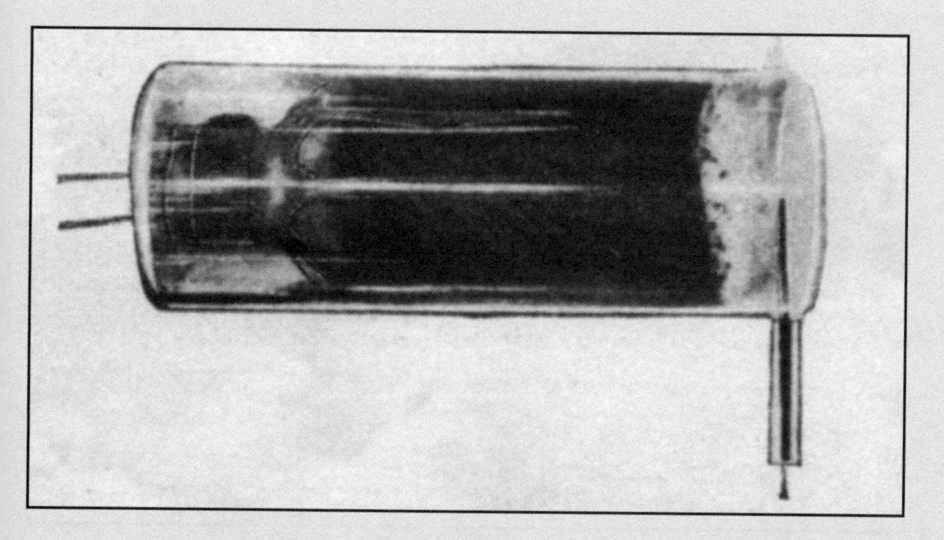

The Farnsworth 'image dissector' tube of 1932 *(USM)*

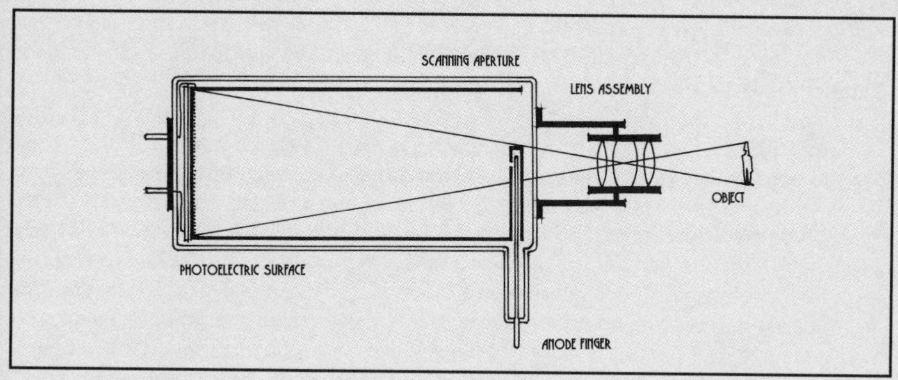

Diagram of the internal construction of the 'image dissector' *(PRJ)*

scanned 20 times a second, he reasoned that this would involve a signal bandwidth of 14.4 megahertz and this could be accommodated only at a wavelength of 1 metre or 300 megahertz.

Comparing this with the then currently available systems, which commonly used raster scans of 45 to 60 lines and frame repetition rates of less than 20, Dinsdale poured scorn on mechanical scanning as ever being capable of producing an acceptable image.

Based on the above analysis, it can be seen that it is only recently with the advent of 'high-definition' television that the pixel rates per millimetre proposed by Dinsdale have become available to the general public so as to achieve an adequate quality of image.

Dinsdale concluded his 'blast' with the observation:

> That television in its present crude and limited form is not acceptable to the general pubic is evidenced by the fact that attempts which have been made both in this country and in England to sell commercial televisors have failed ignominiously. The reason is not far to seek. There is definitely no entertainment value in the present image.

Remarkably, only 4 months later, in May 1931, *Radio News* carried a story about the Farnsworth television system and described in some detail the all-electronic methods that Philo Farnsworth was employing. In particular, this article provided a detailed description of the operation of the Farnsworth 'image dissector' camera and presented a drawing and circuit diagram.

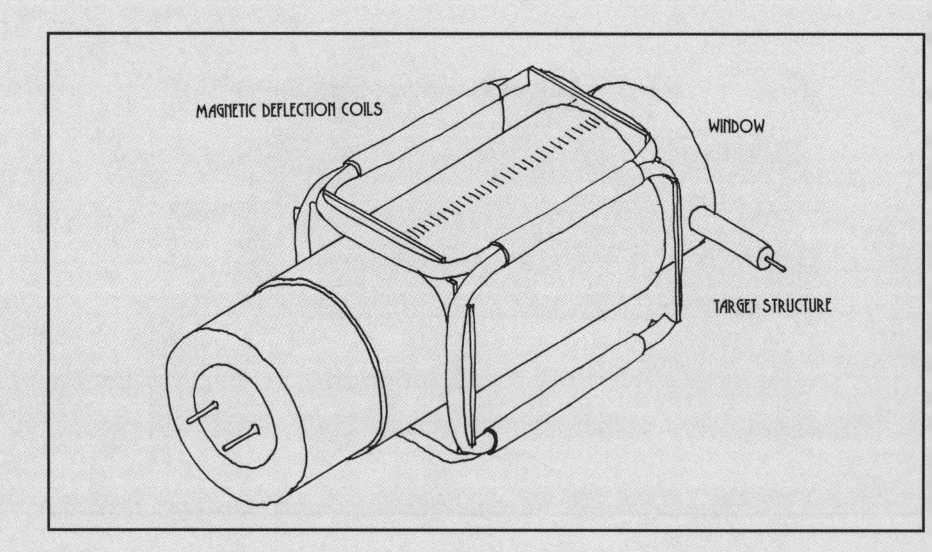

The deflection coils for Farnsworth's 'image dissector' tube *(PRJ)*

As it said in the article (the numbers refer to the diagram which accompanied the article):

> An optical image of a moving object **5** is focused through a lens **3** on to a silvered mirror **6**, this being coated with a material which emits electrons when exposed to light. These parts constitute a sensitive photo-cell of a vacuum type, enclosed in a cylindrical glass tube **1**. The mirror **6** is the cathode. Closely adjacent and parallel to it is an anode **7**, which is maintained 500 volts positive with reference to **6**, by means of a direct current source **8**. The anode consists of a finely woven wire cloth through whose interstices the liberated electrons are projected into the equi-potential space formed by the shield **10**.
>
> Sweeping across the equi-potential space are two electro-magnetic fields which are set up by 'saw-tooth' alternating currents, in two sets of coils placed at right angles around the tube. When one set of coils, diagrammatically represented by **15**, is supplied with a 16-cycle current from an oscillator **16**, it causes a magnetic field to sweep vertically across the tube 16 times per second. When the other set of coils, which is not shown in the diagram but which can be seen in the perspective view, is supplied with a 3000 cycle current, a magnetic field is swept horizontally across the tube 3000 times per second. Their resultant effect upon the electrons in the equi-potential space is to form them into a cathode ray image which successively issues from each tiny element of picture area. This cathode ray is then magnetically focused through the small aperture **11** onto the target or electron collector **13**.

The method of reception of television images in the Farnsworth system appears to have been via a cathode ray tube and, to that extent, was similar to the method proposed and patented by Vladimir Zworykin while working for Westinghouse Corporation in the mid-1920s.

The Zworykin television receptor which he called the 'kinescope' used electrostatic elements to achieve fine focus and image intensity. Although in principle this application of the cathode ray tube in a television display was generally similar to modern television tubes, focusing represented a significant difference. In a modern television tube, focusing is achieved by means of a magnetic coil, as in Farnsworth's camera described above. The diagram opposite shows the main elements of the Zworykin kinescope as it was to be incorporated into the television system developed for RCA, with the encouragement of the managing director, David Sarnoff.

In many respects, the work of Zworykin in producing a television camera that would rival the Farnsworth camera was of more importance than the display unit. It can now be seen that it was to lead almost directly to the next generation of improved cameras that had one vital advantage compared with the Farnsworth model — they had a capacity to store the electric charges of individual picture elements and this in turn allowed the tubes to operate with significantly increased image brightness. In contrast, the geometry of the Farnsworth tube meant that it did not suffer from distortions that were an inevitable consequence of the shape of the tube invented by Zworykin which he called the 'iconoscope'. This tube had the scanning beam set at an angle to the face of the image-forming mosaic and consequently the distorted image had to be corrected electronically.

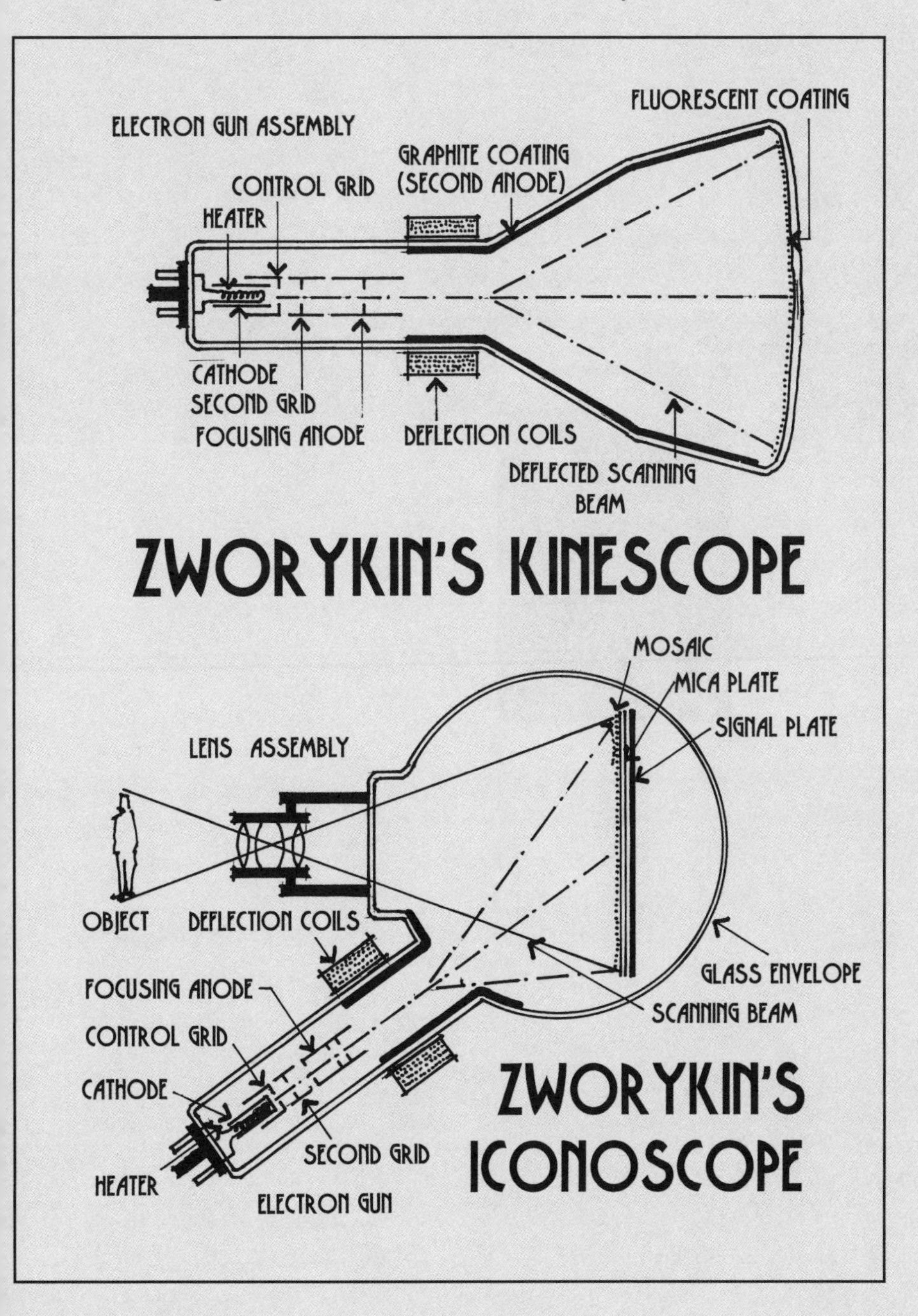

Diagrams showing the internal construction of the kinescope and iconoscope of Vladimir Zworykin, 1930 *(PRJ)*

CHAPTER

9

1918 to 1939

Waves in the air

During the period between the two world wars, there was a rapid expansion in the use to which radio communication was put. Apart from the major developments discussed in the previous chapter, more specialised uses were found for radio in which portability was essential. Such uses involved vehicles, vessels and aircraft, in all of which the driver, helmsman or pilot needed to maintain contact with ground headquarters without access to a conventional wired telephone or telegraph system. During this period, organisations such as the police, fire brigades and the ambulance service all started to make use of the life-saving capability which radio communications could provide.

Early users of radio communications in a mobile situation were the Detroit police force in the United States in 1921, followed by the London metropolitan police force in 1923. In 1928, the Brighton police force began tests with what were later to be called 'walkie talkies', and these new devices were brought into service in 1932.

In addition to the obvious utility of portable radio communications for mobile operations, radio was able to provide long-range communication with isolated rural landholdings, or 'stations' as they are called in Australia. In such remote locations, conventional land lines and telephones were usually not available. Australia, with its huge geographical distances, was in the forefront of a civilian and private-operator communications revolution. In America, Europe and elsewhere, radio amateur communications also demonstrated the value of a network of personal connections, long before the linkages of world communication started to be forged.

AUSTRALIA

Australia is a continent in which distance has a profound impact. Its inhabitants who live away from the cities on the eastern and western seaboard have particular problems to contend with. During the 1920s and 1930s, two events had special significance for the development of Australia and continued a process commenced in 1872 — radio broadcasting and the aircraft helped banish the extreme isolation of the continent.

For all practical purposes, whatever the affections and antecedents of its inhabitants, in the years after the first British colony was established on the shores of Sydney Harbour, Australia had been as totally disconnected from Europe as an orbiting satellite in modern times. From the time Captain Cook landed on Australian shores in 1778, its British colonists may well have seemed to Europeans to be living in another universe, so distant did the other side of the earth appear.

When Cook set out in his converted collier, the *Endeavour*, to circumnavigate the world and discover fresh territory for the British Empire, there must have been many who believed that he would never return. Return he did and with reports of a huge continent populated with a race which was seen as unable to resist the declaration of sovereignty of the British Crown over this land. This presented an immediate opportunity to develop a new and effectively uncontested part of the globe.

In the latter part of the 18th century, the remoteness and inaccessibility of Botany Bay and Moreton Bay provided a marvellous means of removing a running sore from the British body politic — the large number of malcontents and recidivists of all shades and degrees of wickedness. Australia was seen as a convenient and comfortingly distant repository for all the most recalcitrant and uncontrollable criminals of that period locked away in British prisons and 'hulks'. Even after the

A patient is delivered to the aircraft of the Royal Flying Doctor Service *(RFDS)*

practice of convict resettlement had been abandoned, Australia would continue as a far-flung outpost of civilisation until the advent of the electric telegraph.

As discussed earlier, the connection to Europe by telegraph in 1872 began the process of integration of Australia with the northern hemisphere and other parts of the world. Perhaps one can see the World Wide Web as having at last completed that process.

THE ROYAL FLYING DOCTOR SERVICE

From the First World War came the remarkable growth of radio broadcasting and the 'short waves', and also the newest form of transportation, the aircraft. In combating the vast distances of continental Australia, aeroplanes were to have a major impact on the isolation of the inhabitants. In conjunction with radio, it was soon possible for the loneliest outpost to consider itself connected to the rest of the Commonwealth.

In many respects the most potent application of these two technologies used in concert was to come from a perhaps rather unlikely source, the Reverend John Flynn of the Inland Mission of the Uniting Church. In his efforts to develop what would become known as the Royal Flying Doctor Service, he applied a principle that would ultimately underlie world communication as a whole. This was to allow people to be connected via wire and radio wave for safety, for health, for pleasure and for education.

Although it was Flynn who was the driving force in setting up the Flying Doctor Service for the inhabitants of the remote areas of Queensland, the original idea was presented in a magazine article published by the Presbyterian Church. The author of the article was a young medical student, Clifford Peel, who had a profound interest in aircraft and had abandoned his studies to join the Australian Flying Corps in Europe. The article had been written in 1917 as he was travelling to England but was not published in Australia until October 1918. Tragically, Peel had been killed while flying over France, just a month before his article appeared in print. His memorial was to be the life work of the Reverend Flynn who created what he had proposed in his article.

Flynn had been ordained into the Presbyterian Church in 1911 and later was intimately involved in the establishment of the Australian Inland Mission. During the latter days of the First World War, he and his sister raised a considerable amount of money for the purchase of aircraft to start the flying medical service that Peel had envisaged.

At this time, apart from the trans-continental telegraph line running from Adelaide to Darwin, wired connections to any of the inland towns were virtually non-existent. Only on the edges of the continent was a network of telephone cables starting to spread. Flynn realised that having a capacity to attend the sick in the Outback was all very well but, without a means by which such distant people could call for help, the service would be futile. Clearly the answer lay in the rapidly developing technology of radio communication.

The Reverend John Flynn and George Towns in their Dodge set off to explore the inland for radio communications possibilities *(RFDS)*

Alfred Traeger demonstrates his pedal wireless system *(RFDS)*

By 1919 Flynn had decided that the answer to his need for long-range communications was a compact and portable form of transmitter and receiver to be located at every inland homestead and station. However, it needed to be simple enough to be used by untrained operators. As he was to say later, 'Without a wireless transmitting station at every isolated habitation, an aerial medical service would be 75 per cent futile.'

Having a practical turn of mind, Flynn decided that, in order to fulfil his ambition, it would be necessary to learn about radio. Accordingly he set out to obtain a radio experimenter's licence which would make him eligible to become a member of the Wireless Institute of Australia. After much hard work, he received his licence in 1925 and shortly thereafter set off on an expedition into the inland to test his theory about the use of radio for long-range communications. With him and his companion, George Towns, in an open car went a 10-watt amplitude-modulated (AM) transmitter, a standby continuous wave (CW) transmitter and two short-wave receivers. In addition, he carried a broadcast receiver and a four-section metal mast to use as the antenna. This consisted of metal tubing which, when assembled, was about 13 metres (40 feet) tall.

The expedition was generally a success but what became patently apparent was that, if a reliable form of communication was to be achieved over the distances involved, it would have to be a service using CW rather than voice. Technology was simply not up to the task of providing a really compact and long-range voice system at that time.

Mrs Gertrude Rothery uses the first Traeger-designed pedal radio in 1929 *(RFDS)*

During this expedition, Flynn came in contact with another keen electrical experimenter, Alfred Traeger, another radio amateur with a very practical ability to create the sort of equipment that Flynn had been hoping to use. These two would later become involved in a long and very fruitful partnership to establish the Inland Radio Service.

With the early call sign 5AX which later became VK5AX, Traeger progressively developed a compact form of transceiver. This ultimately became the basis of a communications network to bind the inhabitants of the inland into a new form of radio community which still exists.

In his first attempt at developing portable apparatus suitable for use in the inland, Flynn had designed a small CW valve transmitter running at about 10 watts. The valves of this transmitter were supplied with high voltage produced by a Model T Ford ignition coil. The signal that this transmitter provided was known as interrupted continuous wave (ICW). The ragged and unsmoothed alternating current pulses available from the coil were generated by a 6-volt accumulator. At the time, this was quite a common device in the inland and could be charged up by means of a wind-powered generator. When travelling, Flynn would obtain the necessary power by removing a wheel of his vehicle and driving a portable generator with a drive band.

A Royal Flying Doctor Service transceiver for use with morse code, installed at Augusta Downs in 1929 *(RFDS)*

Flynn now had created the first of what he called the 'baby' stations. These were intended to be used in conjunction with the high-power base stations which, logically, were called 'mother' stations.

Following the recruitment of Alfred Traeger, this first transmitter was substantially improved by the introduction of a hand-cranked power supply. This produced reasonably clean high-voltage direct current, with only a minor degree of residual generator whine and voltage fluctuation. In addition, crystal frequency control was becoming increasingly popular at that time and this was also incorporated into the Traeger-designed transmitters. The introduction of crystal control of the transmitter frequency produced a notable increase in stability of the signal. Despite the variable voltage output from the rotary generator,

Base operator from Cloncurry, Maurie Anderson, supervises the erection of an antenna at the Birdsville Hospital in the Outback of Australia *(RFDS)*

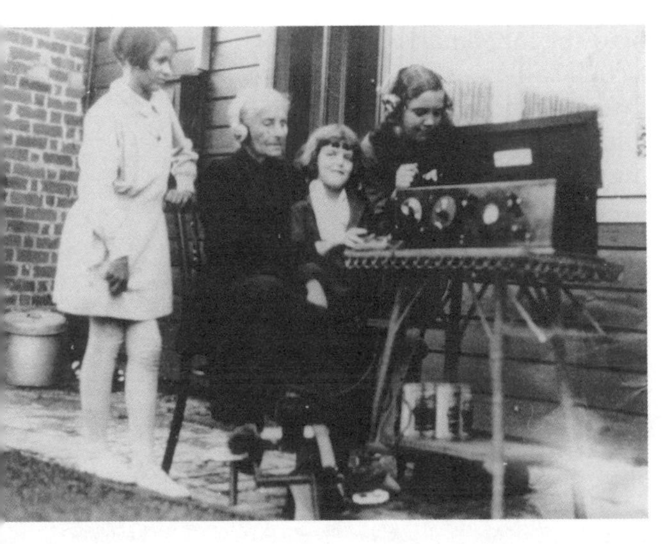

drifting of the signal away from its proper frequency was now prevented. This had been a significant problem in the original Flynn-designed transmitter.

Further field tests with this new apparatus were generally satisfactory to all but Traeger. Upon his return to Adelaide, he set about producing an even better system. This time he developed a foot-powered high-voltage generator producing 350 volts at a power of about 20 watts. This device was to be the basis of inland remote-area radio stations for many years to come. It was eventually replaced by the vibrator style of power supply and, later again, by solid-state systems using low voltage.

By 1929, at last Traeger was reasonably satisfied that he had developed a reliable transmitter and receiver combination. This integrated assembly was frequently referred to as a 'transceiver', a word that would ultimately be universally applied to such combined apparatus. A main station was installed at Cloncurry in north-west Queensland with the call sign VJI, providing a 200-watt, amplitude-modulated voice signal. The first six 'baby' stations were installed at Augusta Downs, Lorraine, Gregory Downs, Birdsville, Mornington Island and Corinda, all using CW to allow morse code signals to be sent to the 'mother' station. This initial network had a superficial area of some 233 000 square kilometres (90 000 square miles), a little less than the area of the United Kingdom (which includes Northern Ireland).

Top Grandma shows the children how to use the Royal Flying Doctor Service transceiver *(RFDS)*

Bottom Alfred Traeger building radio sets in the 1930s *(RFDS)*

Below The 'mother station' at the Hermannsburg Mission as constructed by Traeger and Flynn at the 5CL radio station workshops in Adelaide *(RFDS)*

In these early transceivers, Traeger had made use of the early four-electrode valves, the tetrodes, which could be run with a very low plate voltage. By coupling two bias batteries, it was possible to obtain a plate supply of 9 volts which, with regeneration, provided acceptable reception on the short waves. The transmitter portion of the transceiver used a triode operating in Class 'C' to produce about 10 watts of CW which again was adequate to provide reliable contact with Cloncurry using morse code.

In an emergency, the station owner could call up Cloncurry with a morse signal and the answer would be given with a voice signal. Later, the problems of learning morse code were overcome by the introduction of a mechanical morse code generator. This device, which Traeger had invented, resembled a large mechanical typewriter. This machine made it possible for anyone to send respectable morse signals. Again this was used until voice operation became more common in the latter part of the 1930s as new apparatus became available.

The initial six stations in the 'outback' network of the Royal Flying Doctor Service, as it would ultimately be known, have grown to involve

The *Southern Cross* at Oakland after conversion to its long-distance flying role *(QAN)*

thousands of stations spread throughout the vastness of inland Australia — an amazing development all based on the rather improbable alliance of a Presbyterian clergyman and a self-styled 'agitated ant', both with an interest in long-range radio communications and both with radio amateur licences. This was a partnership in which the vision of Flynn to end the isolation of the inland found a worthy partner in the determined ingenuity and technical ability of a country lad, Alfred Traeger.

THE *SOUTHERN CROSS*

The next event involving portable radio communications is well known to many people, especially Australians. This was the epic flight, begun on 31 May 1928, of the *Southern Cross*, piloted by Charles Kingsford Smith and Charles Ulm. The successful completion of this flight would see another distinguished pair of names carved into the record of Australian exploration.

Designed by the aeronautical firm Fokker, the *Southern Cross* was a commercial aircraft of that period and very well suited to the long-distance journey that the two partners, Kingsford Smith and Ulm intended to make. This was a flight from America to Australia with a series of intermediate stops to refuel along the 11 896-kilometre journey (7389 miles) from San Francisco to Brisbane.

With the initial encouragement of the Australian government, the two adventurers had purchased from the explorer Sir Hubert Wilkins the remains of two Fokker transport aircraft that had been used for Polar exploration. These two aircraft were 'cannibalised' to

The Fokker at the Boeing works, Oakland, prior to conversion to the *Southern Cross (QAN)*

Left The *Southern Cross* at Suva during the crossing of the Pacific *(QAN)*

Right Charles Kingsford Smith sits in the cockpit of the *Southern Cross*, with Harry Lyon, Charles Ulm and James Warner standing in front *(QAN)*

create a new aircraft that was named *Southern Cross*, at the suggestion of Kingsford Smith's initial partner, Keith Anderson.

Originally constructed in Holland at the factory of Anthony Fokker, the Java-born founder of the company, the aircraft that had been purchased from Wilkins were initially assembled in New Jersey in the United States by the Atlantic Aircraft Corporation. Following the purchase by Kingsford Smith and Ulm, they were shipped to the Boeing Aircraft Company in Oakland where they were rebuilt and new Wright engines, paid for by the Melbourne philanthropist and emporium owner Sidyney Myer, were installed.

Given the extent to which Boeing aircraft now pass along the route from America to Australia and back again, this was an appropriate relationship to have been established at such an early stage.

With a nominal maximum cruising range of 6740 kilometres (4187 miles) based on a fuel consumption of 4730 litres (1040 gallons), the Fokker tri-engine transport was intrinsically a safe and appropriate aircraft for the projected voyage over the waters of the Pacific. Because of its size, it was able to accommodate the two pilots and two crew in relative comfort for those times. Unfortunately, the cabin was not entirely enclosed and, as a result, in bad weather the pilots would get saturated and, in addition, the noise of the engines was deafening.

The flight was heavily subsidised by a philanthropic citizen of the United States, Captain G Allan Hancock. The radio apparatus was American, as were the navigator, Captain Harry Lyon, and the wireless operator, James Warner, who had been a shipmate of Lyon.

The *Southern Cross* was a monoplane aircraft with three engines, two in the wings and one in the nose. The latter made forward visibility somewhat difficult and, as a result, landing and taking off tended to be far more difficult than in later aircraft. By modern standards the Fokker was a miniature aircraft, with a wingspan of just under

Left Charles Kingsford Smith, Harry Lyon and Jim Warner standing with Charles Ulm sitting during the parade through Sydney *(QAN)*

Right The *Southern Cross* flying with a temporary solution to the engine-noise problem, silencers, curving over the wings *(QAN)*

22 metres (72 feet). Into these wings and fuselage went 5906 litres (1298 gallons) of fuel, giving the aircraft a range of 5868 kilometres (3645 miles) when travelling at 145 kilometres (90 miles) per hour under reasonable weather conditions. The three engines were Wright Whirlwind J5C of 220 horsepower each and, at a cruising speed of 150 kilometres (94 miles) per hour, rotated at 1600 revolutions per minute.

The flight involved three hops. The first was from Oakland Airport near San Francisco to Wheeler Field near Honolulu, in the Hawaiian islands — this involved a flying time of 27½ hours. The second leg of the journey from Barking Sands (some 145 kilometres (90 miles) from Wheeler Field) to Suva involved a flying time of 34½ hours over a distance of 5050 kilometres (3138 miles). The third and final leg from Naselai Beach, 32 kilometres (20 miles) from Suva, to Brisbane involved a further 20 hours of flying.

In order to undertake this remarkable journey, and to ensure that the intermediate refuelling points were actually reached, the adventurers had access to astro-navigation, dead reckoning and the newly installed radio beacons.

To conserve fuel, they planned to fly at a height of about 270 metres (600 feet) above the sea, and generally this altitude was maintained except where avoidance of tropical thunderstorms was required. On a couple of occasions, Kingsford Smith was forced to fly at an altitude of 2440 metres (8000 feet) to avoid thunderstorms, but even so the aircraft was forced to fly through torrential rain which thoroughly soaked the flyers.

Despite some problems with the radio and antenna system, compounded late in the flight by the loss of the earth inductor compass, what is remarkable about this long-distance flight is that it went as smoothly as it did. A tribute to the fine engineering of both the airframe and the engines — apart from frightening moments associated

The *Southern Cross* on display at Brisbane Airport in 1998 *(PRJ)*

with thunderstorm activity — the flight went almost exactly as planned, with the *Southern Cross* arriving in Brisbane as anticipated.

The only mistake that Kingsford-Smith was able to describe in his memoirs of this historic flight was the failure to oil the earth induction compass. This error resulted in the flyers making their landfall over Ballina some 180 kilometres (110 miles) south of Brisbane. During the final leg of the flight from Suva, for navigational purposes it was necessary to rely on a conventional magnetic compass that was known to be affected by the metalwork of the aircraft.

What perhaps the pilots had not anticipated was that there would be a crowd of 15 000 people to greet them in Brisbane. By the time of their arrival at Mascot in Sydney, the crowd of wellwishers had become a remarkable 300 000. Something else that had not been expected was temporary deafness caused by the noise level of the engines. Despite the size of the crowd, the welcome was completely inaudible to the four adventurers.

Since that historic flight, passenger movements have steadily grown to many thousands each year. Now flights to America in an easterly direction and to Europe in a westerly direction are routine and, where the *Southern Cross* lumbered along at a safe but slow 145 kilometres (90 miles) per hour, the Boeing 747 jets roar across carrying upwards of 300 passengers at a speed of nearly 1000 kilometres (620 miles) per hour. What was in the time of Kingsford Smith and Ulm a dangerous and very uncertain adventure is now routine and, for the most part, safer than driving a motor vehicle.

Now the uncertainties of terrestrial and celestial navigation have been almost entirely removed by the introduction of radio systems of communications and guidance, including the GPS, or satellite-based global positioning system, which is one of the better side-products of the 'cold war'. Another such product, without a doubt, is the the Internet.

CHAPTER 10

1871 to 1939

Comptometry and analysis

From the time of his death in 1871 and for the best part of 70 years, the work of Charles Babbage was consigned to a forgotten corner of history as an aberrant, eccentric and somewhat peculiar manifestation of the Victorian era. However, for those people interested in the science of numbers, his machines and dreams were known. As technology developed over the latter part of the 19th century after Babbage's death, gradually things that he had speculated about started to be incorporated into machines used in science and in the business office.

Initially, the work of the Scheutz family in Sweden saw the creation of difference engines of a rather simpler but workable design compared with those of Babbage. Such machines were used to reduce the burden of calculating large numbers. However, in the longer term perhaps the most important event was the gift to Harvard University of a section of one of Babbage's machines by his son, Colonel Henry Babbage. This action was finally to bear fruit in the mid-1930s.

A Chinese abacus, the original 'calculating machine' *(PRJ)*

A 'difference engine' as made by Georg and Edvard Scheutz in Sweden in 1854 *(PRJ)*

Part of the analytical engine as constructed by Charles Babbage's son, Colonel Henry Babbage *(PRJ)*

HOLLERITH

As described earlier, one of the interesting innovations relating to the science of calculation embodied in the analytical engine of Babbage was the use of Jacquard cards for input of numbers and to set up the program operation. These had been borrowed from the fabric-weaving industry in which they were used to establish the pattern to be incorporated by the looms. This was, in principle, the same system of control of a mechanism as was embodied in the paper tape with punched holes used in the player pianos of the late Victorian era. Through the location of the punched holes in the paper piano roll and their alignment with air-actuated valves, the 'pianola' could entertain families in which the skill of piano playing was not available but a desire to have a fashionable instrument on display in the drawing room was.

The 1880 Census of American citizens took a considerable labour force of clerks 7 years to analyse and compile. Further, the clerical task involved exacting and infuriatingly repetitive work in tallying the totals and this was bound to produce frustration and errors. It was realised that with a growing population the census would be a recurring problem of increasing magnitude but, because it involved data of great importance to a newly developing nation, it could not be abandoned. As a result, the census bureau decided to invite tenders for the carrying out of the next census. Of the various methods proposed, it was that of Herman Hollerith that was ultimately accepted.

Hollerith used as his basic tool of analysis a card based directly on the Jacquard loom control card. The card was 168 millimetres ($6\frac{5}{8}$ inches) by 82 millimetres ($3\frac{1}{4}$ inches) and was set out with a grid of numbers running up to 960. Holes were punched next to the printed numbers, representing the data collected by the census collectors. These cards were then entered into a machine that was able to read the data contained on each card. This was done with an array of metal fingers which completed a circuit. The numerical amounts were then automatically added to the total on a meter assigned to each element of the census. This machine eliminated a time-consuming and repetitive task prone to error but, more importantly, undertook the summation process in about a tenth of the time that the clerks had been able to achieve.

A simple mechanical adding machine from the 1890s *(PRJ)*

The introduction of the Hollerith system reduced the time to analyse and process the 1890 census to 2½ years and provided an answer that was not received with complete enthusiasm by the legislators. The total population had been expected to reach 75 million and the census analysis revealed that the population was actually 62 622 250, based on 13 million census forms returned. To achieve this result, 45 000 collectors had been required in the field, and in Washington a group of 2000 clerks had been required to sort, tabulate and total the data using the Hollerith machines. Following the success established in carrying out the 1890 census, Hollerith formed the Tabulating Machine Company in 1896.

THE MECHANISED OFFICE

With a rapidly growing population and demand for services, the office of the latter part of the last century needed labour for basic clerical tasks that simply was not available in America as was the case in Europe. For this reason, America was a particularly fertile place for the growth of mechanisation as a substitute for labour. As a result, in the period from about 1880 to 1918, some of the most basic and popular office machines were developed there.

Typical of the response to the demand for labour-saving office machines was the typewriter. Because of its speed, accuracy and clarity, this machine could replace several clerical handwriters who, in European offices, spent much time in laboriously copying letters into ledgers. One of the most successful of the early typewriters was produced by the armament firm, Remington, which later was merged with the Rand Corporation to become the well-known firm Remington–Rand.

The Rand Corporation was a father-and-son business set up to serve the needs of offices for filing and document systems and control. Integration with the typewriter company was to create a formidable conglomerate to provide for the needs of business.

Another area of business machines that was a direct answer to the absence of skilled clerks, who, in Europe, were referred to as 'comptometers', involved calculating machines. Initially able to do little more than add and subtract, by the end of the 19th century these machines were developed to perform not only these functions but to multiply and divide also. Typical of the period, the Burroughs calculating machine provided a range of such functions but was overtaken by the printing calculator of Felt and Tarrant. In this later machine, the results of an operation were printed onto paper tape rather than simply appearing in a window as in the case of the Burroughs machine.

NATIONAL CASH REGISTER

Under the inspired leadership of John H Patterson, National Cash Register (NCR) not only developed a most successful line of cash registers, designed to stop petty theft in small businesses, but also invented a system of machine leasing and selling which was to become the model for businesses up to the present day.

During the reign of Patterson, NCR had concentrated on the production and sale of cash registers exclusively and by 1922 it had sold 2 million of these machines. However, despite this success, following the death of Patterson the firm was redirected towards mechanised accounting systems and produced machines similar in general function to the Burroughs machines.

However, before this, in 1911 one of NCR's most successful salesmen and managers, Thomas J Watson, had been discharged by Patterson over some minor incident and had taken over the Hollerith enterprise, by now known as the Tabulating Machine Company. Since the time of Hollerith, this firm had moved into the field of general business machines, using punched cards as data input, and had generally relinquished the specialised field of census analysis to others.

A Facit mechanical calculating machine *(PRJ)*

A Brunsviga mechanical calculating machine *(PRJ)*

INTERNATIONAL BUSINESS MACHINES (IBM)

In the early 20th century, an entrepreneur who might now be called a 'business integrator', Charles Ranelegh Flint, negotiated a business merger in which the Tabulating Machine Company bought out the interest of Hollerith for $1 210 500. In addition, he added to this core two other organisations, the Computing Scale Company and the International Time Recording Company. This new conglomerate was given the name the Computing, Tabulating and Recording Company (CTR). In 1924, this organisation was renamed International Business Machines and is now much better known as IBM.

Thomas J Watson became the General Manager of the Computing, Tabulating and Recording Company in 1911 and was made President in 1914. In a particularly shrewd move that tended to mark all his business decisions, when he joined the company Watson had elected to take a somewhat reduced salary together with a 5 per cent share of the profits of the organisation — ultimately, this would make him one of the highest paid executives in the United States.

Far left A Burroughs manual calculating machine *(PRJ)*

Left An early electronic calculating machine made in Japan in the 1960s *(PRJ)*

Watson applied many of the lessons and practices learned at NCR to the new conglomerate, in particular the notion of leasing of machines rather than selling them. This mode of business, when coupled with backup service or 'refill' as it was usually described, was to keep CTR, or IBM as it became, buoyant even during the disastrous years of the Great Depression from 1929 to 1934.

Of the various business-oriented machine manufacturers and suppliers, it was Remington–Rand that became the great rival of IBM but even then at a relatively low level of competition. By 1935, with the enormous success of its 405 electric accounting machine, IBM controlled 85 per cent of the market in the United States compared with the modest 15 per cent controlled by Remington–Rand.

ANALOG AND DIGITAL DEVELOPMENTS

Looking back at the period before 1940, it is apparent that analytical and calculating machines tended to fall into two distinct groups. Firstly were all the digital machines that could be seen as related to processes involving manipulation of numbers. However, for many years after Babbage, there was a different range of machines that worked on the principle of physical analogy. These machines worked by simulating, in a mechanical fashion, complex physical processes which at that time were not susceptible to numerical analysis.

Typical of these analog machines was the tide predictor of William Thomson, later Lord Kelvin, who at an earlier time had supervised the laying of the first successful trans-Atlantic telegraph cable. This machine was able to indicate the times for high and low tides with basic input data and then provide important local information which would normally have been available only for major shipping centres and ports.

Another machine which was to achieve considerable commercial success in both Australasia and the United States was the 'automatic totalisator' that created a revolution in racecourse betting. Invented by the British-born, New Zealand-educated engineer Sir George Julius, it was first used to take on-course bets in Auckland in 1913. From there it quickly spread to Australia, a nation of gamblers, and then to the United States. Now converted to full electronic operation, it has had a profound effect on the horseracing industry.

A demonstration model of the automatic totalisator, as designed by George Julius, an Australian engineer *(PRJ)*

On a very minor scale, the analog approach can be seen at work in a device commonly used by draughtsmen and surveyors. This is the planimeter which, by the use of a sliding and rotating wheel, can calculate the area of an irregular enclosed area which would be difficult to calculate by conventional means. So useful is this device that even in the age of the computer, it is still routinely available, although by now the numerical answer is provided in digital form and the old vernier readout is no longer required.

One of the largest analog machines produced in this pre-Second World War period came from the brain of Professor Vannevar Bush at Massachusetts Institute of Technology (MIT) in 1928. This enormous machine, weighing in the vicinity of 100 tonnes and known as a 'differential analyser', was laid out on a flat bed and involved a vast array of rods and cogs with circular platens and drive motors. Looking like something out of the mind of Heath Robinson or Rube Goldberg, this machine resembled the toy machinery Meccano.

Interestingly, given this resemblance, following a visit to Harvard in the 1930s by Douglas Hartree, a scientist from Manchester University, and his student associate, Arthur Porter, it was decided to build an analog differential analyser out of Meccano. This small demonstration machine was able to successfully complete a number of significant tasks and produced answers that were within 2 per cent of the output of the full-size machine. Later, a full-size 'differential analyser' was built at Manchester University where it was kept busy for many years. At about the same time, full-scale differential analysers were also built at Pennsylvania and Oslo universities. Of particular importance to the future of computation, in 1935 a special model of the Bush machine was built for the Moore School at the University of Pennsylvania in Philadelphia and set to work.

In the inter-war years, analog machines together with digital machines were used to calculate the trajectory of shells from the newly developing artillery. This set in train another series of events which can be seen as leading inexorably to the development of the electronic computer.

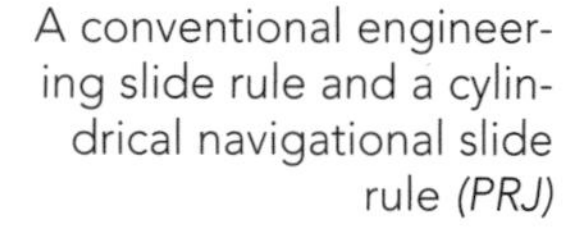

A conventional engineering slide rule and a cylindrical navigational slide rule *(PRJ)*

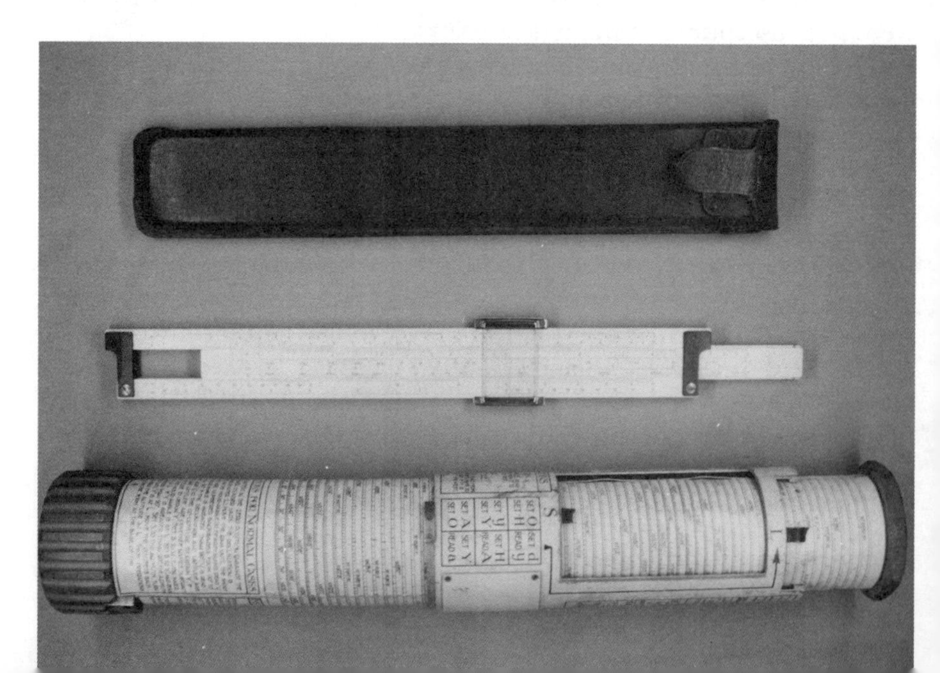

THE HARVARD MARK 1 COMPUTER

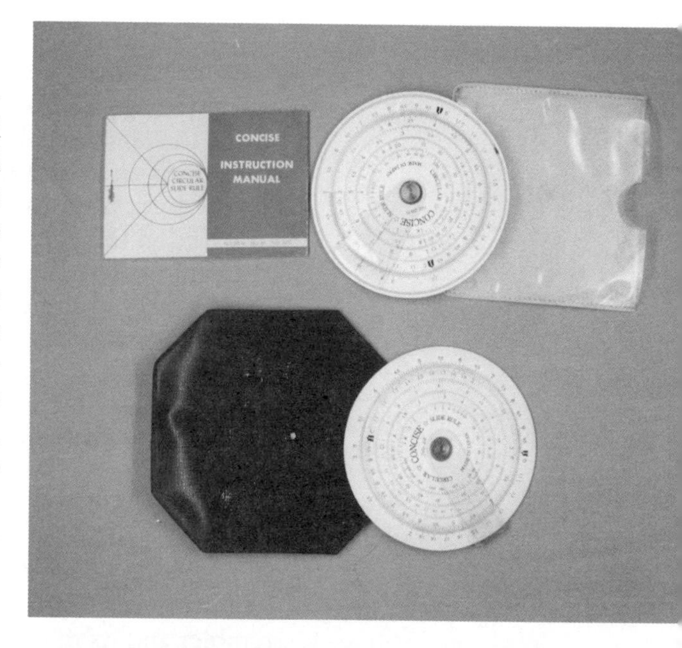

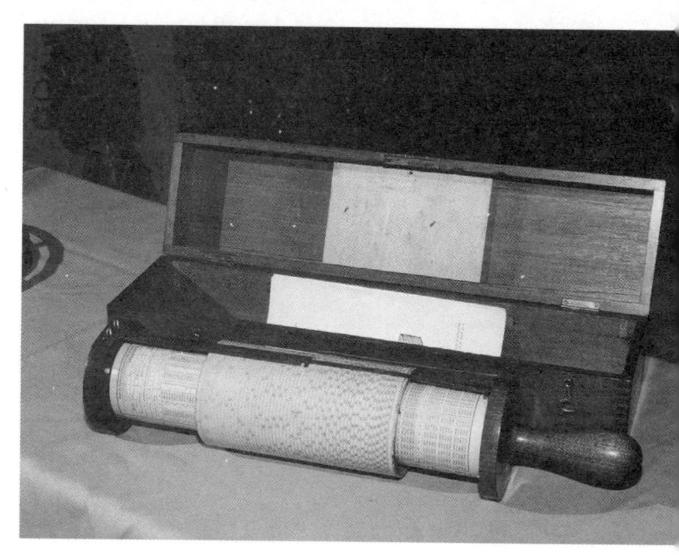

Top The last generation of pocket circular slide rules from the 1960s *(PRJ)*

Bottom A tubular slide rule and carrying case *(PRJ)*

During the early 1930s, a young and single-minded student at Harvard University, Howard Hathaway Aiken, became aware of the small demonstration piece of the Babbage machine stored there. This collection of cogs and wheels which had been donated by Colonel Henry Babbage in 1886 had been gathering dust ever since. Now at last there was someone able to understand what Charles Babbage had been trying to achieve with a machine intended to be powered by a steam engine. More importantly, Aiken could see how the basic principle of the Babbage machine could be achieved using electricity and electric relays, without any of the severe problems of machine tolerance or unsympathetic contractors.

It seems clear that Aiken believed that he had been passed a spiritual baton by the long-gone Babbage in a race to achieve modern computation. Having worked out in general terms what would be required to achieve an electrical analog of the Babbage digital machine that had been called the analytical engine, Aiken submitted a proposal to IBM for sponsorship of the project that he wished to undertake. The machine would be programmed with Jacquard cards which would also be used to input the data; the analysis and counting would be undertaken using electrical relays.

To turn this student project into a useful machine, IBM appointed Claire D Lake, a veteran of the organisation, to assemble a small team to turn the dream into reality. From their collaboration with Aiken came a 5-tonne machine which, originally, was called the 'automatic sequence controlled calculator' (ASCC) but later was to be much better known as the Harvard Mark 1.

In order to achieve synchronisation of all the counting elements of the computer, the Harvard Mark 1 had a drive shaft some 15.25 metres (50 feet) long. The cabinet of the machine was a little longer than this and stood 2.4 metres (8 feet) high and was 600 millimetres (2 feet) in depth. Instructions for this machine were entered via a perforated paper tape. Data were entered via punched cards as developed by Hollerith, and the output from the machine was available through a teleprinter or punched cards. The machine was able to add or subtract numbers 23 digits long in 0.3 of a second, whereas a multiplication of numbers of similar length took 4 seconds. Conversely, a division of this order of magnitude took 10 seconds.

To convert this unprepossessing machine into something that would have the glamour associated with an IBM machine, an industrial designer was employed. The 'shirt front' that was applied by NB Geddes turned the Harvard Mark 1 into an Art Deco masterpiece complete with glass panels and chrome plate. Most importantly, because it was eventually used for war work, some smartly uniformed service men were employed in the computer room to guard this valuable machine.

Commissioned in 1937, the Harvard Mark 1 was finally turned on in 1943 and did some useful work, but it was already obsolete. When it was finally made operational and despite its successful completion of various tasks, it was obvious that it was inherently a very slow machine. In this regard the noise made by the Harvard Mark 1 revealed the source of the problem. This was the thousands of electric relays that it used which were inevitably very slow moving devices.

As described by a contemporary reporter, the machine sounded like 'a room full of little old ladies knitting'. Evidently what was really needed, just as had been the case with television scanning, was the minimal weight of the electron moving through a vacuum rather than the relay trembler and its relatively great weight to be moved every time a calculation was performed.

In the years from 1937 to 1943 and in the face of wartime expediency and pressure, the Harvard Mark 1 had been completely overtaken in technological terms by secret computing machines developed in the United Kingdom for cryptographic purposes and by electronic computers developed in the United States.

As discussed later, these were all electronic machines using thousands of valves (tubes) which were capable of operating a thousand times faster than the Babbage-derived machine of Aiken. If nothing else, the Harvard Mark 1 had successfully proved that the general structure of Babbage's machine had been logically correct and would operate satisfactorily with other means than the originally specified mechanical means. The world was ready for a revolution in computation and the Second World War was about to provide the necessary impetus.

BINARY COUNTING

During the latter part of the 1930s, there were two significant theses written by bright young men at universities in the United Kingdom and the United States which were to have profound implications for the further development of computational methods. The first and in many ways the least known of these was by Alan Turing in 1936 while he was at King's College, Cambridge University. The second was by Claude Shannon at MIT in 1938.

Neither of these documents are matters to be lightly read by the faint-hearted but in the field of mathematics and logic they stand out as fundamental steps in the emerging science of electronic computation, or computing as it is now called.

Turing's paper, entitled 'On Computable Numbers' and based on the work of Alfred Whitehead and Bertrand Russell as set out in their monumental treatise *Principia Mathematica* of 1911, delved deep into the mysteries of mathematical logic. In particular, this paper set out to respond to an unanswered issue raised by the mathematician David Hilbert, brought to Turing's attention in lectures by MHA Newman. Max Newman was later to induct Turing into the highly secret cipher centre at Bletchley Park.

Hilbert, in discussing the assertions of Whitehead and Russell concerning mathematics and the application of truth, had asked: 'Could there be any precise method or approach through which all mathematical problems might be solved?' In answering this question in the negative, Turing invoked the notion of a machine carrying out a defined task and set out to prescribe the most fundamental process that such a machine could undertake. He concluded that such a machine could only undertake tasks that could be specified by a human and described in machine terms, even if inexactly.

This insight led to the notion of what is now referred to as a 'Turing machine' in which the relationship of logical directions as deriving from the thinking process of the human mind could be made explicit in the processes of a machine. These conclusions would later germinate as the foundation of a machine developed to combat the intricacy of ciphers produced by German-operated cipher machines.

Claude Shannon's contribution to the future development of computers lay in his Master's thesis of 1938, in which he applied the formal logic of Boolean algebra to problems of counting and switching in networks. However, the most important idea that was embodied in this paper was one originally applied by the mathematician Gottfried Leibnitz, the great rival of Isaac Newton. This was the realisation that counting could be undertaken with a binary system in which 1 and 0 or the 'on' and 'off' positions of switches could be used to represent the values to be manipulated. This was to be the ultimate 'breakthrough' concept, later applied to all electronic computers.

Although not directly derivative of the work of Shannon, so far as is known, the first application of the new concept of binary counting by machine occurred in Germany. In the mid-1930s, as Germany was once again girding itself for aggressive incursions and war if necessary, Konrad Zuse, a young student at the the University of Berlin at Charlottenburg, had an inspiration. This was that it might be possible to construct a machine which could carry out general-purpose calculations using the binary number system and implemented using electrical relays.

From this idea sprang a machine, known by its creator as Z1, in which a forest of electrical relays did the counting, with input provided by a keyboard and with light bulbs to indicate the results of computations. Importantly this machine appeared to embody what would now be called a central processing unit or CPU and random access memory or RAM. This machine was soon followed by a more sophisticated device that jettisoned the keyboard in favour of data input from punched film. This material was actually 35-millimetre film disposed of from the photographic industry.

This machine was later refined by Zuse during the war years while working for the Henschel Aircraft Company. Fortunately for the Allies, the time frame to develop this machine into a code-breaking device was anticipated to exceed a year and, accordingly, a proposal by Zuse for such a development was rejected.

TYPEWRITERS AND DATA ENTRY

An item of office equipment that has had an intimate association with both telecommunications and the computer from the earliest days is the keyboard. Developed in the middle of the 19th century as a fundamental part of the newly invented 'type-writer' of Christopher Latham Sholes and later manufactured by the Remington armaments firm in the United States, this device has had a remarkably long life in a stable and consistent form.

During the period before the demonstration of Sholes' inventive work in 1867 and its adoption and sale by Remington in 1874 as the 'type-writer', many had tried to create a comparable machine. In all cases the implicit goal had been to produce typefaces by a mechanical process as a replacement for the time-consuming and laborious efforts of the handwriter and calligrapher of the 19th century and earlier. In Europe and the United States, there are records of numerous attempts to produce a machine that would replace the often illegible handwriting of earlier times with something more consistently readable.

The culmination of these earlier efforts is thought to be the machine invented by the Austrian Peter Mitterhofer. In 1864 he created a device that was progresively refined and by 1869 was available in a form now referred to as the 'Murano model'. This machine in a general sense can be seen as fundamentally similar to the typewriters that followed. However, Mitterhofer was a carpenter rather than an engineer. Inevitably his machine lacked the mechanical refinements that were introduced by the Remington Company only a short time later based on the work of Sholes.

One of the most tenacious elements of the Remington 'type-writer', later to become the 'typewriter', was the layout of letters and numbers on the keyboard. As a response to the jamming together of the most frequently used letters and their activating levers, Sholes invented a keyboard that, as a means of achieving typing inefficiency, would have to be considered a masterpiece. What was created was a keyboard in which the most frequently used keys were placed as far apart as possible. From this sprang the QWERTY arrangement of keys with which every wordprocessor operator or typewriter user is painfully familiar.

Ever since Sholes proposed the layout of the typewriter keyboard, it has remained unchanged with the minor exceptions being the introduction of the shift key to allow uppercase and lowercase to be produced and a slight rearrangement of the number keys so that the capital letter 'I' no longer has a dual function to represent 'I' and '1'.

This page from top:
Imperial Model D, c. 1919 *(CB)*
AEG Mignon Model 4, c. 1924, precursor of the 'golf ball' *(CB)*
Underwood Model 3, c. 1930 *(CB)*
Empire portable, British Typewriters Ltd, c. 1935 *(PRJ)*
Creed teletypewriter, c. 1944 *(PRJ)*

An attempt in 1932 by Dr August Dvorak at the Washington University in Seattle to do something to improve the Sholes keyboard, although technically successful, has not achieved any penetration into the office or home environment. Its improvement of about 35 per cent in typing speed is probably insufficient to justify the effort of relearning the keyboad layout, and the Sholes keyboard remains the unchallenged device for typing, keyboarding and data entry to the present day.

Although the keyboard remains as a very tangible reminder of the typewriter and the revolution that it produced in business and private communications, the typewriter itself has almost reached the status of 'dinosaur', albeit quite a small one — perhaps a pterodactyl given its capacity to 'fly' through business letters. In its last incarnation as the Selectric typewriter manufactured by IBM, known more commonly as the 'golf ball' typewriter, a highly reliable and ergonomically satisfactory office machine was created. But for all its advantages, even this machine could not compete with the revolution of the 1970s — the introduction of microprocessor-based word processing. This new office technology would very soon sweep conventional typewriters into oblivion.

With the introduction of the teletype for long-distance high-speed transmission of the written word in the 1930s and as a replacement for the morse key, the beginnings of a new form of business machine were apparent. As computers were developed and as a simpler method of data entry than punched cards or punched paper tape was urgently required, the teletype machine was pressed into service in a new role. From its original form as an escapee from the cable telegraphic networks, the teletype keyboard soon lost its mechanical levers and linkages, using instead key-activated electric switches.

From that time, the teletype has gradually shifted from its earlier independent telegraphic function to a simple adjunct to the new computers. At the present time it is now nothing more than an electronic and intelligent keyboard linked to a computer with a video display unit to present the data or letters and numbers entered. In the period between the development of the teletype and its ultimate conversion to a self-contained keyboard that is now used to drive all forms of computer, there lies an intermediate device, the video terminal and keyboard console as exemplified by the VT 52.

Although most people who live and work with computers have necessarily become skilled in the use of the venerable QWERTY keyboard, it is quite possible that they are the last generation of computer users who will have to go through the agony of learning to touch type with this device. During the last 10 years, voice-to-data-entry software has gradually been refined and in recent times continuous speech dictation capabilities have become available. In the not too distant future it seems probable that most data and text entry will be via voice direct, and the keyboard will be used only for more obscure tasks where voice translation is inefficient, such as in writing computer code. Perhaps the Sholes keyboard may be around for a lot longer than anyone could presently anticipate.

This page from top:
Adler desk model, c. 1955 *(PRJ)*
Teletype terminal Type 33, c. 1965 *(PRJ)*
Televideo terminal Type VT 52, c. 1970 *(PRJ)*
IBM Selectric 'golf ball' typewriter, c. 1975 *(CB)*
Dedicated word processor, c. 1982 *(PRJ)*

CHAPTER

11

1939 to 1945

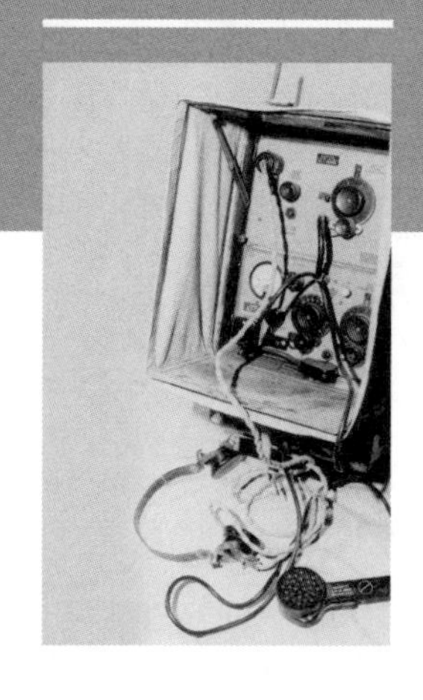

Radio at war

At 11.15 a.m. on 3 September 1939, the sombre tones of the Prime Minster of Great Britain, Neville Chamberlain, announced that the ultimatum and time limit issued to Germany had expired and, as a result, Britain was at war. Poland had now been overrun and the consistent policy of appeasement of the territorial ambitions of Germany's Chancellor, Adolf Hitler, had all come to nothing.

Since the unopposed reoccupation of the Rhineland in 1936 by Hitler's newly revived army, the Wehrmacht, the likelihood of a new war had existed. Indeed the reconstruction of Germany's armed forces in the period after 1932, with Hitler's accession to power, should have revealed his probable intentions. These ambitions had been clearly stated in his book, *Mein Kampf*, written while in prison in the 1920s. Unfortunately, Hitler was allowed to achieve his early goals and this may well have encouraged him to take more outrageous steps later, as Britain and France failed to react to the provocation.

At this stage of the development of world communication, it was extremely significant that the announcement of this fatal event, the start

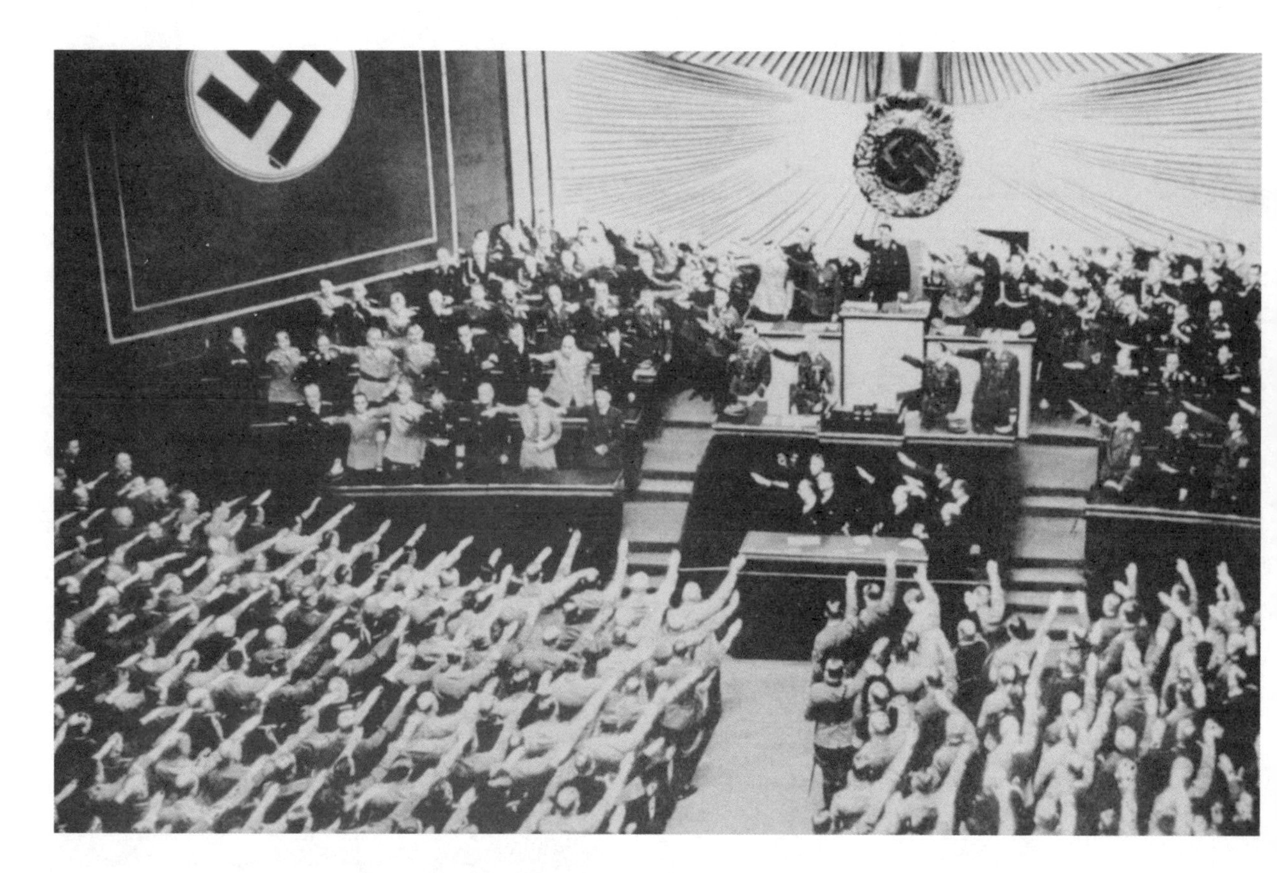

Top Germany's new Chancellor, Adolf Hitler, makes a speech at a Nazi party meeting in the 1930s *(BAW)*

Bottom Hitler with his generals of the Wehrmacht *(BAW)*

of the war, should have come to most citizens via radio. Compared with the period of the First World War, here was a household necessity to be relied on for up-to-date and reliable news. Now the remarkable public hysteria of the First World War that had been directed at potential spies and the use of radio for associated purposes was conspicuous by its absence.

In this war, radio broadcasting was to be used by both friend and foe to inform and persuade, and as the German minister of propaganda, Joseph Goebbels, was to show, it could be applied to good effect as a weapon of war. In this war, the voice of William Joyce, better known as Lord Haw-Haw, was to exhort the Allies to capitulate to the mighty forces of the Third Reich. On the other side of the globe in the Pacific war, the mellifluous tones of a young and beguiling maiden, later known as Tokyo Rose, were used to encourage the Allies to submit to the might of Japan. Ultimately, neither was successful and at the end of the war Joyce was hanged in Wandsworth Prison for treason, even though he was a citizen of the Irish Republic. However, if any indication of the power of broadcasting was needed from this period of conflict, then the voices of these two 'agents provocateurs', and the more sedate tones of the BBC, were potent examples.

As a symbol of hope and sanity in a world made mad by the impact of war, the chimes of Big Ben became very well known wherever the oppressive heel of the Axis forces rested. Seen as both dispassionate and generally reliable, the reporting of news by the BBC became a source of inspiration for those under the power of Hitler. It appears that even the German public, looking for the truth rather than expedient propaganda, would turn to clandestine listening and the voice of the BBC, despite the draconian penalties for such activity. In prisoner-of-war camps, despite the threat of reprisals, clandestine radios were successfully built and used to keep the inmates informed of the progress of the war.

Compared with the First World War, the Second was to be an exercise in rapid movement, involving tanks and aircraft together with long- and short-range communications. In contrast to the stalemate and mud of the First World War, a system of lightning attacks known as 'blitzkrieg' was developed and brought to its devastating zenith in the invasion of Poland. In this campaign, the Polish generals still preserved the fantasy of the First World War that cavalry could be the basis of victory. This time, instead of sharing the mud with the infantry, the cavalry were swept away by the might of mechanised armour, even though of limited power.

Prime Minister Winston Churchill broadcasts to the nation via the BBC *(BBC)*

To this fundamental change of tactics in attack, the German army added a further potent element. This was an assault involving aerial support and coordinated by radio communication with the ground. With the Polish invasion and the announcement of war by Chamberlain, the scene was set for an epic struggle which would take both victor and vanquished to the depths and set in train tensions that would take several decades to resolve on the continent of Europe and in the east, in Russia.

The Polish cavalry charge forward to meet the Panzer tank invasion of the Wehrmacht in 1939 *(BAW)*

STEPS TO A NEW WAR

Despite the immensity of the First World War and the incredible tide of death that it generated, for most citizens not directly involved on the Western and Eastern Fronts it was in a sense somewhat detached from daily life. Apart from the dreadful arrival of successive letters advising of the loss of yet another soldier, sailor or airman, at home things went on more or less as normal for the duration of the conflict. Only in the latter part of the war, with the incursion of Zeppelins and Gotha bombers, were the ordinary citizens of the United Kingdom confronted with the direct realities of war.

In the Second World War, the expectation was that people would have to cope with bombing and these expectations were entirely justified. The counterpoint to blitzkrieg, with tanks on the ground cooperating with aircraft support from the air, was the Blitz, in which bombs rained down on British cities. Later, German cities would suffer the same fate as the armed forces of the Wehrmacht were pushed back on all fronts.

This was a war of movement and communications but it was also a secret war of codes and ciphers. Most particularly, however, it was a citizens' war, as whole nations became embroiled in the horror of bombing and the threat of gas attacks.

The steps to this war were many and complex but it seems clear that Chancellor Hitler served as a focus for his Reich's dissatisfaction with the outcome of the First World War and the stringency of reparations demanded by the Allies. Blame for the outcome of this conflict was conveniently directed at international financiers and particularly the Jews who were to become the war's most tragic victims. Whatever the motivation, following his ascension to power in 1932, Hitler began a determined process of rearmament, at first covertly with the collaboration of Russia and factories far to the east but latterly without any concern for dissembling.

The British Prime Minister, Neville Chamberlain, and Mrs Chamberlain leave Number 10 Downing Street *(BAW)*

THE WAR IN EUROPE

In 1936, Hitler took his first bold and potentially dangerous step when his army was sent to retake the Rhineland. With the failure of either France or Britain to react to this overt aggression, Hitler's boldness increased. In 1938, the 'Anschluss' (joining together) of Austria with Germany was completed and then the invasion of the Sudetenland of Czechoslovakia was successfully accomplished. This was soon followed in 1939 by the invasion of Czechoslovakia itself, after the shameful acquiescence of the Allies at Munich. From this debacle, Chamberlain, or 'Regenschirm' (umbrella) as he was scornfully known in Germany, returned, proclaiming 'Peace in our time' — peace for a while but in truth in a 'fool's paradise'.

It has been suggested that one of the subtle factors that emboldened Hitler to follow the path of aggression was an apparently innocuous but later notorious event at Oxford University in 1934. This was the adoption by the Oxford University Union of a motion proposed by CEM Joad in a student debate that 'This House will in no circumstances fight for its King and country'. While by no means certain, such an apparently trivial action may indeed have served to confirm Hitler's perception that the British had been intellectually crippled by the First World War and were now so decadent that all spirit for a fight had evaporated. This may well have persuaded him that his determination to achieve an expansion of Germany to the east and into Polish and Russian territory would not be opposed.

Evidently, until the push for further living space, 'Lebensraum', and the assault on Poland in 1939, this was a correct perception. However, Poland was to be the 'final straw that broke the camel's back' and at last the British bulldog would stand firm and accept no more goading.

Perhaps the supreme irony in this over-ambitious last step to conflict was the use of tanks and aircraft to spearhead the attack. The tactics of blitzkrieg were largely a British invention and had come out of the experience of the First World War, championed by a tank soldier, Sir John Fuller, and a military historian, BH (Basil) Liddell Hart.

The theory put to such devastating effect by its chief exponent, Heinz Guderian, relied on the notion of applying concentrated force in a limited area of the opponent's front line and, once a breakthrough had been effected, bursting through in an expanding torrent of armour supported from the air.

The German invasion of the Lowlands and Northern France and movements up to the escape of British troops from Dunkirk *(PRJ)*

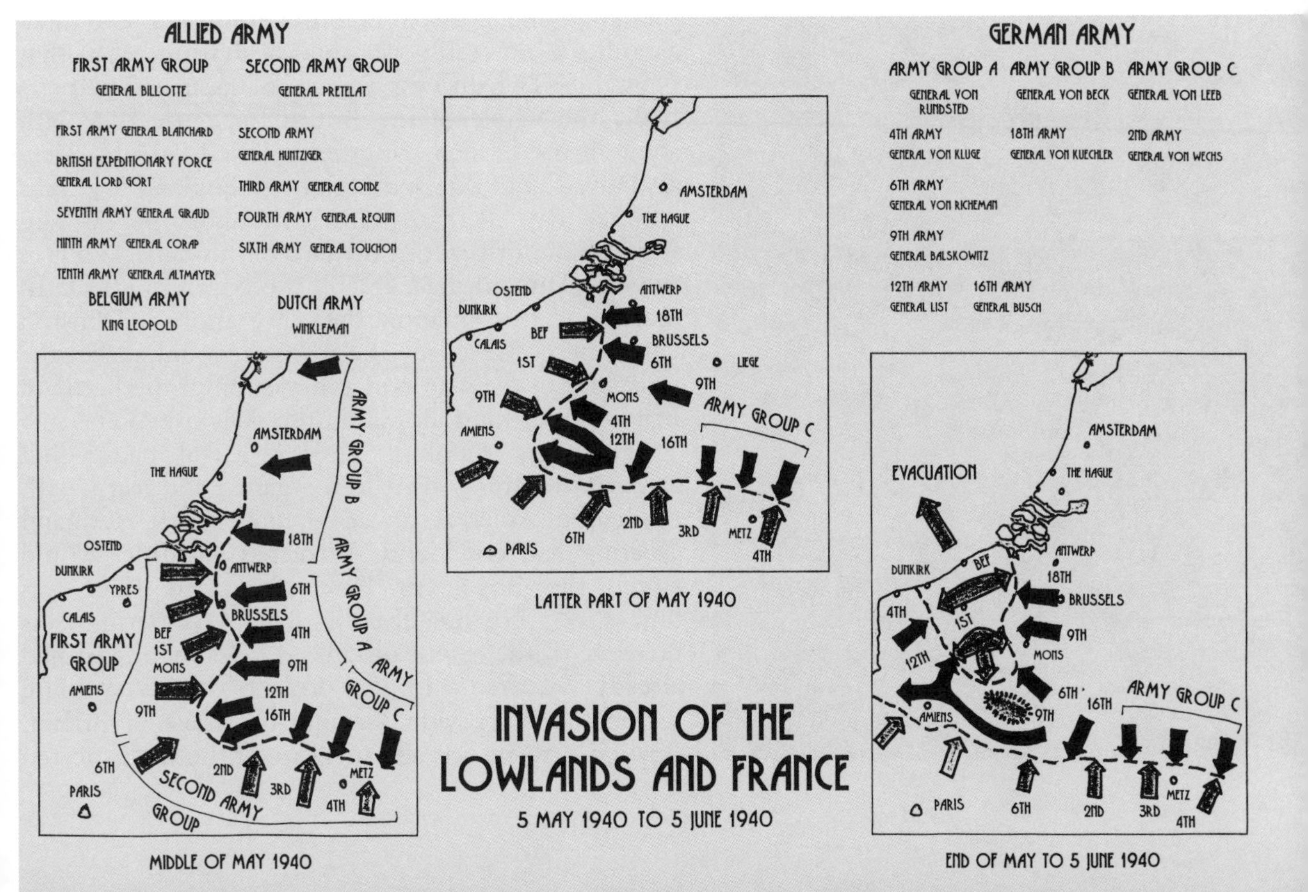

The escaping British Expeditionary Force streams out across the sands of Dunkirk to the fleet of waiting ships *(BAW)*

The fleet of ships, small and large, carry the troops back to England in 1940, avoiding a total military disaster *(BAW)*

The Second World War in Europe can also be seen as a conflict in which a cumulation of fundamental mistakes by Hitler and his henchmen were finally to bring the Third Reich to its knees.

In 1940, when the British Expeditionary Force and the French army had been squeezed into a pocket around Dunkerque in northern France (known more commonly in English as Dunkirk), for reasons that have never been adequately explained Hitler 'blinked' and paused before authorising the final assault. The result of this hesitation was that Operation Dynamo was able to be launched and virtually the whole of the British force together with some of the French army were carried back to safety at Dover and the south-east of England to fight another day. Perhaps Hitler thought that Britain would look for peace in the face of this debacle but in the end Dunkirk had exactly the opposite effect. In reaction to this situation, the exhortations of Winston Churchill conveyed by the wireless would stiffen the resistance of the nation in anticipation of the Battle of Britain that soon followed in the skies over Kent.

Following Dunkirk, after a useful pause that allowed the British army to reorganise and rearm with the help of America, the aerial conflict that Hermann Goering had confidently predicted to Hitler would sweep the Royal Air Force from the skies was unleashed. Fortunately for Britain, Goering was unaware of the extent of radar development that had already occurred and the protective early warning screen that it provided around the coast of Britain. Beyond that, he was also unaware of the sophisticated system of fighter interception control that had been developed since the end of the First World War. Lastly, he was up against a phlegmatic and extremely determined person, Air Chief Marshall Dowding, usually referred to as 'Stuffy' from his reserved manner.

As noted earlier, Dowding had been intimately involved in the application of wireless communication to aircraft during the early days of the First World War. Since that time he had taken a keen interest in the methods of radio control of aerial combat. More importantly, he had been a major supporter of the introduction of radar for aerial defensive purposes. He was probably as well placed as any airforce officer to deal with the problem of containing the assault of the Luftwaffe while husbanding the limited resources of aircraft and pilots. In the end, despite pressure from other officers not in possession of top-secret information provided under the codename 'Ultra', he coolly maintained control and ultimately ground the German airforce to a standstill. As a result, the projected sea invasion from France by the German army, codenamed 'Seeloewe' (sea lion), was called off in late 1940, never to be reinstated.

A military radio of 1925, the 'Cork' set (receiver section) *(PRJ)*

The transmitter section of the 'Cork' set of 1925 *(PRJ)*

Perhaps the most extraordinary error made by Hitler was his decision to launch Operation Barbarossa, the invasion of Russia in June 1941, given the disastrous attempt by Napoleon Bonaparte in the previous century. There can be no doubt that Hitler, as an avid student of military and political history, was aware of the French army campaign against the Russians and the bitter defeat that resulted from this incursion and from which the armies of France never fully recovered. Why he should have imagined that he would be able to do any better seems in retrospect quite inexplicable. But fathoming the mind of someone who, at least in his latter days, was clearly unbalanced and dependent on medical assistance and drugs was not an easy task even for the strategists of the Second World War, with the advantage of greater temporal proximity and access to material which provided a direct insight into Hitler's mind and methods.

British troops using the 'Cork' set in the Far East in 1926 *(PRJ)*

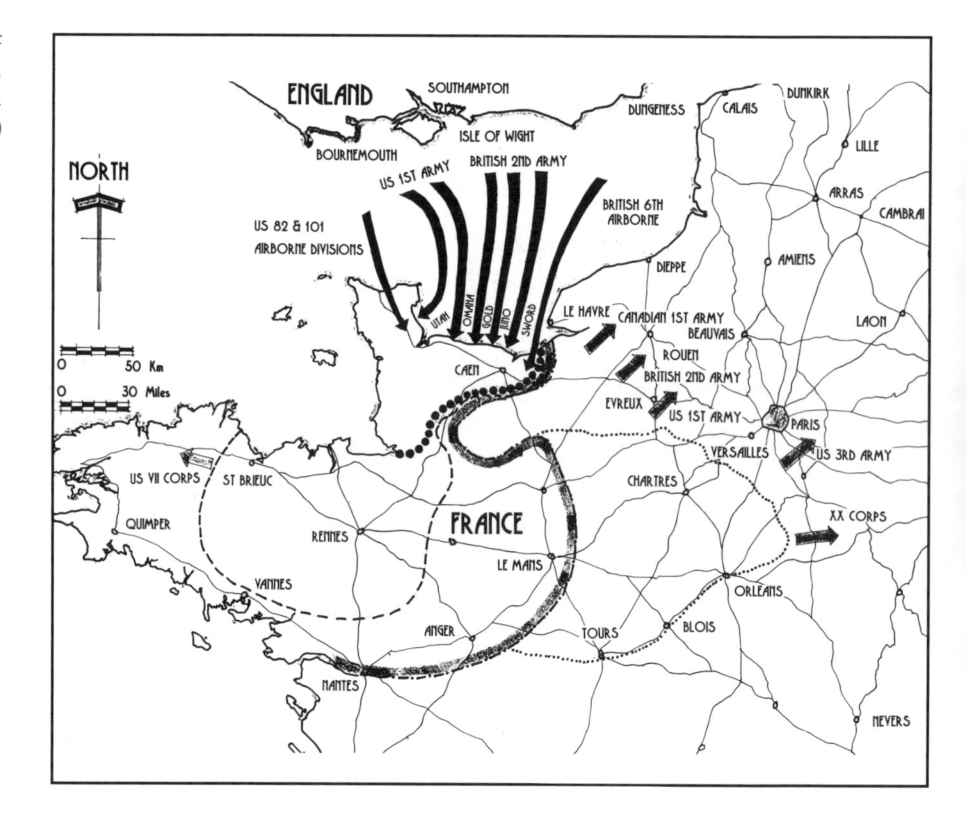

The Allied invasion of France and the advance to the north-east in 1944 *(PRJ)*

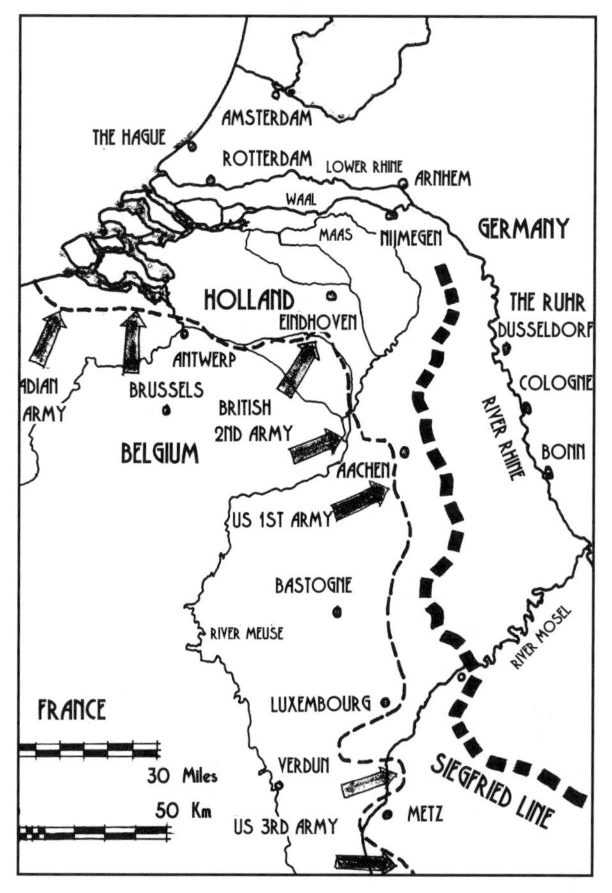

The Allies move north and east, pushing the German army back towards the Siegfried Line *(PRJ)*

The vitally important and revealing material that became available to the Allies, based on the efforts of the Government Code and Cypher School at Bletchley Park, was again something that was completely unknown to the German military. The senior officers of the German army, airforce and the navy all seemed to have a remarkable level of faith in their cipher machines. In this they reckoned without the presence in the Allied intelligence-gathering force of people like Max Newman, Alan Turing, Gordon Welchman and others. Such dedicated and brilliant minds had by the start of the war taken the first major steps to unravelling the most arcane of modern mysteries, the Enigma-based codes which were distributed among the highest levels of the Allies as 'Ultra' intelligence.

As the tide of war gradually began to swing in favour of the Allies after 3 hard years of combat, a carefully orchestrated concert of misinformation successfully persuaded the German High Command and the Dictator that the Allies would attempt to reinvade Europe by the shortest route via Calais. As the success of D-day, 6 June 1944, was to demonstrate, in warfare the most important element for success is surprise. By a variety of subtle and not-so-subtle means, the American, British and Canadian forces managed to get ashore with remarkably low levels of casualty and death. This was in marked contrast to the First World War debacle of Gallipoli — perhaps something had been learnt in the hard school of battle.

In order to convey the general outlines of the war in Europe, the shifting fortunes and lines of attack are best conveyed by maps. The maps in this chapter show how the initial thrust of the German army, the Wehrmacht, through Holland, Belgium and the north of France was accompanied by a swing through the Ardennes into southern France. Where, in the First World War, this southern swing had been absent and the opponents had finished up on a line of stalemate through Flanders to the coast of the English Channel, this time the Wehrmacht bundled up the Allies, pushing them back to the coast of France at Dunkirk in a campaign of a mere 6 weeks. Here the French army capitulated and the British army was rescued by the fleet of small ships organised by Admiral Ramsey from his headquarters at Hell Fire Corner, in the cliffs above Dover.

Britain was now placed in a lonely and perilous position but nevertheless managed to stave off the might of the Luftwaffe. With the supply of American aid in the form of lend-lease, Britain continued to survive to see the Americans enter the war following the attack on Pearl Harbour by the Axis ally of Germany, Japan.

The flow of goods and armaments that crossed the Atlantic under this policy of aid from the United States at first was severely mauled by the submarine fleet of the German navy, the Kriegsmarine. Gradually a combination of newly developed technology, radar, and a capacity to read the German Enigma traffic arrested the high level of depredation of Allied convoys.

With the entry of the United States into the European conflict, a decision was made to turn the fight to the enemy and an invasion force of American and British troops landed in North Africa. This force was intended to supplement the British forces in Egypt and, after a see-sawing campaign along the edge of the Mediterranean through Tripoli (now Libya), Tunisia and Algeria, firstly against the Italian army and then with the German troops of Rommel, a victory at El Alamein changed the course of the war. After this engagement, Churchill observed that, before it, the Allies had never had a victory and after, they never had a defeat.

Following the success in North Africa, the Allies pushed their way up through Sicily and Italy to later merge with the successful breakout from the D-day landings in Normandy in 1944. On the Eastern Front, after the defeat of the German army at Stalingrad, the intervention of the Russian winter ensured the ultimate destruction of the German forces, as it had for Napoleon and his French army. The Wehrmacht was then driven all the way back to Berlin, where the Russian army joined up with the Allies coming up from the south and west. At this stage, Hitler and his wife and former mistress, Eva Braun, committed suicide in the underground bunker next to the Chancellory in Berlin and a terrible chapter in European history finally drew to a close.

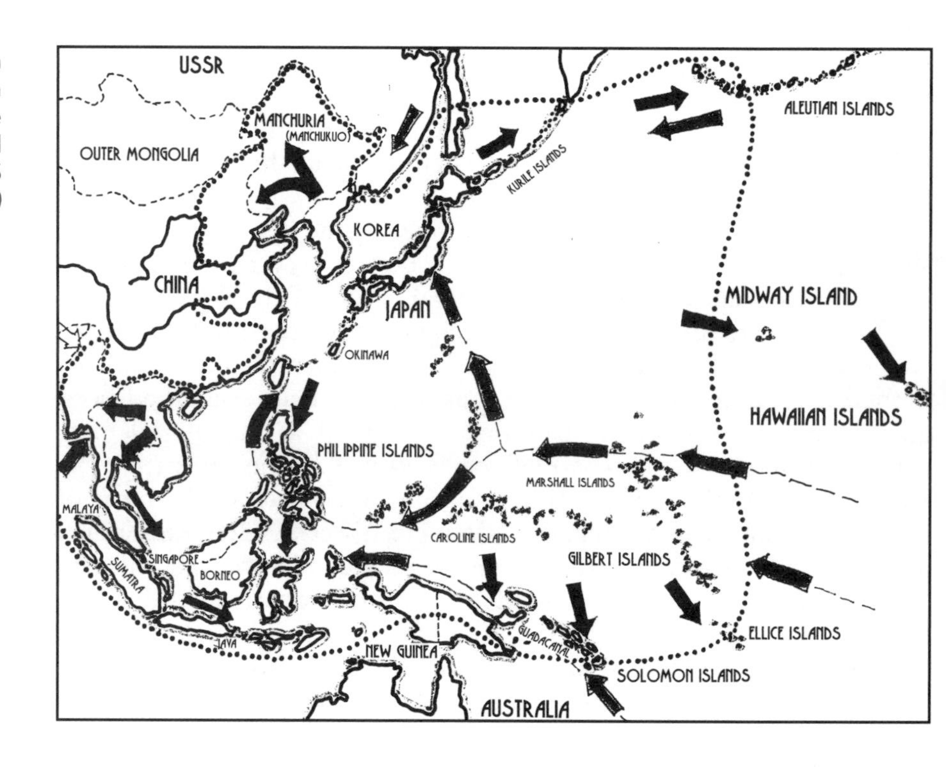

Japanese military invasion and expansion through South-East Asia and the Pacific, 1938 to 1943, and retreat to defeat in 1945 *(PRJ)*

THE WAR IN ASIA

Opened up to European access in the middle of the 19th century, Japan maintained a generally feudal society well into the 20th century. However, once the decision to allow access to American, British and other trading interests was made, it was quickly followed by a decision that Japan should remake itself in the mould of the Europeans.

The navy to use as a model was British and the military model was German. The wireless communication came from Marconi and its use is said to have contributed to the defeat of Russia in the war of 1906. However, far more importantly, Japan saw itself as an imperial power and sought to obtain its own version of 'Lebensraum'. The available space was seen as lying in Korea and beyond in Manchuria and, ultimately, China. In the 1930s, Japan began an active policy of invasion and colonisation.

The so-called South East Asian Co-prosperity Sphere as it was later described by Japan became a thinly disguised vehicle to take over the territory of neighbouring states and obtain the raw materials necessary for a newly developing industrialised society seeking to achieve equality of status with the great powers.

The aggression of Japan and its outward movement of conquest can be seen in the map of Japanese expansion. The initial invasion of Manchuria and its conversion into the vassal state of Manchukuo under the titular leadership of the deposed Chinese emperor Pu-Yi was soon followed by a southern movement into mainland China. The rape of Nanking and the invasion of Shanghai were to leave indelible crimson stains on the psyche of the Chinese, whether at home or abroad. Moreover, they are stains that appear to remain as vivid today as they were at the time of the events in 1937. But then, too, for Allied prisoners of war under the heel of the Japanese, the memories may have faded somewhat, but there remains very little forgiveness for the brutality that was experienced.

The debacle of the unconquerable fortress of Singapore was to see many Australian and British servicemen taken to a slow death by starvation and beating on the infamous railway from Malaya to Burma and anywhere else that expendable labour was required by the conquering Japanese. But for every Australian or Briton lost in the labour camps, perhaps ten times that number of Chinese and Asian labourers lay dead. It has been suggested that for every sleeper of the Burma Railway, a corpse can be assumed to lie close by.

Japan's great mistake of the Second World War was to vastly underestimate the huge industrial resources of the United States and to believe that from a comparatively minuscule industrial base it would be able to wage war and achieve ultimate victory. From the initial surprise attack and victory at Pearl Harbour in 'a day of infamy' as Roosevelt described it, Japan would reach the limit of its imperial expansion in about 2 years. From then on, the Allies would progressively beat the Japanese army back on all fronts until finally the war was arrested by the dropping of two atom bombs on Hiroshima and Nagasaki and the receipt of Japan's unconditional surrender.

ARNHEM

By 1939, radio communications were a well-established element of military, naval and air strategy and tactics and innumerable examples of its successful use could be described. However, there were still a number of very basic lessons to be learnt about the use of radio communications in a mobile role. Once again it was from the disasters of war that some of the most important insights were developed. An examination of one action that was indeed disastrous for the Allies is far more revealing than the successes — the operation usually referred to as the battle of Arnhem but known officially as 'Market Garden'.

Radio communication and its failure was to play a very significant part in this unfortunate action. However, before considering how this arose and its impact on the operation, it is necessary to have a clear idea of the state of the military campaign in Europe during September 1944. Only then can a full appreciation of the importance of Arnhem and its impact on future developments in radio communication be obtained.

The advance towards Germany

The tremendous success of the D-day landings in June 1944 was followed by an Allied advance northwards and eastwards that in many respects resembled the blitzkrieg the German army had launched in 1939. However, it was a blitzkrieg on three fronts rather than one and involved three Allied armies, two American and one British. By late June 1944, on the continent of Europe the Supreme Commander, General Dwight D Eisenhower, had under his command nearly 2 million men, almost half of whom were American.

Field Marshall Montgomery of Alamein broadcasting via the BBC *(BBC)*

Part of the Allied forces, the Twenty First Army Group under General Montgomery, advanced northwards through France and Belgium and, on 25 August, Paris was liberated. The Allies continued their advance and by the latter part of August 1944 the German army was in a state of near rout.

Destruction of the French railway system by the Allies before the Normandy landings meant that the transport of goods was almost entirely reliant on the road system. Supplying the needs of the Allied army, in the north some 70 000 strong, became a logistic nightmare. Without supplies and particularly fuel, the advance slowed to a crawl. The German army received a breathing space and immediately regrouped to arrest the uncontrolled retreat. German resistance stiffened and the opportunity for a quick end to the war was lost.

The airborne army

By the latter part of August 1944, the First Allied Airborne Army under American leadership had been kept in reserve in England for a number of months. As the only Allied reserve forces available for the battle for Europe, by August 1944 it was eager and impatient to help in defeating the German army.

In early September 1944, General Montgomery persuaded General Eisenhower to undertake a more concentrated attack on the German forces. This would be in the form of a blitzkrieg type of attack. However, compared with the action against Poland by the German army in 1939, this offensive would open with an airborne attack involving massive surprise. The aim would be to secure the bridges across the Waal, the Maas and the Neder Rijn in the Netherlands and open the way into the industrial heartland of Germany, the Ruhr.

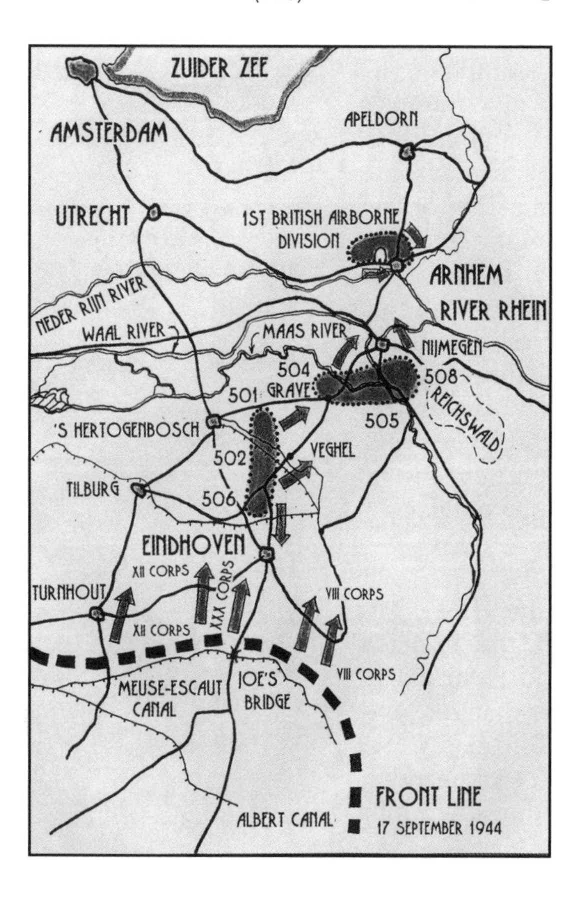

Operation 'Market Garden' and the push to secure the bridges at Nijmegen and Arnhem *(PRJ)*

The planning of this operation, codenamed 'Market Garden', was undertaken with incredible speed in just 7 days. From the air, it involved a combined parachute and air landing operation in which two American airborne components would be applied together with one British airborne component. This part would be the 'Market' of the codename.

On the ground, the XXX Corps with tanks, artillery and infantry would try to punch through the corridor opened by the parachute troops so as to reach and take all the bridges including that across the Neder Rijn at Arnhem. This part was the 'Garden' of the code name.

With an eye to the political realities of a joint American–British venture under the overall command of a British general, the British forces were given the most risky element of the plan to undertake. This was the capture of the Arnhem bridge some 106 kilometres (66 miles) to the north of the front line of the Expeditionary Force.

In planning the 'Garden' element of the proposed assault, it was clear that there would be a series of major problems to contend with if the operation was to be a success. The principal issues that had to be considered were:

- The bridges at Grave, Nijmegen and Arnhem had all to be taken and held.
- The level of German resistance to the attack would be critical.
- The road corridor used for the advance was narrow and extremely vulnerable.
- The logistics of supply for the advancing forces were quite intimidating.

From the outset it was clear that the airborne troops at Arnhem were very much at risk. This was recognised by Lieutenant-General FAM Browning, commander of the British First Airborne Division, in his famous response to General Montgomery that 'Market Garden' might involve going 'a bridge too far'.

Phase 1: The Arnhem landings

Initially all went well and the armada of aircraft and gliders streamed away from England across the Channel and landed a large proportion of the British contingent of the operation at the drop and landing zones west of the main town of Arnhem. However, this satisfactory situation was to change all too soon as the newly stiffened resistance of the German army began to be felt.

A three-pronged attack was launched towards the centre of the town, at which time major radio communications problems occurred. The first prong, consisting of the 2nd Battalion under Lieutenant-Colonel John Frost, moved swiftly along the northern side of the river and, without too much difficulty, reached and held the northern approaches to the Arnhem bridge. The other two prongs of the attack were soon held up by Panzer troops who attacked from the north-east. While Frost held the end of the bridge with limited forces, the rest of the attack was stopped in its tracks.

Phase 2: The southern landings

While the British troops were held up and attacked both within and on the western edge of Arnhem, the landings of the US Airborne Divisions were proceeding with few problems initially. The southern 'stepping stone' between Eindhoven and Veghel was established but all too soon things began to go wrong. Arriving at the bridge across the Wilhelmina Canal at Son, the paratroops were just in time to see it blown up by retreating German forces. Later, a Bailey bridge was built across the Son by sappers working feverishly through the night.

South of Nijmegen, the US paratroops also landed with little trouble near the Groesbeek Heights and moved off to the north to secure the bridge across the River Waal at Nijmegen. Resistance to capture of the bridge was very determined but, following a valiant attack across the river in collapsible British boats, the US paratroops stormed towards the north end of the bridge as British troops and tanks attacked from the south to capture the intended target.

Phase 3: The advance of XXX Corps

While the landings at Arnhem and places south had been taking place, the ground troops and armour of XXX Corps had commenced their advance to the north from the start line at the Meuse-Escaut canal. Delays were caused by the demolition of the bridge at Son but, ultimately, after some very hard fighting, the forces reached Nijmegen in time to join the American paratroops in capturing the vital bridge.

Although the British tanks were able to rumble across the bridge, to the north of Nijmegen the advance was brought to a complete halt. The road ahead now ran as an elevated corridor across marshy ground with the German forces established in flanking positions. This was an untenable situation for a tank advance and the northerly thrust was arrested just 12 kilometres (7 miles) south of Arnhem.

Phase 4: Collapse, retreat and aftermath

With the hold-up north of Nijmegen, it became apparent that time had run out. Even before the enforced halt had occurred, Frost and his remaining force of 110 men had been overrun. With the exception of the few surviving members of the battalion who now were forced to surrender, the rest of the assault force at the bridge had been completely annihilated. It was now clear that unless a strategic withdrawal was executed, the remaining British force on the northern side of the Neder Rijn would suffer the same fate. At this stage, one of the great disasters of the First World War was to play a part: Gallipoli.

A similar withdrawal operation to that achieved at Gallipoli was now arranged and the tattered remnants of the 10 000 men who had landed at Arnhem moved across the river in boats brought up from the south before the flanking German forces realised what was happening.

Despite this small success, the campaign as a whole could be seen only as a tragic waste of valuable and skilled fighting men. Of the approximately 10 000 men that had landed west of Arnhem, some 1400 had been killed and 6000 taken prisoner. This was a complete disaster and comparable with the 'Charge of the Light Brigade' in the Crimean War. An operation planned in haste and based on what the Polish Major-General Sosabowski had described as 'arrogant optimism, completely unjustified', it had led to a withdrawal as painful as Gallipoli or Dunkirk.

A communications disaster

As discussed earlier, the problems of radio communications associated with the Arnhem end of 'Market Garden' were to have substantial and damaging effects on the way the operation was conducted. With the benefit of hindsight, some of the lessons of 'Market Garden' were probably self-evident. However, perhaps a military debacle was needed to drive home the message that adequate power and efficient antennae are essential, particularly at high frequency, if reasonably reliable communications are to be maintained over even moderate distances while operating in a mobile mode.

To fully appreciate why the problems with radio communications occurred at Arnhem, it is necessary to consider how radio had developed in the British army up to that time. By the end of the First World War, trench radios had advanced to the stage that they operated reliably on the long waves below the AM broadcast band and over distances of several kilometres. Over the next 20 years, the frequency of operation was increased and by the start of the Second World War a band of frequencies between about 2.5 megahertz and 8 megahertz was in use by the army.

By comparison, the American army developed the use of very high frequencies, VHF, for mobile operations of dispersed troops up to about 12 kilometres (7 miles). With its late entry into the war, America was able to spend considerably more time in developing this new form of VHF radio communication.

A significant difference between the British army radio system and that of the American army could be seen in the requirements for efficient and appropriate antennae. For maximum efficiency at HF as used by Britain, an antenna had to be many metres long, whereas with the American system using VHF it was possible to use antennae measured in centimetres with almost 100 per cent efficiency.

In battle, the antenna system of portable radio apparatus becomes a major concern, in particular its physical size. Even trailing a whip antenna a couple of metres long becomes a major nuisance if not a convenient aiming point for the enemy. For this reason, making mobile whip antennae as short as possible, consistent with achieving reasonable communications, is an inevitable requirement of an army on the move.

As well as this consideration, local conditions can have a major impact on the range over which a signal will be received, with higher frequencies being more heavily attenuated by intervening vegetation than lower frequencies. In addition, immutable physical laws ensure that the shorter the antenna at high frequency the less efficient it becomes as a means of radiating a signal. Moreover, this was a matter that the British army was well aware of. As a radio handbook of the period makes quite clear, the inevitable consequence of a shortened antenna relative to the length required to achieve resonance at the frequency of operation is attenuation of the radiated signal.

By comparison, with antenna requirements at the high frequencies, the reduction in antenna size possible for a system operating at, say, 45 to 55 megahertz in the VHF band is quite dramatic. A quarter-wave whip antenna at 52 megahertz is about 1.5 metres (5 feet) long and this can radiate its energy at 100 per cent efficiency. By comparison with this, a 2-metre (6½ feet) whip operated even at, say, 5 megahertz is achieving only a very small percentage output of energy for the input of energy from the transmitter. To be 100 per cent efficient, it, too, would have to be a quarter wavelength long or 15 metres (50 feet) in length. Such a size is quite impractical for a whip antenna on a moving vehicle or a tank.

Apart from the problem of antenna inefficiency at HF, the power output of the British Number 22 transceivers used in conjunction with Number 18 sets in this operation was very low. The Number 22 set produced in the vicinity of 1.5 watts for CW operations and less than that for RT or voice operations. By comparison, the Number 18 set produced a tiny $\frac{1}{4}$ of a watt in the CW mode and accordingly its range was particularly restricted. By comparison, $1\frac{1}{2}$ watts is about the same level that a handheld CB radio transceiver produces today.

At Arnhem, the inherent problems of low power and short antennae became very obvious in frustrated attempts of the signallers to maintain contact over modest distances. Such difficulties had occurred previously and, given the distances from the drop zones to the west of Arnhem, radio communications were likely to be a problem. To go into battle with obviously inadequate communications seems in retrospect quite extraordinary. However, as noted earlier, the whole operation was set up in 7 days and, in a sense, the problems that occurred were largely endemic to the system then in use.

A military high-frequency transceiver of 1938, the Number 11 set — high-power CW at 4.5 watts and a range of about 32 kilometres (20 miles) maximum depending on the antenna *(PRJ)*

As a comparison, the VHF portable pack radios used by the American troops in the same operation further south at Grave and Nijmegen achieved a reliable range of between 13 and 16 kilometres (8 and 10 miles). Communication problems were not reported for these landings, and other problems that occurred were largely associated with the strength, determination and resistance of flanking German troops.

Apart from the radio failures on the ground near Arnhem, another major problem occurred in the provision of communications with supporting aircraft. The supply of incorrect crystals, used to determine the frequency of operation, appears to have been the cause of this problem. This communications function was to have been provided using type SCR 193 radios netted to the Second Army. Requests for air support should have been relayed via the Second Army and the Second Tactical Air Force Control Centre to the 83 Group of the Royal Air Force. To allow direct ground-to-air contact, VHF radios, type SCR 522, were also provided. Unfortunately it appears that the 1st Airborne Division was never able to contact the RAF or obtain much-needed air support. No doubt the combined nature of the operation and the mixing of American and British troops and radio resources all helped to create this element of the communications debacle.

In retrospect, it seems highly probable that the failure of HF radio at Arnhem was to be one of a number of significant factors that persuaded the British army to convert to VHF operation for company, battalion and brigade communications at a later time.

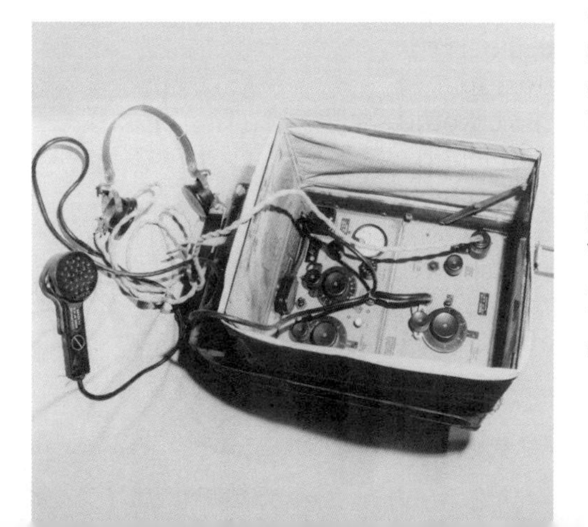

Left middle The Number 19 transceiver of 1941 covering the HF band, installed in tanks and vehicles, included a short-range VHF communications set operating at 230 megahertz *(PRJ)*
Left bottom The Number 18 transceiver backpack set of 1940 operated between 6 and 9 megahertz with a maximum range of 16 kilometres (10 miles) with a long rod antenna *(PRJ)*

Afterthoughts

In a number of respects, the Arnhem operation could be seen as a British version of the application of the principles of blitzkreig, in which a concentration of force at a weak point creates a 'breakdown' through which an expanding torrent of tanks and men can burst. However, in the lowlands of Holland, the conspicuous difficulty with the application of this principle was the constraint of topography. If the tanks could have spread out in a torrent after crossing the Nijmegen bridge, the story of Arnhem might have turned out quite differently.

Apart from this, in the use of parachute-borne troops, mobility was critical. Lightly armed paratroops could be useful only where the opposition was pushed off-balance and the momentum of attack maintained. Once this movement was arrested and troops placed in a defensive position, the end was almost inevitable, particulary where the opponents had available tanks and artillery and desperate and determined troops fighting at the edges of the Fatherland.

In the context of a fast-moving battle, reliable communications represent an essential element. In principle, although radio had the capacity to provide such a facility, in the 'Market Garden' operation it was a conspicuous failure for the British troops. In the face of developments which were unanticipated by the military planners, failure of communications materially contributed to the creation of a major military 'shambles'. Further, it exaggerated the damage that an inflexible response had already promoted.

As observed by the military expert Karl Marie von Clausewitz in the 19th century, a peculiar difficulty of war is the uncertainty of the data upon which one can act. Arnhem certainly demonstrated the truth of such a perception. Evidently, the lightly armed British forces were unable to maintain mobility and the momentum of attack when opposed by a large and unexpected concentration of German tanks and soldiers near Arnhem. Because of the serious breakdown of radio communications with the ground-based headquarters, rapid redeployment of troops to meet this situation did not occur and the initiative generated by surprise was lost. Beyond this they were not able to respond adequately to the unexpected high level of German resistance. In this respect, the failure of radio communications at Arnhem can now be seen to have contributed very materially to a sorry chapter in military history.

Top to bottom
Wireless set Number 22 of 1942 was a general-purpose HF transceiver with a maximum range of about 32 kilometres (20 miles) using a long rod antenna *(PRJ)*
The American BC1000 VHF backpack transceiver of 1944 operated in the frequency band 44 to 48 megahertz, with a range of between 5 and 8 kilometres (3 and 5 miles) using FM radio telephone *(PRJ)*
The remarkable US-designed 'Handy Talky', the BC 611 F high-frequency radio telephone, operated on HF frequencies and was crystal-controlled *(PRJ)*
Inside the BC 611 F showing the battery compartment and the array of miniature valves *(PRJ)*

CHAPTER

12

1939 to 1945

Codes, ciphers and Colossus

Few people today have not heard of the 'great secret' of the Second World War. This involved the ability of the Allied (British and American) intelligence services to read the encoded messages of the Axis (Germans and Japanese). This secret was so sensitive that it did not come to public attention until nearly 20 years after the end of the war in 1945. Even then, the exposure of the secret was at first quite obscure, one of the first discussions of the subject appearing in a book written by a Polish author.

Somewhat later, the subject was more clearly revealed, although still rather obliquely, in the notable book by David Kahn entitled *The Codebreakers*. Then, in 1974, the end of the 30-year limit on secrecy that had applied since the war's end, there appeared a book entitled *Ultra Secret* by FW Winterbotham. In a readable and ultimately very popular description, Winterbotham explained how the British had succeeded in reading German ciphers and had distributed this information to a number of senior and specially chosen officers in the fighting forces.

'ULTRA' INTELLIGENCE

Winterbotham's 1974 book did much to explain how the intelligence material collected by the codebreakers of Bletchley Park was used and disseminated as a potent weapon to defeat the Nazis. Much of the specific information relating to the technical aspects of the deciphering process came to light only after 1980. In addition, the remarkable contribution of Poland and its extraordinary team of young codebreakers was then revealed. Their attack on the secrets of the German codes had begun prior to 1930 and had achieved a significant level of success. This was in spite of the introduction of machine-generated codes that appeared initially to be incapable of decipherment by conventional means.

As a final provocation and what would be the immediate prelude to the start of the Second World War, the Nazi forces were assembled on the border of Germany and Poland. Now, in the few days of peace that remained, the threat of invasion was all too clear. Realising the value of the work that had been done by their cipher experts, the Polish cipher bureau decided to turn over its resources to its British and French counterparts.

The mansion house at Bletchley Park *(PRJ)*

Seen from the perspective of a later time, there can be little doubt that this generous decision had profound repercussions. The ultimate benefit to the Allied war effort is very difficult to estimate but it is certain that it played a very significant part in the ultimate outcome of the hostilities.

By the commencement of the Second World War, the system that was used by the Germans to carry out ciphering of written messages involved a systematic jumbling of the letters using a machine to simplify and speed up the process. The Enigma machine which was used for this task employed a cipher key that repeated so infrequently that even a fast modern computer, carrying out a process of testing all alternatives, would take so long to find an answer as to make the translation of the message too late to be useful. The encoding process as undertaken by the Enigma machine was thought by the German forces to be unbreakable. In a technical sense and in the absence of electronic computers, it was probably a justifiable assumption. However, as it turned out there was more to deciphering long repeat key messages than a simple process of testing every possibility in an exercise of brute force.

In fact, it was through a combination of techniques that the messages were broken. Part of the technique was based on the errors and misdemeanours of the cipher clerks who carried out the enciphering process. Setting up the Enigma involved a number of critical steps which were laid down in a field manual distributed to the cipher clerks at regular intervals. Through laziness or boredom, often such steps were not properly carried out. In particular, the use of formal letter headings on a repetitive basis in fresh messages provided the cryptographers at Bletchley Park with a most important chink in the armour of the Enigma-generated cipher. This process was described at a later stage by a noted member of the Bletchley Park code-breaking group, Gordon Welchman, who was nearly prosecuted by the British government for his indiscretion.

Apart from this, however, what the Germans did not fully appreciate was something that was very effectively stated by the mathematical genius Alan Turing, the author of the historically important paper on 'computable numbers'. This paper was, in large measure, responsible for setting the foundations of modern computer programming.

What this paper showed was that, if it was possible to build a machine to produce a complex cipher structure, then it was also possible to build another machine which could be designed to unravel the complexity. In the case of the German encoding system based on the Enigma machine, it was the Polish code-breakers in the 1920s, using a combination of mathematics and simple machinery, who first demonstrated that the proposition of Turing was correct.

One of the original huts at Bletchley Park *(PRJ)*

What is now known by all cipher users is that there is, in reality, only one absolutely safe and unbreakable method of encoding a message. With the emergence of systems such as those embodied in PGP (pretty good privacy) and seen on the Internet, a remarkably high degree of security is offered. However, with sufficient power, even the security of PGP may be possible to breach. In the end, to be absolutely sure that what one wishes to hide cannot be read by an unwanted third party, it is to the 'one-time pad' system that one must turn.

Beyond even the complexity of ciphers created by machines in which extremely long keys can create something akin to true randomness, there lies the ultimate security of the 'one-time pad' cipher. In this system, truly random arrays of numbers are printed onto pairs of paper pads (one-time pads).

One of each pair of pads is distributed to the field operators and one is kept at the home station. In the field, the letters are converted to numbers based on the position of each letter in the 26-letter alphabet. The random numbers are then added to the letters of the message in sequence, converting the plain text to cipher text. When a page of the pad has been used, it is immediately destroyed, hence 'one time'. At the receiving end, using the same sequence of numbers as used in the field, the random numbers are substracted from the cipher text revealing the plain text.

The randomness of the cipher key is transferred to the cipher text and the resulting message cannot be broken into by statistical methods or by repetitive testing with a computer.

Despite the invulnerability of the cipher text generated with the one-time pad, in a war situation this method is impractical. The number of pads required and the problem of their distribution create

The Enigma enciphering machine with its case open and the plugboard (Stekker) visible at the front and the cipher wheels at the back *(PRJ)*

insuperable logistic problems which the Enigma machine was intended to overcome. However, Enigma was to prove vulnerable to attack due to human error in its use. Where absolute security is essential, the 'one-time pad' remains the only completely reliable method of enciphering a message.

THE ENIGMA MACHINE

Until the advent of machine enciphering and computers, complex, clever and secure ciphers usually involved a considerable amount of effort by ciphering clerks, and mistakes and errors were common if not inevitable. Also, despite the need for conscientious work to avoid penetration of the ciphers, clerks always seem to be badly paid. For these reasons, there were early attempts to mechanise the chore of converting plain text to cipher and vice versa.

In the period immediately following the First World War there were several attempts to produce more sophisticated and secure ciphering machines. In the United States the work of Edward Hugh Herbern led ultimately to ciphering machines being made in Sweden by the Hagelin Company, and these were later adopted by the American army. By comparison, the work of Walter Scherbius in Germany led to the adoption of the Enigma enciphering machine by the Wehrmacht during the 1930s.

The Enigma enciphering machine, as used by the German army, looked like a strange cross between an old-fashioned typewriter, a manual adding machine and a telephone switchboard. However, as a relatively complex electromechanical device, its purpose was simply to convert the letters of a message typed into it to an insanely complex 'jumble' of letters which, in theory, could not be understood by any third party without access to a similar machine.

However, beyond having access to another Enigma machine, the recipient of a coded message had to set it up in a particular manner, which was set out in instructions distributed to the operators in a very organised, efficient and secure manner. The administrative and logistic problems of providing instructions to the operators in the field in war conditions were relatively small compared with what would have been required if a 'one-time pad' system had been used.

In detail, the Enigma consisted of a keyboard, much as one finds on an electric typewriter, connected to a panel of lights which were battery operated, with the intermediate wiring carried through a system of rotary switches and a plugboard. Each of the switches, or rotors as they

Left A boxed set of cypher wheels for the Enigma enciphering machine *(PRJ)*

Right Inside a cipher wheel showing the 'jumble' of wiring connecting contacts on one side to contacts on the other *(PRJ)*

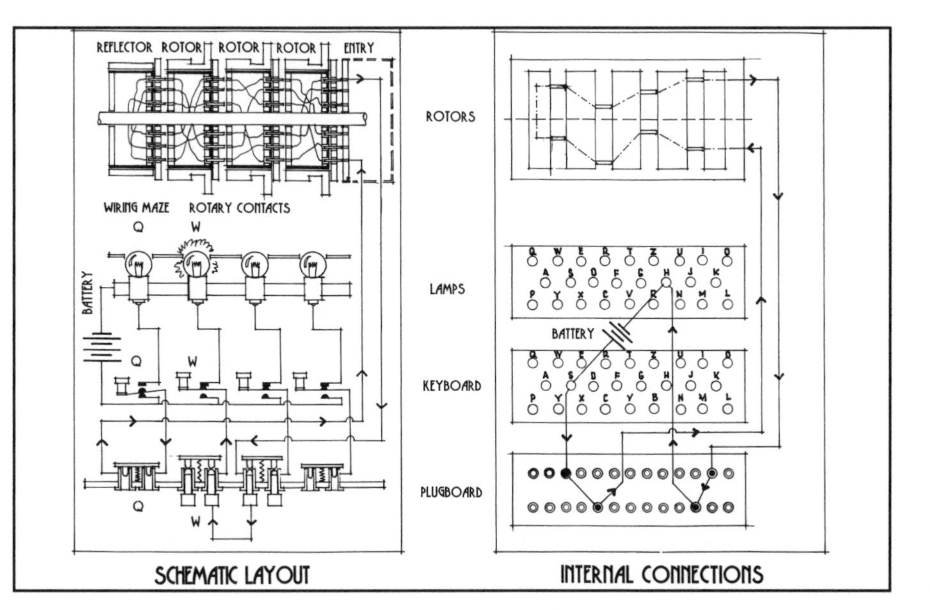

Diagram of the Enigma enciphering machine *(PRJ)*

were called, had engraved upon them a random alphabet and, in the case of the Enigmas used by the German army, there were three sets of rotors per machine in use at any given time. Later, two further rotors were supplied so that three out of five rotors could be used.

The diagram shows how the current from the battery flowed through a key when it was pressed, wound its way through the jumble of the wiring in each of the three rotors, was reflected back through an end rotor and finally emerged to light up a bulb in the display board, having passed through the final complication of a plug and patch board. If the first letter of the message was 'A', for example, somewhere on the array of lights another letter would light up, for example 'Z'. If one returned the machine to its basic setup position and then pressed the letter 'Z', the bulb representing the letter 'A' would then light up. A message encoded by one machine could be decoded by another machine at the receiving end by simply typing in the code and copying down the letters which lit up at each key press, assuming that it had been set up in the same way as the sending machine. This important property of 'reciprocity' made for a very flexible and rapid method of coding and decoding messages. However, inclusion of the reflector rotor in the Enigma machine produced one subtle but serious cryptographic fault — no letter could ever be encoded as itself. This feature was not emulated in other ciphering machines developed for the Allies or subsequently.

THE SECRET LISTENERS

From the very earliest days of wireless, it was quite obvious that as a system of communication it was inherently vulnerable to eavesdropping. One of the reasons that the British army was initially so reluctant to become involved with the new method of communication was the realisation that there could be a problem in preventing the enemy from hearing what was being said as messages were passed.

As Sir Oliver Lodge commented in 1903: 'Wireless or open methods cannot compete in point of secrecy or certainty with closed or cable methods, and can only compete with them in point of speed and accuracy by aid of great improvements and new inventions involving little less than discoveries.'

Many of the very painful lessons of the First World War served as models of what to avoid in the Second World War. In particular, the Germans, dismayed by the vulnerability of their signals which was revealed rather incautiously in the late 1920s by Sir Alfred Ewing, turned to machine-generated ciphers. By comparison, the British consolidated, centralised and coordinated radio signal interception and decipherment under the control of the Government Code and Cypher School.

Radios commonly used by the 'Y' service, the secret listeners *(PRJ)*

Unlike the Germans, the British mobilised its legion of radio amateurs, not only into the armed services but also into a clandestine corps of 'listeners', known as the 'Y' Service, which provided much of the raw material for decoding at Bletchley. Soon after Hitler came to power in 1933, the new Nazi administration, fearing that free-spirited radio amateurs were likely to represent a security risk, shut down all the radio clubs and, more seriously, made very little use of the talents of the technically expert amateurs. They came to regret this, as Field Marshal Goering was to acknowledge at a later stage.

Apart from the Polish and British success in breaking the ciphers used by the Germans, the immensely important part played by those who intercepted the radio signals which carried the enciphered information has now become clear. A recent book by Nigel West, entitled *GCHQ*, has helped to dispel much of the deliberately obscuring camouflage surrounding the monitoring of the radio frequency spectrum. The covert listeners who undertook this task were, during the war years, gathered together in a variety of listening establishments. The headquarters of this operation was originally known as the Government Code and Cypher School located at Bletchley Park. Later, this name was changed to the Government Communication Headquarters (GCHQ).

THE 'BOMBE'

Although the capacity of one machine to analyse the complex output of another was first applied to cipher problems in Poland in the 1920s, it was brought to its most significant application in Britain after the commencement of the Second World War.

During the 1920s, with an escalating level of German belligerence and an increasing feeling of isolation, the Poles took an ever-greater interest in the ciphered signals of their dangerous neighbour. Early on, the Poles were able to intercept an Enigma enciphering machine and make a copy of it without being detected. However, having the machine was only a small part of the problem of dealing with the

The British 'Bombe' used to break the Enigma codes to produce the Ultra intelligence *(PRJ)*

Enigma code, and other devices were needed to help in the process of cryptanalysis. One of the most important of these pieces of apparatus was a large electromechanical device known as a 'Bomba', which was used to test alternative settings of the Enigma ciphering machine.

For a while during the 1930s, the Poles were able to decipher a good deal of the German traffic. However, the end came quite suddenly when two extra rotors were introduced in 1938, increasing the alternatives from three to five. Where previously the Poles had been able to carry out the decryption task with six Bombas, now sixty would be needed.

At this stage, realising that the Germans were very close to launching an invasion, the Poles gave all their work to the French and the British at a meeting in Warsaw. Ultimately, their foresight and generosity in providing this valuable material to the Allies was to pay dividends in winning the war. However, in the short run, it did nothing to delay or defeat the new German military tactics of blitzkrieg, employed to devastating effect during the next few months. What was critical to the success of the blitzkrieg was the application of air power, as applied via Stuka dive bombers used to support the rapidly deployed armada of tanks on the ground. The blitzkrieg technique as developed for the German army by its most vigorous exponent, Heinz Guderian, involved close control of operations by radio from a command vehicle located close to the head of the column of tanks. In addition, messages enciphered with the Engima machine were also sent back to headquarters from this same vehicle. Radio and enciphered messages represented the vital element in keeping control of the rapidly moving and advancing tank squadrons and the success of the German campaign in Poland.

With the defeat of the Polish army, much of the constructive work of cipher-breaking passed to England and specifically to the new secret centre at Bletchley Park. Here the required number of what from now on would be called 'Bombes', constructed by the British Tabulating Machine Company of Letchworth, were set up and the job of breaking the Enigma-generated code began in earnest.

Despite the precedence of the 'Bombe' in the story of Ultra intelligence, it is another machine that is the first to have been replicated at the revived Bletchley Park. When it is realised that this machine may ultimately be acknowledged as the world's first operational electronic computer, this inversion of the anticipated order of work is not hard to understand.

However, work is now progressing apace to replicate the 'Bombe'. The skeleton of this vitally important machine was completed by 1997. This fascinating piece of apparatus that was so crucially significant in unravelling the complexity of Engima messages will soon be heard in operation once again.

ENCIPHERED TELETYPE

At a later stage of the war and quite separate from the system which generated the Enigma codes, there was another, supposedly even more impregnable, system developed in Germany. This method was used to transmit high-level traffic via both radio and telephone lines and was based on two teletype encrypting machines. The first of these was, in effect, a supplementary device to be attached to a conventional teletype machine and was made by the firm of Lorentz. It was known as the 'Schlusselzusatz 40' or SZ 40 and had two sets of five encoding wheels. Subsequently, a second version was produced, known as the SZ 42.

Later again, a composite teletype and enciphering machine was produced by the firm of Siemens and this was known by the Germans as the T 52 or, more commonly, the 'Geheimschreiber' (secret writer). Both these machines employed multiple rotating drums with adjustable projecting pins which were used to encode the plain text typed into the teletype. The machines in turn produced encoded Baudot using the method devised by GS Vernam in 1917.

Baudot, which was the normal output from a teletype machine at that time, is frequently referred to as a 'code' because it uses a system of marks and spaces (tone or no tone) to represent the letters of the alphabet and the numbers. For computer-literate people, this will be recognised as a binary form of code as used by computers in the machine language with which they are given instructions. In general principle, Baudot is very similar to ASCII code and is not dissimilar to morse code.

In terms of hiding the meaning of a message, Baudot by itself is no more difficult to read than ASCII or morse code. All one needs is a teletype machine of identical construction and the messages can be read. To encipher the Baudot code of one machine so that it could not be simply translated by another machine, a set of rotors with projecting pins was used to add binary cipher elements to the binary code that resulted from the plain text as it was typed into the teletype machine.

Left The Lorentz online teletype enciphering machine (SZ 40) *(PRJ)*

Right The Siemens online teletype enciphering machine, the Geheimschreiber (T 52) *(PRJ)*

These pins were located on the rims of the rotors and, as they turned, the plain binary Baudot code was encrypted by the addition of binary elements supplied by the pins. This process involved the creation of a long repeat cipher key, determined by the number of rotors and pins.

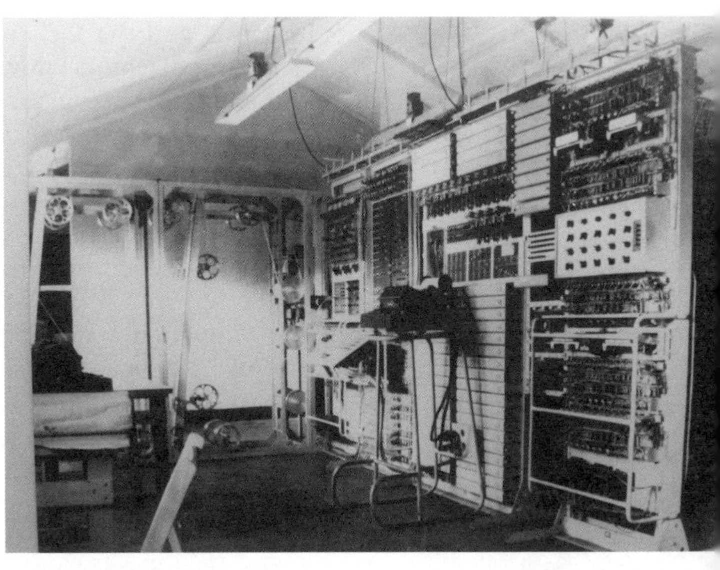

The electronic computer for breaking the German online enciphered teletype, the Colossus *(PRJ)*

THE FIRST ELECTRONIC COMPUTER

Needless to say, the complexity of the output from the teletype enciphering machines required an equivalent level of complexity in analysis for messages to be extracted from the 'jumble' of enciphered bits. As a result of some very significant inventive thinking by Alan Turing, Max Newman and others at Bletchley Park, a specification for a machine was developed which was intended to deal with the problems set by online encoding of teletype. The first product of this inspired work was the creation of an electromechanical device which was constructed by technical staff of the British Post Office and came to be known as 'Heath Robinson', because of its resemblance to one of the strange machines drawn by the artist Heath Robinson, who produced illustrations for the daily newspapers in wartime England.

The principal designer of this first machine was CE Wynn-Williams of the Telecommunications Research Establishment, and to a large extent it relied on conventional telephone and telegraph technology of the period. In particular, and an important portent for future developments, it incorporated about two dozen thyratron vacuum tube switches which were able to emulate the switching function of the mechanical rotors (rotary switches) of the Lorentz and Siemens machines. The 'Heath Robinson' machine was to prove very effective, if rather prone to mechanical troubles, making use of punched paper tape to provide the data input which was driven over rollers and past the reader unit at high speed. Those who operated these machines recall that sometimes during a run the paper tape would break and the air would be filled with small pieces of paper, rather like confetti.

The replica of the Colossus as built at Bletchley Park during 1995 *(PRJ)*

With the advantages of a machine basis of analysis established, it was seen as appropriate to extend the technique, and further development led to a new and far more powerful machine. At the suggestion of Alan Turing, Tommy Flowers of the Post Office Research Station at Dollis Hill was called in and began to build the new machine, assisted by other electrical wizards from his department. With the benefit of hindsight, the machine that was built can be seen as effectively the world's first true electronic computer, involving initially an array of about 1500 valves (EF50) and thyratrons. In a second version, this figure was to grow to about 2500 vacuum tube devices.

These vacuum tube switches (thyratrons) emulated the relatively slow rotation of the mechanical rotors of the Lorentz enciphering

Looking behind the racks of the Colossus with the decoding machine, Tunny, in the course of construction also visible *(PRJ)*

machine and the Geheimschreiber but at a blinding speed for 1944. This later machine, known as 'Colossus', was capable of employing conditional Boolean logic as applied in any modern computer. For this purpose a system of toggle switches and a plugboard were used.

The main work of Colossus was to analyse the settings of the rotors in the German machines. However, in order to produce plain text from the cipher text, another device was required. This was known as 'Tunny', derived from the cover name for all of the enciphered teletype traffic received from the Axis, 'Fish'. The Germans called this system the 'Sagefisch' (swordfish) and the codebreakers at Bletchley Park adopted a dangerously similar range of names, such as Tunny, Sturgeon and so on, all grouped under the general name of 'Fish'.

What the Tunny machine was able to do was to emulate, with the use of conventional Post Office Type 3000 relays, the action of the Lorentz and Siemens machines, once the settings of the rotors had been established using the Colossus. Tunny received the stream of the enciphered Baudot and converted it into a stream of plain text German which then could be taken off for translation and distribution. Here was the source of the 'golden eggs' that Churchill referred to as coming from 'geese that never cackled'.

The Colossus was 3.6 metres (12 feet) wide and 2.3 metres (7 feet 6 inches) high. In front was an IBM electric typewriter which had a conventional transverse carriage. This operated with such violence that the stand on which it was mounted would walk across the floor if not restrained. It was usually 'roped' into position, adding to the 'string and sealing wax' appearance of the whole construction. More importantly, however, the apparatus worked and worked brilliantly.

Unfortunately for historians, at the end of the war, because of its secret application, the Colossus was dismantled completely and few photographs of the original installation remain available. Only a few years ago, that would have been the end of the story of Ultra and the extraordinary secret of code breaking in the Second World War. However, with the reopening of Bletchley Park, a new and fascinating endeavour has commenced. This is the replication of the Colossus which was recently completed and is now able to duplicate the method of code breaking used during the war years.

The Bletchley Park computer and microcomputer museum *(PRJ)*

IMPACT OF ULTRA

During the first months of the Second World War, knowledge of the contents of German codes and, in particular, Hitler's most secret orders allowed British troops to be spirited away from the disastrous shores of Dunkirk, thus saving the nucleus of a new British army. Soon after, in the Battle of Britain fought in the air over southern England, knowledge of German intentions carried in their ciphered radio messages allowed the Royal Air Force to conserve its resources and ultimately triumph over a large and potentially overwhelming opponent.

Later in that same conflict, the Americans were presented with reasonably clear indications of Japanese intentions to attack the Pearl Harbour naval base in Hawaii extracted from enciphered messages. Despite this, the failure to send the information to the garrison and naval commanders in sufficient time resulted in a major disaster for the American navy. The surprise raid launched by the Japanese air fleet dealt a shattering blow to America's fleet lying unprepared below. Complete disaster was averted only by the fortunate absence of American aircraft carriers which had put out to sea a few days previously.

Following the attack on Pearl Harbour, the Americans determined to inflict retribution upon the architect of this successful Japanese assault, and many months later the opportunity arose. Disregarding the potential breach of security that the action could bring, a flight of aircraft was arranged to ambush Admiral Yamamoto in midair. This was carried out by long-range American aircraft off the coast of Borneo on the basis of deciphered material that had been received in Japanese radio transmissions. Yamamoto's plane was shot out of the sky and the Admiral died never knowing that it was the penetration of supposedly secure Japanese ciphers that had sealed his fate.

By far the most important outcome of all the secret work of the Second World War was that it led directly to the first generation of electronic computers in the period after 1945. Once again the swords of war were converted into the ploughshares of a new revolution: computer science.

CHAPTER

13

1945 to 1975

Solid state and silicon chip

Only a few years after the end of the Second World War, a revolution occured. It was a revolution that would displace nearly 40 years of technological development and experience with the vacuum tube or valve. Further, it was a revolution based on discoveries that scientists of the Victorian era, when the world communications network first began, would have found quite inexplicable.

Such scientists would have had difficulty grasping the concept of electrons moving in a vacuum, as demonstrated by the development of valve or tube technology during the period after 1912. Now the incredible secret lurking in the heart of certain crystalline substances would be revealed but this would involve a knowledge of quantum mechanics. This understanding and description of the inner workings of the atom and its accompanying electrons was developed during the 1920s by the Danish scientist Niels Bohr. For that reason, its insights would have been completely unknown to and unimaginable by Victorian scientists.

In the period before 1920, the crystal detector had been a common device in radio communication. At that time, its fundamental method of operation had been a complete mystery. Before any conception of what was involved in crystal rectification could be developed, an understanding of the processes of electron energy states at the atomic level would be essential. Over the latter part of the 1930s, while preparations for war went on apace, this fundamental knowledge was being developed. This, in turn, served as a prelude to research work at the Bell Telephone Laboratories in the United States that ultimately led to the development of the first transistor in 1947.

What is now taken for granted by anyone using any piece of electronic apparatus in which solid-state devices or silicon chips are used is that atoms can move around in certain forms of solid semi-conducting material. Further, compared with the older thermionic vacuum tubes, such solid-state devices can be used for amplification of weak signals over a vast spectrum of frequencies as well as for counting and memorising numbers for a variety of complex purposes. The first vital characteristic of a solid-state device is that it is able to operate with only a single source of power, unlike a valve or tube, which needs a heated cathode in the vacuum to throw off electrons. The second vital characteristic is that such devices can be made microscopically small as most recent generations of silicon chips and integrated circuits have demonstrated. Where the first computers, developed both during and after the Second World War, used several thousand valves and took up the space of several large rooms, today's solid-state microcomputers have millions of active devices crammed onto a piece of silicon the size of a couple of postage stamps. Despite this minuscule size, the modern microcomputer is far more powerful than the first-generation computers, and can be tucked into a container somewhat smaller than the average hardcover book.

Despite the apparent novelty of the solid-state devices which were developed during the war years and which ultimately would lead to the solid-state amplifier, the transistor, developed through the research of William Shockley, Walter Brattain and John Bardeen, solid-state diodes had been in use as early as 1905. In the form of the humble crystal detector, the cat's whisker, the galena-based diode was used extensively in the first generation of radio broadcast receivers in the 1920s. When valves were developed as amplifiers, the crystal detector was largely forgotten. The development of radar at centimetric wavelengths in the Second World War forced a search for a more effective detector than the valve.

As described in an earlier chapter, the answer to this search was the development of a new generation of solid-state detectors. At this stage, however, the mode of operation still remained obscure and further research during and immediately after the war was required to formalise the mechanism that allowed this odd diode-type behaviour to occur in solid semi-conducting substances.

THE TRANSISTOR

During the mid-1930s, a new and aggressive Director of Research was appointed to lead the Bell Laboratories, the research arm of the American Telegraph and Telephone Company. In a short space of time, Mervin Kelly came to believe that all the hot and fragile valves and unreliable electromechanical relays used in his generation of telephony would by some new means have to be replaced. He decided that with appropriate research effort a new device or devices based on the development of radically different technology would become available to solve the problems of unreliable telephone exchange operations. His clear intention was that the Bell Laboratories would be at the leading edge of research in discovering such new technology and devices, whatever they might turn out to involve.

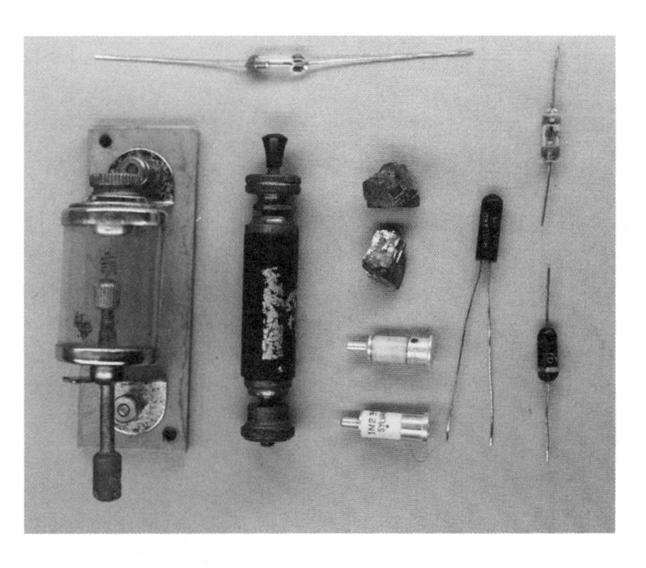

Left to right:
Cat's whisker galena detector, Perikon detector, Sylvania and Westinghouse radar detector cartridges, galena crystals and solid-state diodes *(PRJ)*

In order to initiate this new research and considering that it was likely to arise out of the new understanding of physical processes at the atomic level in solid materials, Kelly appointed a number of bright young graduates. One of these new employees had a particular knowledge of quantum mechanics as it had developed by that time. This same young man also had a background in electrical engineering and, with such qualifications, not surprisingly was selected to lead the new Bell Laboratories research team. With such an array of skills, William Shockley was particularly well suited to this task. Shortly after he was recruited to the Bell Laboratories in 1939, he was joined by another older scientist and experimenter, Walter Brattain, who was also to play a significant part in the discovery of the transistor.

Before the start of the Second World War, these two, together with other suitable graduates in solid-state materials, began to look at the operation of copper oxide rectifiers which were commonly used at that time for producing direct current from alternating current sources. Despite this common application, the underlying physical operation of the copper oxide rectifier was quite unknown. Determining how such devices worked was seen as a doorway to some substitute for the valve and relay of telephonic installations.

In 1939, Shockley and Brattain tried for the first time to imitate the vacuum tube using solid material, by adding a third electrode to a copper oxide rectifier, but without any success. Despite this failure, it is clear that Shockley was sufficiently convinced that a solid-state amplifier could be developed that he made notes on the subject in his research notebook.

At this stage the war intervened and both Shockley and Brattain were moved away to other more important tasks involving the detection of magnetic fields and the associated problem of detecting submerged submarines. However, the work continued at Bell because the need for a new form of rectifier at microwaves became an urgent necessity with the development of the magnetron used in radar. Russell Ohl, Jack Scaff and Henry Theurer began work on purifying silicon so that more consistent and reliable diode detectors could be produced, and from this work came the chance discovery of an important aspect of solid-state materials. Known as the P-N junction, this structure, which was created in a sample of purified silicon by accident, was found to have particular rectifying characteristics. Later, this junction was to assume crucial importance in the search for the solid-state amplifier.

An Ivalek cat's whisker crystal set from about 1950 covering the medium-wave band *(PRJ)*

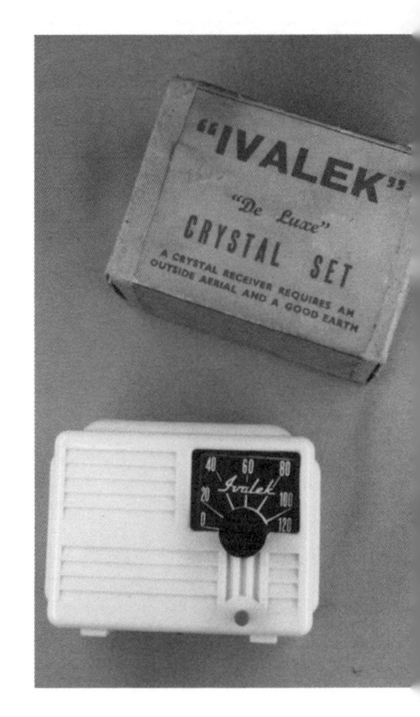

Transistors in various packages from glass OC series through metal-cased to ceramic; silicon chips including a Z80A CPU, a 486 DX 33 MHz CPU (Intel); operational amplifiers and solid-state memory chips (*PRJ*)

During this period, much of the research on solid-state devices moved to Purdue University in the United States where, in retrospect, it can be seen that these academic researchers missed discovering the transistor by a 'cat's whisker'.

After the war, Shockley and Brattain returned to Bell Laboratories in 1946 and were shortly joined by another very talented researcher and theoretician, John Bardeen. Much of the research work at Bell Laboratories before the war and at Purdue University during it had involved the materials germanium and silicon. Inevitably, these materials continued to be the basis of research in the period after 1946.

After considerable work, Shockley proposed a new type of theoretical device which would now be called a 'field effect' transistor. Despite the theoretical basis of the device being seen as adequate, initially it could not be made to work. Bardeen later developed an explanation of this stubborn behaviour. However, in working with this experimental device, an important discovery was made that became the first major step towards the transistor as we know it today.

Brattain and Bardeen had been working together experimenting with wire connections to the surface of pieces of exceedingly pure germanium into which a tiny admixture of another element had been made. This process of adding an impurity into the germanium was known as 'doping' and represented the method by which a solid-state semiconductor material was able to be created, in which mobile electrons or 'holes' (places where electrons were absent) were available to carry an electric current. The experiments with wires led Brattain and Bardeen to try an arrangement with a physical structure having a close analogy to the old cat's whisker crystal detector.

They found that by using two metal wires set very close together and touching the surface of a piece of the 'doped' germanium, it would amplify a signal fed into it or go into continuous oscillation if feedback was provided. This first solid-state amplifier was discovered in December 1947 and was soon patented as a distinctly new device.

This began the revolution that would see the vacuum tube or valve displaced in all but specialised applications in less than 20 years. At about this time, the new device was given the name 'transistor', which had come from the fertile mind of JR Pierce (who worked at the Bell Laboratories from 1936 to 1971). He had derived the name from the conjunction of two words, 'transfer' and 'resistance', from an analogy with the 'transfer-conductance' of the vacuum tube.

The 'point contact' transistor went into commercial production in October 1951 at Western Electric, the manufacturing arm of Bell Laboratories. However, the 'point contact' transistor proved to be a very temperamental device and, although it certainly started the solid-state revolution, it was not until it was replaced with a far more reliable device in the form of the 'junction' transistor that the battle for recognition was finally won.

The 'junction' transistor was developed by Shockley as early as January 1948 in response to the unsatisfactory and 'noisy' performance of the point contact transistor. However, due to an injunction from the Bell Laboratories hierarchy, Shockley was prevented from publishing his work and the device did not appear until 1951.

On the basis of these various strands of revolutionary work and discovery, it is not surprising that Shockley, Brattain and Bardeen are now almost always seen together as a triumvirate in early photographs of the researchers. This is despite the fact that, in reality, it was Brattain and Bardeen who developed the very first workable transistor. In addition, the good early working relationship of the three researchers was subsequently soured by personal rivalries which a Nobel prize, jointly awarded in 1956, did something to mend.

The first 'point contact' transistor of 1947 in its original 'lash up' form with two sheets of gold leaf touching the surface of a 'doped' germanium crystal *(PRJ)*

Interestingly, the 'field effect' transistor (FET) that had been conceived of as a theoretical device by Shockley in 1946 was ultimately developed by his semiconductor research team some years later. The FET remains an extremely important variant of the basic transistor family, having characteristics far more like the triode valve than do those of the junction transistor. It was not developed for quite a long time because, shortly after its initial conception, it had been realised that it bore a close resemblance to a device that had been invented 20 years earlier and patented. This was a primitive FET which came from the mind of a German immigrant, Dr Julius Lilienfeld. He had filed a patent in Canada in October 1925 and then subsequently in October 1926. This patent had later been granted on 28 January 1930 with the US patent number 1,745,175. For this reason, a patent for the 'field effect' transistor was not sought at the time that Shockley defined it initially and for that reason the first transistor patent was awarded to Brattain and Bardeen. Despite this, their names are always coupled with that of Shockley in the invention of the transistor. Given that he was undoubtedly the major driving force in transistor research going back to before the war, this does not seem at all unreasonable.

Despite its revolutionary advantages, the transistor was initially a very expensive device costing something like eight times the price of a triode valve able to do a comparable job. In addition, it was initially promoted as a simple substitute for the valve before its characteristics and potential were fully appreciated. This led to attempts to apply it to intrinsically unsuitable electronic tasks. Perhaps more importantly, it was an unfamiliar device vying for attention in the well-established field of vacuum tube technology. In order to understand and apply the new solid-state devices, electronic engineers and scientists were required to relearn their basic skills. This engendered a good deal of professional resentment and militated against the immediate acceptance and application of transistors. Perhaps more significantly the

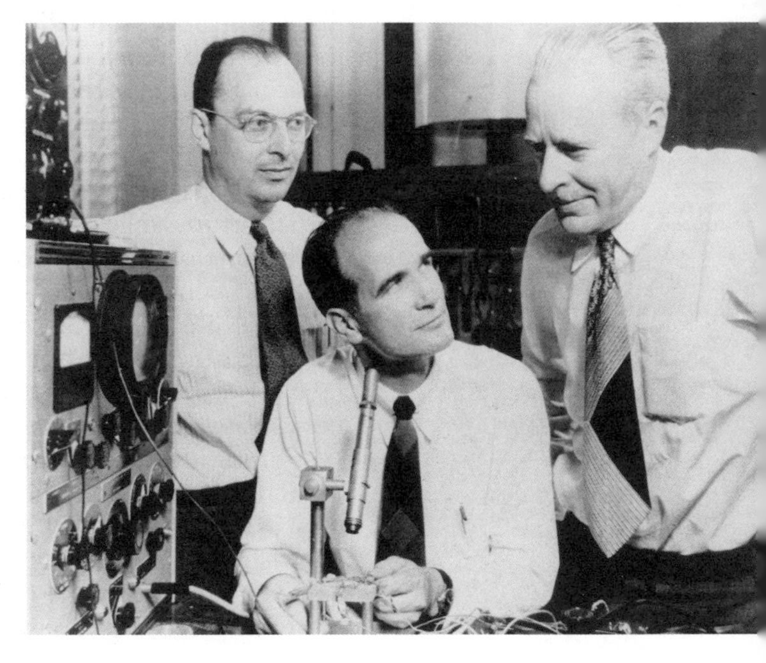

John Bardeen, William Shockley and Walter Brattain in 1948 shortly after the discovery of the transistor at Bell Laboratories *(LUC)*

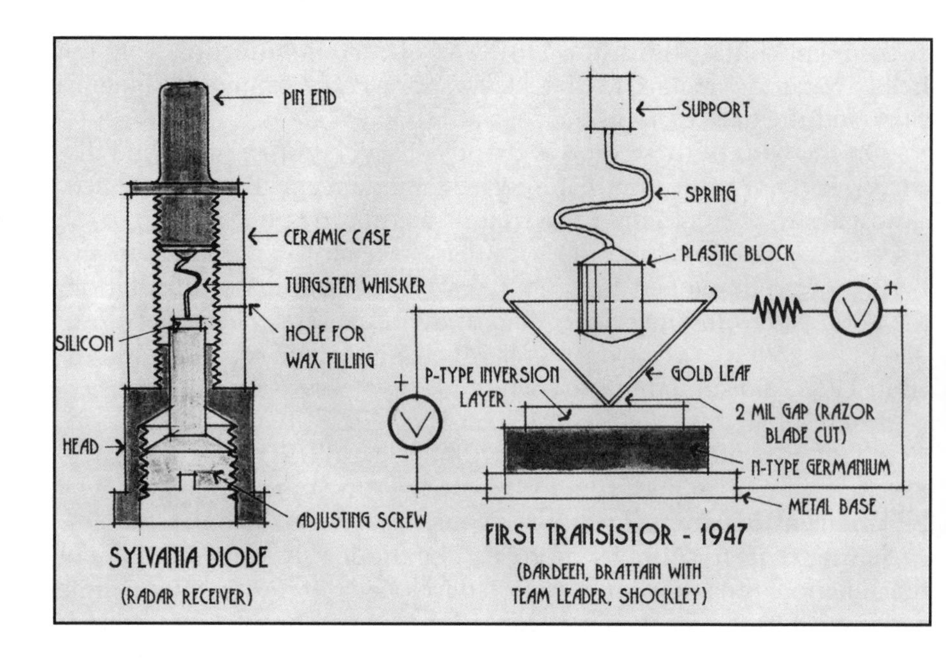

Steps in solid-state development 1 — diagrams of the Sylvania diode for radar detection and the first 'point contact' transistor *(PRJ)*

device also created a degree of fear that it might well lead to a loss of jobs. In the longer term, this fear was probably justified and one that has escalated in importance as other potent solid-state devices have appeared since 1947.

Despite this opposition, once the transistor was perceived to be a reliable device, it began to penetrate the market in a number of different ways. A development that was to have a dramatic impact in consolidating the impression of reliability was the development of the silicon junction transistor. All the earlier successful Bell Laboratories transistors had been based on germanium, but they had been very susceptible to heat and thermal runaway. By comparison, the silicon transistor was far more robust and heat-tolerant. When it appeared, having been developed by Texas Instruments in 1954, it was immediately seen as a logical replacement for the germanium transistor, other than for special purposes. In particular, this new robust transistor was seen by the military as particularly suitable for a number of applications in difficult environments.

The Regency 1 radio *(ET)*

Although there was an increasing demand for transistors in the general market and the commercial uses for transistors proliferated, it was particularly the military that provided the demand for the new devices. As the volume of transistors dramatically increased, there was a corresponding and continuous decline in the cost of production, reflecting conventional economies of scale in mass production. This desirable result converted the wildly expensive transistor into something that could be used for general applications and as a substitute for the common and, by now, relatively inexpensive device, the vacuum tube or valve.

During this period, the outbreak of the Korean war was to add a particular emphasis to the military concern with new systems and devices. The transistor was small, light and more robust than the valve and therefore particularly suitable for use in the harsh environment of combat. Apart from the obvious application to portable field communications radio, the transistor found an appropriate use in proximity fuses of artillery shell. Once again, one could see that warfare had produced the impetus required to establish a new technology, just as had happened with the valve in the First World War. The rapid reduction in the cost of the transistor soon saw its use extend into the consumer market for entertainment goods.

One of the first and most spectacular public, as opposed to military, early successes for the transistor was in the field of popular radio entertainment. At last it was possible to construct a small, light and inexpensive-to-operate radio which could be carried around in the pocket. The first of these new radios was commissioned by Texas Instruments, using four of their new transistors, and was built by Regency as the TR1 in October 1954. It was an instant success and did not take very long to cross the world to Japan where, shortly after, Sony Corporation was established in unimpressive premises. This company went on to make its first pocket radio which was an immense success and started the avalanche of Japanese-manufactured solid-state radios and other electronic goods.

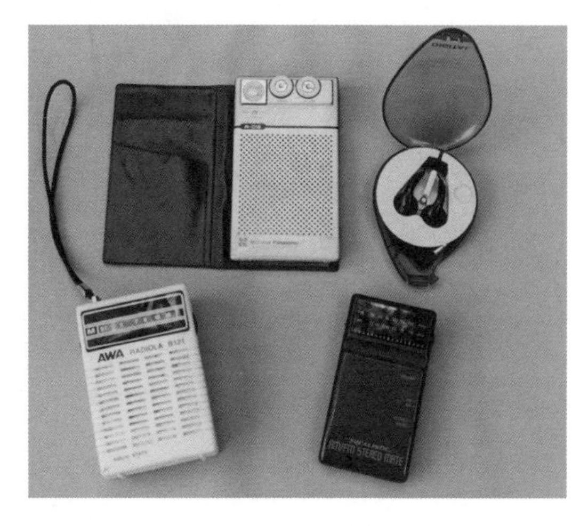

Pocket transistor radios, including the AWA Radiola B121 (AM), the National Panasonic R-012 (AM) and a Tandy FM stereo and AM receiver of 1995 (Model 12-189) with earpiece store *(PRJ)*

Whether the Regency was intended to appeal to such a client or the attraction was simply a fortunate accident is not clear, but a significant early purchaser was Thomas J Watson, the president of the burgeoning computer firm IBM. Perhaps unsurprisingly, very soon after Watson's purchase of 100 of the new transistor radios for his senior managers, IBM had produced a computer using 2500 transistors instead of 1200 conventional valves, saving enormously on space and operating cost. Known as the 1401, here was the start of a new revolution in the use of transistors which saw the technology become universal in computers in less than 5 years.

For some time, the transistor remained a difficult device to fabricate,

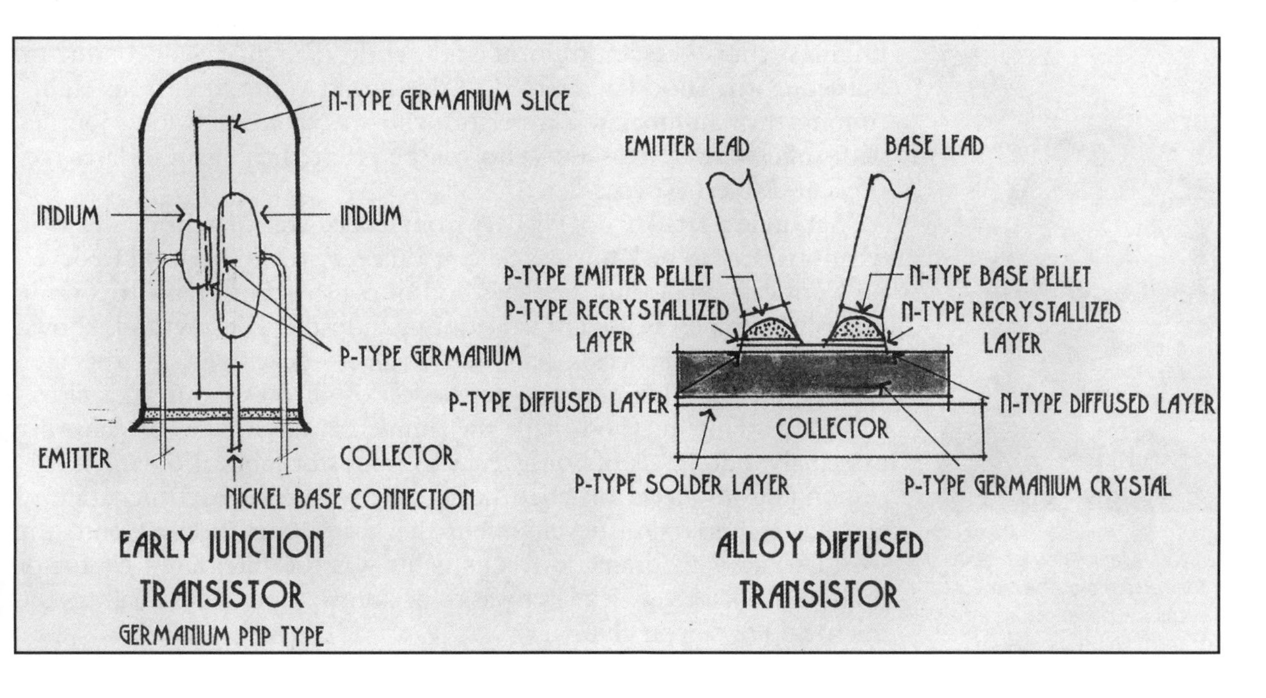

Steps in solid-state development 2 — early junction transistor and the alloy-diffused transistor *(PRJ)*

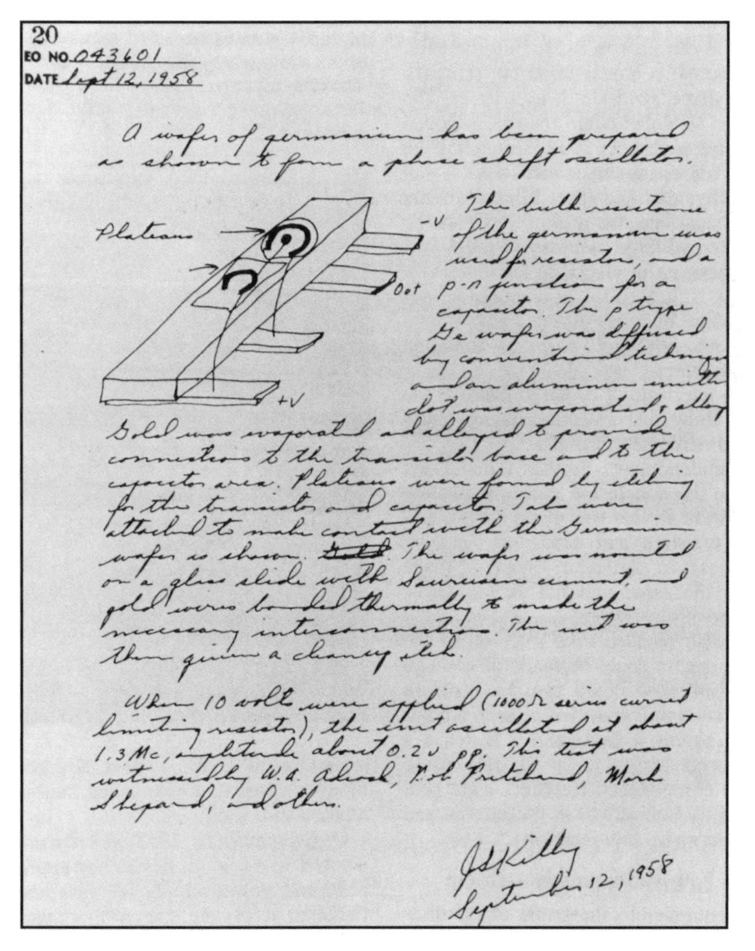

20
EO NO. 043601
DATE Sept 12, 1958

A wafer of germanium has been prepared as shown to form a phase shift oscillator.

Plateau

-V

+V

The bulk resistance of the germanium was used for resistors, and a p-n junction for a capacitor. The p type Ge wafer was diffused by conventional techniques and an aluminum emitter dot was evaporated & alloyed. Gold was evaporated and alloyed to provide connections to the transistor base and to the capacitor area. Plateaus were formed by etching for the transistor and capacitor. Tabs were attached to make contact with the Germanium wafer as shown. The wafer was mounted on a glass slide with Sauereisen cement, and gold wires bonded thermally to make the necessary interconnections. The unit was then given a cleanup etch.

When 10 volts were applied ($1000\,\Omega$ series current limiting resistor), the unit oscillated at about 1.3 Mc, amplitude about 0.2 v pp. This test was witnessed by W. A. Adcock, Bob Pritchard, Mark Shepard, and others.

J S Kilby
September 12, 1958

A laboratory note on the construction of an integrated circuit by Jack Kilby of Texas Instruments in 1958 *(LUC)*

suffering from contamination by minute amounts of inappropriate materials that destroyed its performance. One of the most elusive of these contaminants turned out to be copper which, before its identification as the culprit, had been referred to mordantly as 'deathnium', due to its capacity to destroy newly made transistors.

By 1953, Shockley had left the Bell Laboratories in order to set up his own business manufacturing transistors. Known as the Shockley Semiconductor Laboratory, the new organisation began to develop a range of refinements to the transistor construction process. In this it was mirrored by many old and new organisations which, during this period, sought to develop the technology of transistors and market them as the basis of a range of new consumer products. In particular, important developments would come from former associates of Shockley who resigned en masse after a disagreement with the new technology tycoon.

As a basis for this his new business, Shockley had decided that he would exploit the process of 'double diffusion' that had been developed at Bell Laboratories prior to his departure. Somewhat later, eight of his brightest employees left to form a new entity, Fairchild Semiconductor. Subsequently, Shockley referred to this group as the 'traitorous eight'. Among their number was a person who was to travel a long journey with solid-state devices and who will be referred to again in later sections — Robert Noyce.

Jack Kilby of Texas Instruments at the time of his proposal for an integrated circuit device *(LUC)*

Set up in a factory not very far from its erstwhile parent company in what was later to be known as Silicon Valley, Fairchild went on to develop a critically important new method of constructing transistors using photolithography as used in the printing industry. This method, known as the 'planar' process, had earlier been developed at the Bell Laboratories and involved the deposition of silicon vapour on a silicon substrate. This method had the immense advantage of allowing extremely high levels of consistency in transistor fabrication and a corresponding low level of failed devices. It was to be a cornerstone of another technological development that was about to burst onto the world scene with dramatic results. This was the integrated circuit or monolithic device, which later was to be known more colloquially as the microchip or silicon chip.

INTEGRATED CIRCUITS

Although the idea of the 'integrated circuit' had a very long pedigree in radio engineering, having been the basis of an early German triple valve in a single glass envelope, it took some while for the same approach to be used with transistors. What was ultimately to demand the development of such a new arrangement was the rapidly escalating complexity of devices using transistors. Despite the rapid reduction in the size of transistors, circuits were still being hand-fabricated with conventional wires and solder connections and this inevitably produced errors and lack of consistency. Even the introduction of printed circuit boards still involved a person with a soldering iron to place and connect each transistor. 'Dry joints' were a major problem and remain the bane of circuit-board construction even today — microcomputer 'motherboards' have been replaced more because of this problem than almost any other.

The first Texas Instruments integrated circuit based on the Kilby design of 1958 *(LUC)*

The complexity that results from the need to connect multiple discrete components came to be described in the late 1950s as the 'tyranny of numbers', and it was apparent that, unless it could be overcome, ambitions for ever-greater miniaturisation were bound to be frustrated. Appreciation of the problem and its solution came from two different organisations, both of which had grown up in the transistor era after 1947. The first of these was Texas Instruments which had created the Regency radio, and the second was Fairchild Semiconductor, based on the group that had broken away from William Shockley's firm.

During the latter part of 1958, at Texas Instruments a researcher, Jack Kilby, had run into the problem of multiple connections in complex devices and, in mulling over what could be done, was struck by an inspiration. He had previously worked on the design of hearing aids at a firm called Centralab, having joined this small organisation as a graduate in 1947, and so was well aware of the fundamental problems of complex devices at a miniature level. Centralab was quite a small company with limited prospects and so, after a long apprenticeship in electronic research, Kilby left and in 1958 joined Texas Instruments where the opportunities were greater.

In the period immediately after 1954 and with a licence from Bell Laboratories, Texas Instruments had begun the mass production of transistors and later switched to the production of the new silicon-based transistors. By 1958, when Kilby came to Texas Instruments, transistors were being created by forming them onto a base material and diffusing the necessary additional parts of the transistor onto this base. In considering the problem of connections, during July 1958 Kilby came to the conclusion that it would be possible to make other devices at the same time as the transistor, all set on a common substrate material. Although in one respect this was a rather extravagant notion, given the cost of production of silicon of sufficient purity for transistors and resistors and because capacitors are able to be far more cheaply produced by other methods, nevertheless the idea was, in principle, a major advance.

The men from Intel — Andrew Groves, Robert Noyce and Gordon Moore *(LUC)*

This notion of an integrated circuit was soon converted into a working model of the device in which a transistor, resistors and capacitors were all formed together on a common base of silicon. When connected up to a power supply, this device was able to produce a continuous sine wave as a result of deliberate feedback. The Kilby device was submitted to the patent office in February 1959.

Evidently this was the first and perhaps most vital principle of monolithic construction but, in one respect at least, Kilby did not solve the problem of connections adequately. In order to join together the components, he had in mind very fine discrete gold wires. Naturally this required the intervention of a human hand and a method of fusion analogous to soldering. With this came all the problems of unreliability and repeatability of conventional point-to-point construction.

Some months later while mulling over exactly the same problem, Robert Noyce reached the same general conclusion but with a fundamentally important different technique in mind. His company, Fairchild Semiconductor, which had been formed by the eight escapees from Shockley's organisation, had decided to exploit the new planar system of transistor construction, using silicon as opposed to germanium. This process, which involved setting a thin layer of insulating silicon dioxide over the top of the exposed embedded components by a process of diffusion, had been discovered by Jean Hoerni and quickly patented.

Given this new process of transistor construction, Noyce had available an intrinsically superior structure in which to create embedded connections. What he now conceived of was a design in which not only the components but also the connections could be laid out on a silicon substrate, all in one process. By this means the intervention of the human hand was avoided and the problem of errors removed. Now the various devices and their connections could be drawn up and set into position using the photolithographic process that had been developed to create the new generation of transistors. This was the basis of all silicon chips from now on.

The integrated circuit of Noyce was defined in January 1959, many months after the inspiration of Jack Kilby but, despite this, it was the first to receive a patent — an interesting reflection of the impact of bureaucracy in the Patent Office. Naturally this situation created a patent war

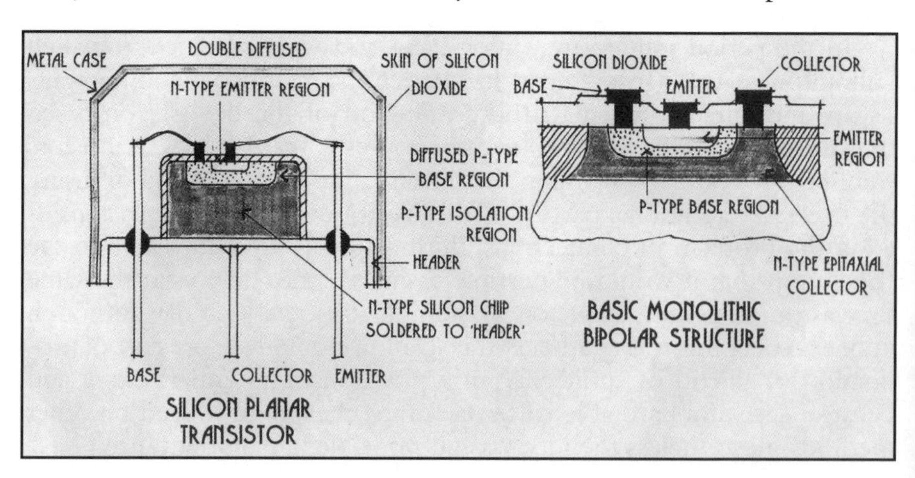

Steps in solid-state development 3 — the silicon planar transistor and monolithic bipolar transistor construction *(PRJ)*

with Texas Instruments which had submitted Kilby's device to the Patent Office before it had even received the Fairchild application.

At a later stage, Noyce sold out his shares in the company he had helped to found, Fairchild, and with Gordon Moore created a new organisation in 1968, International Electronics. Later this name was shortened to one that is instantly familiar to the microcomputer user, Intel. With this move came a concentration on the application of integrated circuits to the computer industry in which the basic notions of integration were to be applied in successive generations of chips up to the present.

The patent war rolled on for the best part of 10 years and finally was settled in favour of Noyce following an appeal to the Court of Appeal in 1966. However, by that stage the decision was all but redundant because, meanwhile, Fairchild and Texas Instruments had signed cross-licensing agreements. In addition, by this stage the silicon chip was so commercially successful that the question of patent rights was largely academic. For the industry and the professions, Jack Kilby and Robert Noyce are now seen as the joint inventors of a critically important element of the modern electronic technology.

Various solid-state calculators from 1968 on — the Sharp EL 120, the Sharp EL 8104 and Sharp EL 347 *(PRJ)*

CHIPS WITH EVERYTHING

It did not take the electronics industry very long to appreciate the usefulness of the silicon chip and one of the first applications of this new device was in small electronic calculators. Although developed by a number of small and adventurous firms, this new area of electronics was soon overtaken by the large players in the industry. One of these was Texas Instruments and this was to be highly significant because it represented the next step to the present-day state of technology.

Jack Kilby, who had developed the idea of the monolithic integrated circuit, was given the job of developing the Texas Instruments version of the calculator. This was undertaken by a team using the same architecture developed by Von Neumann for the EDVAC computer. In the inclusion of input, memory, processing and output, it was also reminiscent of the analytical engine architecture of Babbage a century earlier.

By April 1971, Texas Instruments had available its new calculator, the Pocketronic — a rather optimistic name given that the device weighed in at more than 1 kilogram (2.5 pounds). Further, given its price tag of $150 in 1971 dollars, one would have needed not only a fairly strong pocket to accommodate the new device but also a deep one. However, what came from this new development at Texas Instruments ultimately was to have highly significant ramifications. Among other things, a whole generation of electronic and general scientists and engineers threw their slide rules into a drawer, never to use them again.

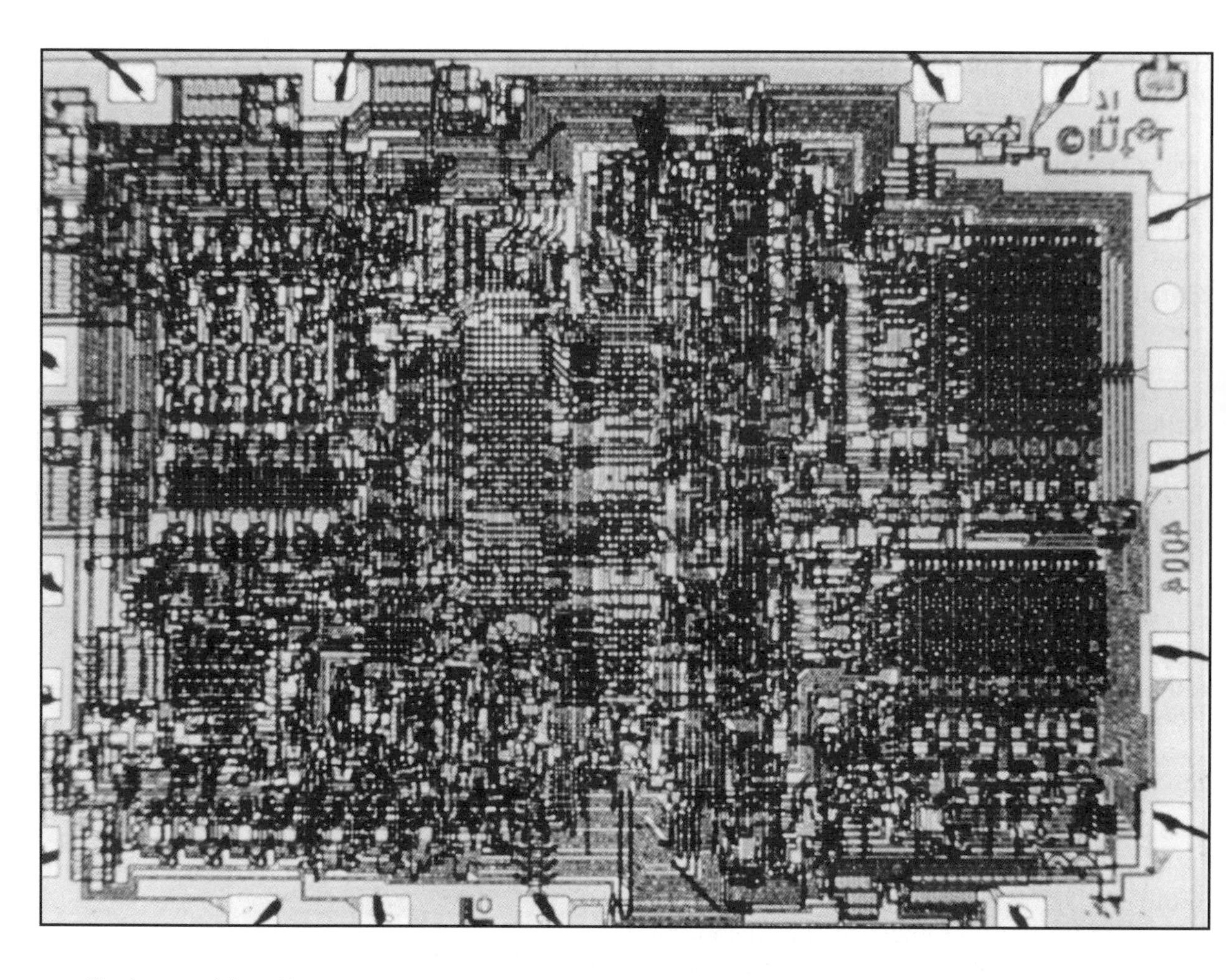

The layout of the 4004 microprocessor as designed at Intel by Federico Faggin, Ted Hoff and Stan Mazor *(EA)*

Ultimately, the market for electronic calculators became saturated and in the turmoil a Japanese entry into the more conventional desktop calculator market was abandoned. This would have been of little significance except that the Japanese firm concerned, Busicom, had earlier contracted with Fairchild to have a set of integrated circuits produced for its intended machine. Now this work was aborted. However, during the development of these chips, Ted Hoff at Intel had come to the conclusion that the proliferation of chips was a thoroughly wasteful and inefficient way to proceed and had developed the notion of a single chip, able to undertake a variety of different tasks. From this idea came the first microprocessor, the 4004, developed in 1971. This device involved a central processing unit (CPU) set on a single chip and able to be programmed to undertake a variety of different tasks. It was able to process four bits of information at a time and was quickly followed by the 8008 which could process eight bits.

However, apart from the recognition of the general usefulness of the chip and the applications that were being developed with solid-state technology, there was an event in that period which would create ripples of tension that continued to resonate for several decades. This was the flight of the first satellite carried into orbit by a ballistic missile, coming from the new technology of Soviet Russia. Sputnik and its infantile 'bleeping' radio signal alerted an already sensitive United States that its position as a world power was under significant threat. If the paranoia developed by the detonation of the nuclear devices of Russia had been serious, Sputnik was to create a furore of anxiety and mutual accusation which escalated yet again when more Sputnik satellites were launched. The McCarthy era will long be remembered as one of the most unpleasant aspects of this time. However, the impact of this concern was a vastly increased pressure for technological advancement. Silicon integrated circuits were precisely what the military required to guide newly developed missiles to their distant destinations.

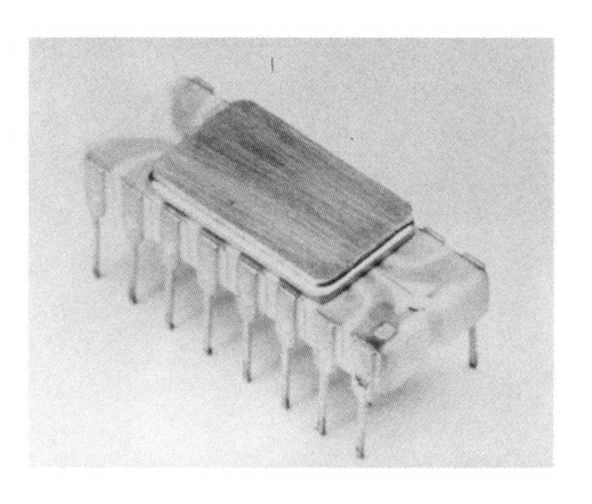

The first microprocessor, the 4004, in a sixteen-pin, dual inline (DIL) package *(EA)*

The intense proliferation of the use of the silicon chip during the 1960s led to a frightening degree of diversity that quickly started to slide into chaos. New special-purpose chips would be demanded and fabricated and, once their application was established, would no longer be required and the manufacturing process would be discontinued. What was required was a general-purpose chip that could be made to do a variety of tasks. The microprocessor earlier referred to was the answer to that problem and appeared at precisely the right time to answer a new demand.

CHIPS AND MACROCOMPUTERS

As earlier noted, the application of transistors to computers came out of a fortunate decision by Texas Instruments to manufacture a miniature radio, the Regency. This device inspired IBM to use transistors in their computers in 1955 at a time when memory was provided by a complex array of toroidal magnetic cores knitted together with sensing wires.

Memory would later be the basis of one of the most important uses of the new integrated-circuit chips. As assembled into progressively more extensive arrays, the new monolithic memory was ultimately far cheaper and far more reliable, and did away with a complex structure with limited capacity in terms of the amount of memory that was available. By comparison, on the basis of the work of two Fairchild engineers, Robert Noyce and Gordon Moore, who went on to create the now very well known new firm, Intel, 1000-byte chips had been made by 1968 and, by 1983, this type of device had grown to accommodate 256 kilobytes of memory. Today, 64 megabytes is commonly used even for entry-level machines. However, with the fall in the price of memory chips, for some memory-intensive applications such as computer-assisted design (CAD) and desktop publishing, far more memory than that is frequently used.

However, all this new application of silicon chip technology to computers was confined to the big machines of the major suppliers like IBM. What was to have the most profound impact was the application of the previously mentioned 8008 microprocessor and its cousin the 6502 microprocessor made by Mostek. Two enterprising young men, one with a radio amateur ticket, were to see the possibilities of this new device and begin another revolution that is with us still. Steve Jobs and Stephen Wozniak would create the first commercially viable microcomputer, the Apple.

VALVES, RECEPTION AND TRANSMISSION

It must seem quite strange to a generation of solid-state users that, until the 1960s, virtually all equipment involving electronics was based on a device which used a vacuum trapped inside a glass or metal bottle, the valve (or tube as it is called in the United States). For the best part of 60 years this device was the basis of a complex industry supplying every facet of demand and one that still has the power to invoke warm feelings of nostalgia among those who were brought up with it.

THE VALVE (TUBE)

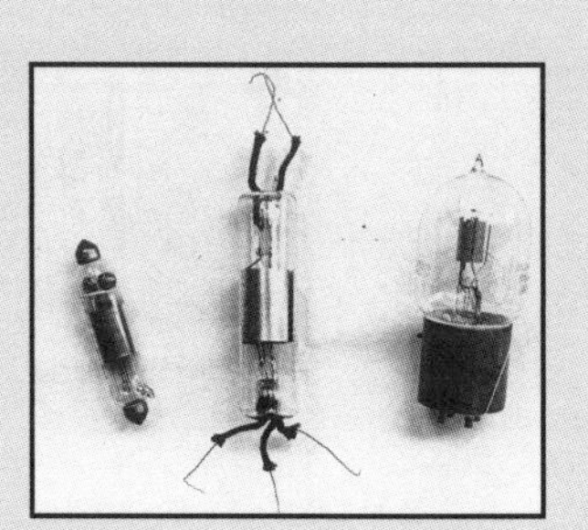

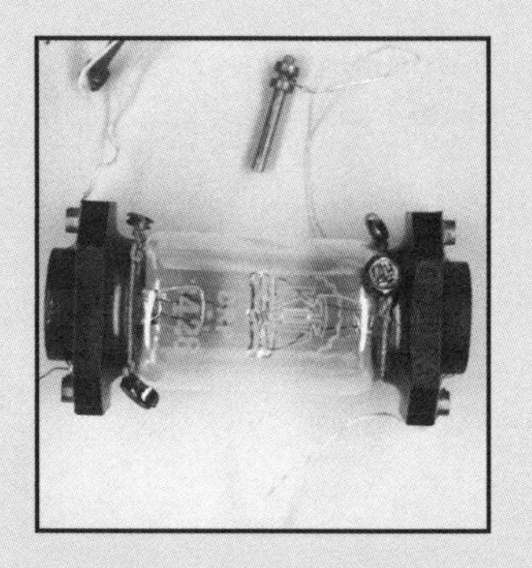

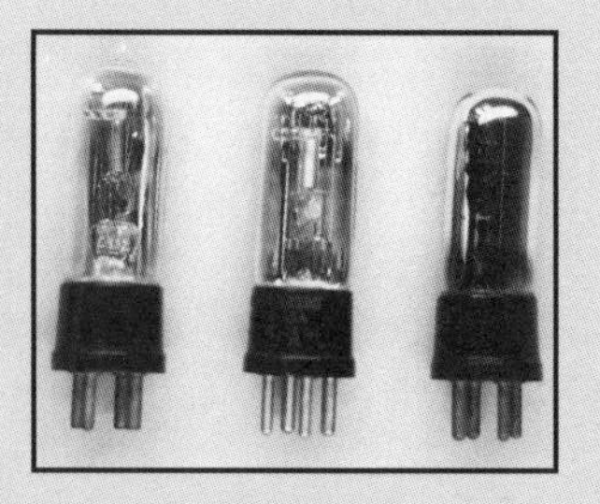

In the First World War, the valve was rapidly developed once its potential was fully understood. This took an appreciation of the need for a 'hard' vacuum achieved by removal of the residual gases once the initial vacuum had been formed. Until the introduction of the 'getter' inside the envelope of the valve, used to fix the residual gases by a process of heating and 'flashing', the valve remained a temperamental though potentially highly sensitive device. The problem with the early 'soft' valves was that achieving the optimum operating condition inevitably involved specific adjustments for each valve in use. Such optimum conditions were also inevitably inconsistent from one valve to the next. Selection of the best devices was by test and, even then, the characteristics of the particular valve could change once in use.

Early valves made use of a tungsten filament, heated to white heat, to generate the necessary cloud of electrons. Because of this, these valves were known as 'bright emitters' and competed quite successfully with the ordinary light bulb as a source of light. A very inefficient and short-lived form of cathode, this was soon replaced by a filament containing a small percentage of thorium and providing a prolific shower of electrons when heated. During the early 1920s, such thoriated tungsten filaments were replaced in receiving valves with filaments doped with strontium and barium able to emit electrons at significantly lower temperatures than the thorium-treated material. Appropriately, these newer types of tube were described as having 'dull emitters'. With the advent of valves designed to be run from the alternating current mains, a new type of cathode was developed in which an internal resistance wire heated up a tube coated with an emissive material and insulated from the heater. By this means the modulating hum of the 50 or 60 hertz mains was prevented from interfering with the signal being amplified by the valve.

From the diode of J Ambrose Fleming and the later triode of Lee de Forest, over the next 40 years valves became more complex in electrode structure and, ultimately, came in a huge variety of configurations to suit particular uses. Specific knowledge of the physical changes that could be made to construction and electrode spacing and the associated impact on the

From top to bottom:
Marconi V24, AWA Expanse B, Marconi VT *(OTC)*
Siemens and Halske Type A valve of 1916 *(PRJ)*
Early telephone amplifier tubes made by STC in Australia, the SY4102D *(PRJ)*
Early broadcast reception tubes of 1924, the UX120 series *(PRJ)*

performance of the valve meant that these variables could be controlled in a fundamentally predictable manner. In particular, control of gain and valve size were factors that saw major development and change.

From a device very much akin to the parent light bulb in 1905 both in terms of shape and luminance, the valve by the end of its life in 1965 had been reduced to the size of the average human thumb. In the process it shed its separate base in favour of integral pins set in the glass envelope and the internal heater became almost invisible except in low light conditions. Metal envelopes were also used for various special tube types.

The diodes and triodes earlier described were, in the late 1920s, joined by a new valve type known as the tetrode. With an additional grid set between the control grid and the anode or plate, the inherent capacity of the three-electrode valve to go into self-oscillation due to inter-electrode capacitance (known as the Miller effect) could be prevented. A typical example of this new tube as supplied to the markets of the world was the type 224.

Later again, the tetrode was also found to be unstable under certain conditions where secondary emission from the screen grid could produce oscillations. It was discovered that a further electrode set close to the plate and tied back to earth would overcome this problem. Known as the 'suppressor' grid, the valve that embodied this configuration was called the pentode. More complex multi-electrode valves were subsequently developed, the best known involving two or more valve structures within the common glass envelope. Typical of this latter class of valves were those developed to be used in the mixer stages of superhet radios such as the pentagrid converter, also known as the heptode in the United Kingdom, and the triode hexode valve. During the late 1930s and after the Second World War, a drive to produce miniature, low-filament-current valves suitable for portable applications saw the emergence of the series of valves which included the 1T4, 1S5, 1R5 and 3S4. These valves operated at a filament voltage of 1.5 volts and could be arranged in series so as to operate from a 7.5-volt supply when the usual configuration of a four-valve superhet was in use. This supply was usually incorporated into a single multi-voltage battery pack with voltages set at 45 volts for the plate and 7.5 volts for the filament supply.

What killed off the valve, except in certain specialised situations, was the problem of power to operate the heater before any electron flow could be induced. In many respects this was a more important consideration initially than the reduction in size that solid-state devices could achieve. Later, when silicon integrated circuits appeared, even this size issue could no longer be suppressed.

Despite the demise of valves in consumer products and in general, valves still possess some particular advantages compared with discrete solid-state devices. Before the advent of modern techniques designed to overcome inter-modulation distortion in junction and field-effect transistors, the best of the valve radios were superior to the equivalent transistorised replacements. The method of overcoming this problem in transistors was the use of front-end mixing using ring modulation diode arrays, with or without a radio frequency amplifier stage.

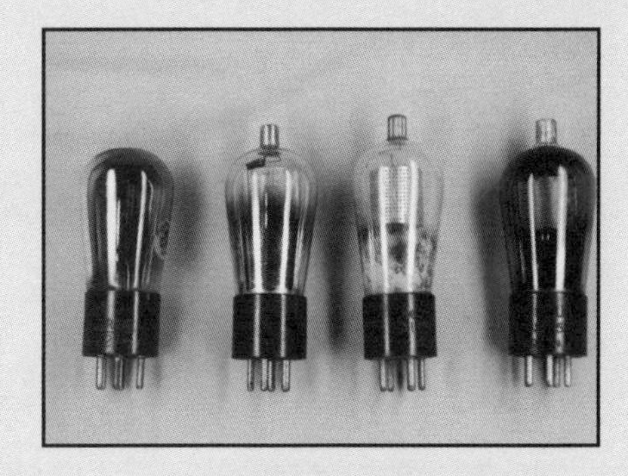

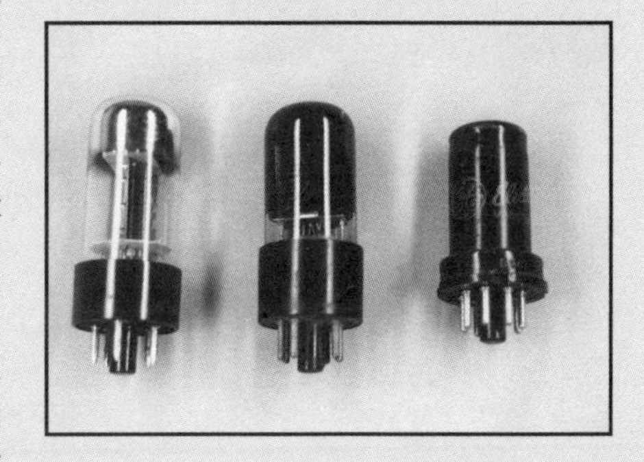

Top Two telephone repeater valves by STC in the 4019 series from 1926 and receiver valve of the same period *(PRJ)*

Centre (from left to right) A UX 112, UX 222, de Forest audion triode and an Arcturus blue-glass triode from 1928 *(PRJ)*

Bottom Compact valves from the 1930s from with 6V6 power valve in the middle and metal shell 6K3 valve on the right *(PRJ)*

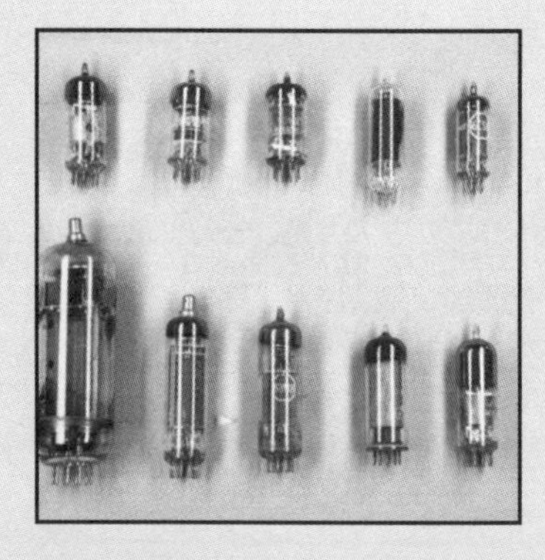

Above left to right: Minature glass valves from 1944 to 1965 *(PRJ)*

Classic power amplifier valves — an 807 *(left)*, a 4-125A *(middle)* and an 813 *(right)* *(PRJ)*

Power amplifier valves from the end of the valve era — the 6146, a 6DQ6 'sweep tube' for use in valve television and a 4CX 250K ceramic valve *(PRJ)*

In this same context, the possibility of electromagnetic pulse energy induced in the antenna and generated by nuclear explosions was much better catered for with a valve front-end in a receiver than a solid-state device. It was also reported that, during the Gulf War, the high level of electrostatic voltage developed by flying sand had a very serious impact on the reliability of military radio communications systems. There was also some suggestion that the American army discovered that elderly Russian valve radios were withstanding this problem far better than the more modern solid-state radios.

Apart from such esoteric attributes of the valve, there is no doubt that, in limited applications, it remains a very useful device. Most notable is in amplitude modulation (AM) broadcasting where the high-powered valve amplifier remains very common and is revived by replacement of the heater at the end of its service life.

Radio amateurs, too, have discovered that it is generally cheaper and easier to get a valve amplifier to produce the legal limit of 400 watts peak envelope power (United Kingdom and Australia) than to use solid-state devices, however attractive such technology may be in other respects.

In passing, one particularly esoteric device developed to provide high-power, pulse radio frequency (RF) energy for radar was the magnetron. It is now the basis for all the microwave ovens that are used to warm up meat pies and 'television dinners' throughout the Western world. In common with other high-power valve-based amplification systems, the microwave oven has lurking inside it the main disincentive to using valves — a formidable high-voltage, high-capacity power supply. This is an absolutely lethal device and should not under any circumstance be allowed into the care of the unwary or electronically ignorant. In particular, such devices should be kept well away from the inquisitive, electronically inclined person with a screwdriver.

With a very high voltage, but more seriously a capacity to produce many milliamps of current at that high voltage, the microwave oven power supply is, in terms of output power, similar to what is used to electrocute criminals in the United States, although applied as AC rather than the DC required to operate a thermionic device.

Current high-power transmission valve with ceramic and metal body, the 3CX5000H3 *(PRJ)*

In this, the technician's tag — 'It's the volts that jolts but the mils that kills' — is relevant. This is a good reminder for anyone game enough to service a valve amplifier while in operation. In addition, the 'one hand in the pocket at all times' rule makes basic good sense. An early radio amateur who fell foul of such a power supply, built to drive an early television set, was the noted writer Ross Hull. He had gone to America to work with the American Radio Relay League and became snagged in his own deadly apparatus. He is commemorated in Australia with an annual radio competition. Tragically, shortly before his death, he had written an article about safety in the 'shack' and then proceeded to disobey some of the rules of safety that he himself had suggested.

Although for many brought up in the era of valves the friendly glow of heater and filaments evokes warm feelings of affection, the counterpoint is the problem of heat generation. This was a factor that clearly contributed to the short life of many components in apparatus based on valves. By comparison, it is remarkable how long-lived are solid-state devices which, until recently, usually operated at room temperature. An enormous number of early solid-state television receivers are still operating happily, albeit with peculiar colours on the screen associated with reduced output from colour guns in the cathode ray tube.

Despite the advantage of cool running in early solid-state devices, with the development of high-speed, multi-transistor central processing units in modern microcomputers, the problem of heat has started to reappear. This has an impact on fatigue of components due to 'turn on' and 'turn off', and the 'mean time between faults' in computers has started to reappear as a significant issue. One of the counterpoints to this problem is that, despite the inevitable power consumption, a considerable number of computers on business desks are left running overnight with the monitor switched off.

Top A valve transmitter of the Royal Navy, *c.* 1922 *(PRJ)*

Bottom The receiver in Royal Navy ship's installation of 1922 *(PRJ)*

VALVE RECEIVERS

The availability of the valve, firstly in the form of the triode but later with an increasing number of electrodes, allowed more sophisticated and complex receivers to be developed between 1916 and 1965. Following the simple regenerative radios based on the work of Edwin H Armstrong and the neutrodyne receivers of Louis A Hazeltine, for many years the superhet (supersonic heterodyne) receiver reigned supreme. Having a capability to separate the increasing number of signals in the medium-wave and short-wave bands while retaining a high level of sensitivity, the superhet remained as the basis of virtually all receiver design until almost the end of the valve era.

Early superhets generally had a relatively low intermediate frequency (IF) at around 40 kilohertz which allowed good adjacent channel selectivity to be achieved. However, a significant problem in the medium-wave band was the propensity to produce images of other stations located only 80 kilohertz away from the desired station as an 'image' underlying the main signal. A progressive elevation of the IF to about 455 kilohertz almost completely overcame this problem in the broadcast band, but it persisted on the short waves. The solution to the 'image' problem in the high frequency (HF) bands was to introduce double conversion, with a first mixer stage providing output to an IF stage at, say, 1.7 megahertz, and then a further mixing stage to a second IF at 455 kilohertz.

In addition to such problems associated with the process of mixing, the earliest superhets had separate tuning controls for the antenna input and the local oscillator. This meant that a difficult operation was imposed on the user and this was a major problem for the unskilled. It was solved by ganging together the tuning capacitors and by adding appropriate adjustable capacitors to achieve proper frequency tracking of the two circuits. This process, known as 'oscillator padding', allowed the two circuits to follow different frequencies over the tuning range of the radio but at a more or less constant separation, being the intermediate frequency.

In the first superhets, it was usual to provide the local oscillation by means of a triode. Later, this technique was changed to involve the production of both the local oscillation and the mixing in a common pentode. The apparently economical method known as the autodyne had a number of practical problems, most serious being its unsuitability for use on the short-wave bands which began to be very popular in the early 1930s. At this stage, the availability of the triode hexode and the pentagrid converter saw the abandonment of the autodyne for many years. It reappeared for a while in early transistor superhets after 1950 when the new three-electrode transistor, roughly equivalent to a triode, became available.

The emerging interest in the short-wave bands also confronted the listener with a major problem of signal 'fading' and this, in turn, demanded a solution. This was automatic gain control, which not only solved the fading problem but has remained a part of radio receiver design ever since to mask the differences of signal strength on local broadcast stations.

In the period immediately before the Second World War, demands of the amateur radio fraternity in the United States saw the development of some very high quality radios which were pressed into service during the war years. Perhaps the best known communications receiver of this period was the HRO with its removable coil boxes to cover the various sections of the high-frequency part of the spectrum. In Australia, a superior copy of this radio appeared as the Kingsley AR7, a thoroughly desirable radio even by today's standards, with impressive frequency stability and capacity to reset to a given frequency. Both these radios featured an unusual dial in which the changing frequency was reflected in an incrementing digital presentation on windows on the edge of the dial. The small problem of this arrangement was that it required the operator to read off the digit in the window and compare it with a chart on the coil box to obtain the frequency — somewhat clumsy, but extremely accurate compared with linear or circular dials of that period.

A Hallicrafters S 20R of about 1938 *(PRJ)*

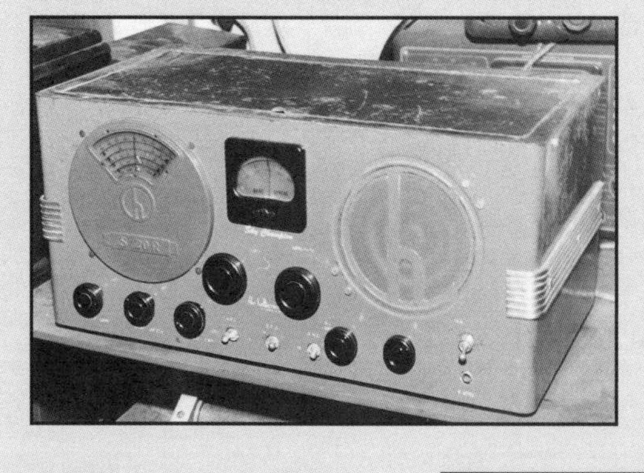

The introduction of single-sideband (SSB) reception in the immediate postwar period, later enthusiastically embraced by the radio amateur community, also led to a significant change in circuitry. This was the introduction of the product detector stage, compared with the earlier simpler beat frequency oscillator required to resolve continuous wave (CW) signals. Despite the cumbersome name, the product detector was in reality simply a form of modulator to allow proper mixing of the local oscillation with the SSB signal, given the absence of the carrier that would be available in an AM or CW signal.

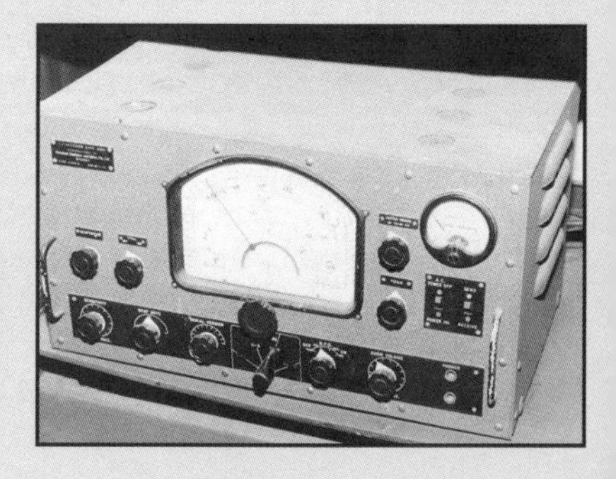

Right A high-frequency communications receiver by STC, the AMR 300 *(PRJ)*

Far right The Australphone transmitter and receiver by AWA *(PRJ)*

By the late 1950s, frequency stability and adjacent channel interference had come to be problems that required solving. In particular, the introduction of single-sideband transmission of audio produced particular demands on frequency stability that earlier radios could not respond to adequately. Both with single-sideband transmission and the use of narrow-band data transmission, the frequency stability of the local oscillator stage needed to be contained to a few hertz rather than kilohertz as was adequate before the Second World War.

A Marconi transmitter for the short-wave bands of 1927 *(AWA)*

To solve such problems, it was necessary to employ more complicated circuitry involving crystal control and crystal filters at the intermediate frequency in superhet-type radios. At this stage, the physical characteristics and size of valves began to represent a severe problem as the increasing number of functions needed to be accommodated. This was solved in the following decade by the progressive introduction of solid-state devices and the development of synthesised frequency techniques.

In this later period, the application of the phased lock loop to frequency determination and use of varactor diodes to control tuning produced a revolution which only solid-state devices could make feasible.

One of the most successful early responses to the new demands of radio receiver operators was met in the Wadley loop method of triple mixing. This was very successfully applied in the Racal RA 17 with a formidable valve count but with advantages that make it a very attractive radio even now.

At the amateur level of activity, it is interesting to reflect on a product of this late stage in the valve era designed and promoted by the magazine *Electronics Australia*. This was the Deltahet, a discrete front-end based on the Wadley loop technique and intended to be used with a conventional communications receiver set to a fixed elevated frequency. This served as the final intermediate frequency amplifier and audio output stage.

As a valve receiver, many Deltahets were built and functioned very well. Unfortunately, a later solid-state version of this radio was less than successful, mainly as a reflection of the continued use of point-to-point wiring using tag boards rather than circuit boards. Repeatability of the solid-state Deltahet was probably beyond the capacity of the average hobby constructor and the position was made far worse by the inclusion of a broad bandwidth, intermediate frequency, reminiscent of the form of a television video intermediate frequency strip. In addition, a sweep generator was required to set this radio up and, as a result, many non-working Deltahets are reputed to be lurking in junk cupboards around Australia and in New Zealand.

The Barlow Wadley solid-state commercially produced radio demonstrated what was possible with this circuit using a single circuit board. As a result, it remains a remarkably high-performance radio for its time and therefore, for the collector, a highly desirable specimen.

VALVE TRANSMITTERS

The Caernarvon radio telegraph station was equipped with valves in 1920, replacing the Alexanderson alternator that had been installed only a year or so before. In this, it was one of the earliest major installations involving high-output power from valves. Designed by HJ Round, this new continuous wave transmitter contained an array of 54 relatively low powered valves operated in parallel. The output from these valves working in concert was 100 kilowatts. With this available, a highly reliable long-wave link to Australia was possible and, accordingly, a news service direct from Britain to Australia began at the end of 1921.

From this time on, developments in valve technology were rapid. By 1926, with the acceptance of the short waves rather than the long waves as the basis of long-range radio communication, the Marconi Company developed a valve-based transmitter which was to have a remarkably long service life. The SWB 6 was a four-stage transmitter using a newly developed valve, the CAT-2. Two of these valves were operated in a class C push–pull configuration as the output stage, which in turn were driven by two similar valves in the two previous stages.

With an input power of 14 kilowatts supplied from an 8-kilovolt high-tension supply, the CAT-2 valves required their filaments to be supplied with 20 volts at 100 amps. Originally intended to be used for the medium- and long-wave bands, the valves were designed to be cooled with distilled water. However, on the short waves, and at the service frequency of 11.66 megahertz, water proved to be unsuitable and instead the coolant used was kerosene. Later, the original CAT-2 was replaced, in the SWB 6, by an upgraded valve, the CAT-9. In this form, certain examples of this very successful transmitter were still being operated 25 years later.

In the early 1920s, the new phenomenon of broadcasting also became possible with the use of high-power valves. Initially, transmitters in this service tended to have a small number of stages, with a master-oscillator stage driving the power-output stage directly. Modulation was by means of the system devised by Heising, involving choke-coupled series drive to the final power stage. This produced an efficiency of less than 50 per cent. In order to increase the power of these broadcast stations, at a later stage it was common to add linear amplification, despite the inherent inefficiency of this method.

In Australia, radio broadcasting stations in the broadcast medium-wave band produced power levels between 100 watts and 10 kilowatts. Cooling was frequently achieved using liquid, although later the introduction of air-blast cooling simplified transmitter construction. The counterpoint to this advantage was the excessive noise produced with air cooling.

After the Second World War, Heising modulation was generally replaced by high-level anode modulation and this allowed efficiencies of around 70 per cent to be achieved. In addition, advances in valve techniques by this time allowed electronic noise levels to be held to 60 dB below 100 per cent modulation. However, with a 3 per cent harmonic distortion level over the range of audio frequencies broadcast, 30 to 10 000 hertz, the output of such broadcast stations could not be described as exactly 'hi-fi'. However, the use of Class B modulator stages made for far more efficient energy use in the newer valve transmitters and represented a benefit that offset the low fidelity of the signal produced.

Before the Second World War, most high-power transmitter use on the short waves involved keying of a carrier for the transmission of high-speed morse. Later, this was supplemented by the use of teletype and, for this, frequency shift keying (FSK) was introduced. The necessary swing of 30 to 42.5 hertz was achieved in the RF exciter stage, a simpler method than switching a carrier off and on, as applied to morse code transmission. This had required the load to be shed into a ballast system.

After the Second World War, the introduction of independent sideband transmission increased the efficient use of the short waves, particularly in the trans-oceanic telephone service. However, the problem that accompanied this new technique was the need to control cross-talk between the two sidebands. In addition, as for early radio stations, amplification of the signal had to be accomplished by means of linear amplification.

Despite the huge impact that solid-state devices had on the use of valves after about 1965, high-power transmission valves remain an area in which this old and highly reliable technology is paramount, particularly where the power level exceeds about 10 kilowatts. No doubt, with time, this limit will gradually increase as advances in solid-state construction are made but, for the foreseeable future, the valve seems certain to retain an assured place in the armoury of transmission devices used for broadcasting.

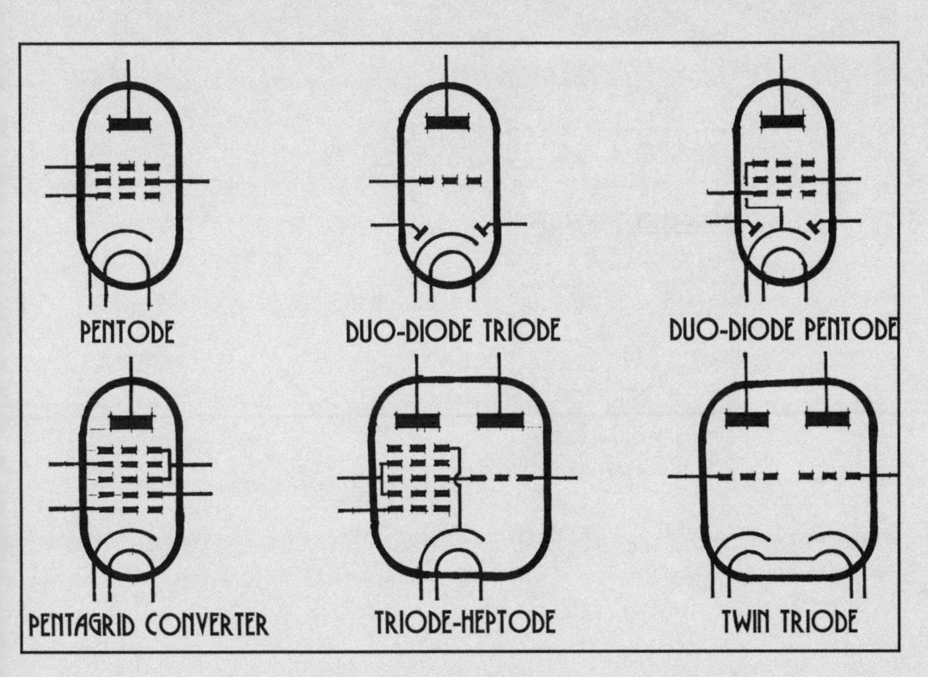

Common outlines and connections of typical receiving valves *(PRJ)*

CHAPTER

14

1945 to 1975

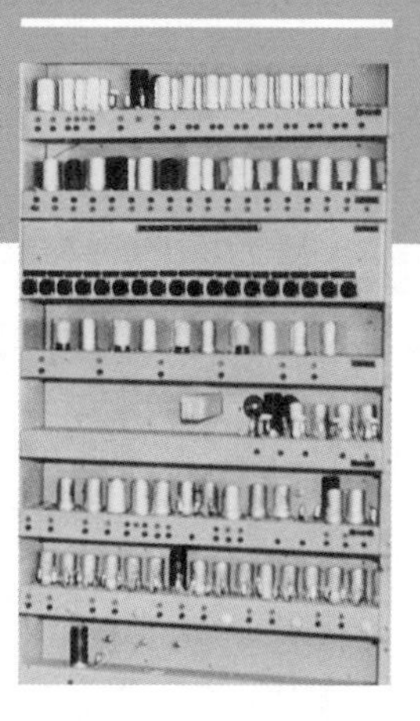

Creation of the computer

A current compact edition of the Oxford Dictionary defines a 'computer' as 'an electronic machine for making calculations, storing and analysing information fed into it, and controlling machinery automatically'.

On this basis, the machine developed at Bletchley Park during the latter part of 1943 could be adequately described as a computer, given its function of assisting in analysing complex German codes and printing out data to a teletype keyboard. What fundamentally distinguishes this machine from later and better known computers is that, at the end of the war and on the instructions of Winston Churchill himself, all traces of it were very effectively obliterated.

Most of the parts for Colossus had come from conventional British Post Office equipment sources and so it was possible to return most of the parts to store for future telecommunication purposes. Fortunately for future historians, not all the records were destroyed. In more recent times, photographs and descriptions of the Colossus have been found in archives in the United States. These had been obtained during inspections by visiting American military and expert personnel in the course of the Second World War. These have provided a basis for the reconstruction that can now be seen at Bletchley Park and which is able to replicate the tasks that the original Colossus carried out in 1943 and 1944.

COMPUTERS IN THE UNITED STATES — ENIAC

In the United States before its entry into the war following the attack on Pearl Harbour, new field guns and other forms of artillery were in the process of development. For each new gun created, it was necessary to prepare what was known as a 'firing table' which allowed accurate gun-laying over the range of the particular weapon.

At the Moore School in Philadelphia, the analog computers of Vannevar Bush were set to work to provide the basic data for these tables. This work was supported by a progressively growing army of women 'computers'. These women were employed by the Ballistics Research Laboratory, which had been created in 1938, and they undertook a task that, to a modern-day reader, could only be seen as mind-bendingly boring. The differential analysers of Bush produced the data and the women operated desktop electromechanical calculators to massage the results into the necessary tabular information.

The testing facility of the Ballistics Research Laboratory was located at the Aberdeen proving ground, approximately 100 kilometres (60 miles) to the south-west of Philadelphia on the edge of Chesapeake Bay. This made the Moore School at the University in Philadelphia an ideal location at which to establish a centre to analyse the data coming from the test-firing of weapons.

In 1942 there were approximately 100 women working at the task of computation, but over the next 3 years this number grew to around 200. Despite this large number of people carrying out the analytical work, each firing table would take about a month of time of one of the women 'computers'. Mary Mauchly, who supervised this team of 'computers', was married to an academic member of the Moore School, Associate Professor John W Mauchly. With a long-standing interest in computation developed from involvement with weather forecasting, Mauchly considered the work his wife was involved in and decided it was based on an inefficient process.

Although Mauchly was not directly involved in the war work, he now proposed a machine to assist in producing the firing tables. In August 1942 he put together a paper entitled 'The Use of High Speed Vacuum Tubes for Calculating', in which he proposed an electronic computer which could do in perhaps 60 seconds what a human 'computer' would be able to achieve in, say, 20 minutes. For some inexplicable reason, this memorandum disappeared into the filing system of one of Mauchly's superiors at the Moore School and had to be reconstructed from secretarial notes in March 1943.

Mauchly's paper, when replicated and distributed, caught the eye of Lieutenant Herman H Goldstine. He had joined the Ballistics Research Laboratory in 1942, having gained a PhD in mathematics from

John Mauchly and J Presper Eckert receive recognition in 1973 for their work on ENIAC of 1945 *(EA)*

Chicago University. His support allowed Mauchly, together with a very enthusiastic electrical engineer, J Presper Eckert, to carry out the work of producing this electronic computer, later known as ENIAC — 'Electronic Numerical Integrator and Computer'. In Mauchly's design, it was intended to have 18 000 tubes and was anticipated to cost US$400 000 in 1942 dollars.

Importantly, as a response to the general unreliability of vacuum tubes at that time, Eckert decreed that the 18 000 tubes were to be run continuously and at two-thirds the conventional heater voltage. This resulted in a substantial increase in tube reliability, 10 000 hours being achieved rather than the conventional 3000 hours.

The first contract for the construction of ENIAC was let in May 1943, the machine ultimately being constructed under the general supervision of the director of research at the Moore School, John Grist Brainerd. Eckert was nominated as chief engineer and Mauchly as principal consultant. Goldstine was the technical liaison officer between the army and the university team for what was to remain, during the war years, a 'classified' project. This secrecy was to have considerable repercussions in the way that the machine was revealed to the public and, in the process, caused much animosity later on.

Although it was a cooperative effort, it is quite clear that Eckert's engineering input was of critical importance. In particular, his knowledge of valve technology ensured reliability. However, perhaps more important was his insistence on consistency based on a number of basic electronic modules that were used throughout ENIAC, all being produced to a very high standard of engineering theory and workmanship.

Despite the exciting possibilities offered by ENIAC, it proved to be intrinsically rather an inflexible machine. In order to undertake a new job which required a different set of instructions, extensive rewiring of the machine had to be done by hand, a most tedious process. In addition, ENIAC had a very small numerical capacity, a somewhat ironic counterpoint to its very large number of vacuum tubes. However, it was capable of undertaking 5000 operations per second and to that extent provided a significant improvement in output compared with the human 'computers' in their firing-table computations. Among other tasks, it was used to assist in the massive calculations required to design the 'trigger' mechanism for the nuclear bomb, developed as a part of the 'Manhattan Project'.

Despite the 'classification' of ENIAC, it was officially demonstrated in the early part of 1945 and this was reported in the *The Times* of London on 14 February 1945. After a quite short working life, the machine was switched off in October 1955 and taken to the Smithsonian Institute for display purposes.

VON NEUMANN ARCHITECTURE — EDVAC

Appreciating the fundamental limitations of ENIAC, Mauchly and Eckert began to develop the design of a new machine that was referred to as EDVAC. After conceptual work on this new machine during 1944, an important, if accidental, meeting took place when the celebrated mathematician John Von Neumann met Lieutenant Goldstine at a conference. Although not revealed in his discussion with Goldstine, Von Neumann was at this time working intensively on the development of the nuclear bomb, then referred to by the codename 'Manhattan Project'. This work involved massive calculations associated with the design of the nuclear firing mechanism. When Goldstine revealed the work that had been undertaken to create an electronic computer, Von Neumann was intensely interested. This led to him becoming a consultant to Mauchly and Eckert. This association, in turn, led directly to the development of the 'stored-program computer'.

Later, the far more flexible computer known as EDVAC, standing for 'Electronic Discrete Variable Automatic Computer', was created. In its development, Eckert had proposed the use of mercury delay lines as a method of data storage and as a substitute for the rather extravagant electron tube storage that had been incorporated in ENIAC. The delay-line technology had been developed by William Shockley as part of the war work that had taken him away from solid-state devices before his return to that field in 1946.

Von Neumann's involvement with this project soon led to a realisation that both instructions and work to be undertaken could be stored in the computer. Further, a program of work could be read into the machine with punched cards or paper tape. Babbage would have been delighted, if not entirely surprised.

This new approach to electronic computing was the basis of a machine with a much smaller vacuum tube count compared with ENIAC and a far greater capacity to handle numbers. The architecture of this new machine, now referred to as 'Von Neumann architecture', was remarkably reminiscent of that used in Babbage's 'analytical engine'. Compared with both ENIAC and Babbage's machine, both of which had used decimal numbers, EDVAC was based on the use of binary numbers.

In June 1945, much to the dismay of his associates — Mauchly, Eckert and Goldstine — Von Neumann wrote an article entitled 'A First Draft of a Report on the EDVAC'. This report had an immense impact and was rapidly copied and distributed, although initially it was intended for only a small internal circulation. To the annoyance of his colleagues, this publication very effectively destroyed any capacity to patent the concepts embodied in EDVAC. Although Von Neumann is now generally acknowledged as the developer of this architectural concept, not surprisingly his actions effectively destroyed the harmonious relationship with Mauchly, Eckert and Goldstine.

The earlier machine, ENIAC, was finally activated in November 1945, effectively too late to be useful in the ultimate conduct of the war. Despite that, observers described it as an impressive sight when in operation, with its battery of lights flashing. No doubt the job offers to Eckert from both Von Neumann, now resident at Princeton University, and IBM, were a reflection of the perception that a successful machine had been achieved. In the event, such offers were resisted and, together with Mauchly, Eckert set up a new business in March 1946. This was to result in the creation of their version of an electronic computer known as UNIVAC.

COMPUTERS IN BRITAIN — EDSAC

Given all the development work that had occurred at Bletchley Park, it is a sad reflection on postwar Britain that so much of the major development work in computing ultimately passed to the United States. Perhaps this was a function of postwar lethargy or perhaps a response to the intense secrecy established in regard to the war work. Whatever the reason, it took some time for the most potent ideas developed in association with Colossus to percolate into the field of electronic engineering.

After the end of the war in 1945, Alan Turing was involved in the establishment of a computer development program at the National Physical Laboratory and in the creation of a machine that was called Pilot ACE. The other stalwart of the Colossus program, Max Newman, went to Manchester University as professor of pure mathematics and there became involved in computer research and development. In this he was soon joined by Professor FC Williams, later Sir Frederic, who joined the university as the professor of electrical engineering. Their work would lead to arguably the first stored-program computer, the Manchester Baby Machine (also known as the Mark 1), which successfully commenced operations on 21 June 1948.

Dr Maurice Wilkes at Cambridge University with EDSAC, which first operated in May 1949 *(EA)*

However, the principal person involved in Britain's successful early entry into the stored-program computer field was Maurice Wilkes. He had been one of the recipients of Von Neumann's EDVAC report on computer architecture. Following this, Wilkes had gone to the United States to attend a course at the Moore School. Due to university dithering, the funding for this exercise was not finally approved until close to the end of the course. The result was that Wilkes was able to attend only the last 2 weeks, although, given his background, he managed to assimilate the remaining material quite easily. It should be noted that, at this stage, Von Neumann's conception of EDVAC remained largely a theoretical model and it appears that the machine was never completely finished.

On his return to the United Kingdom, Wilkes developed a machine which also used 'mercury delay lines' as for EDVAC. In recognition of this, the British machine was known as EDSAC — 'Electronic Delay Storage Automatic Computer'. On 6 May 1949, a papertape-based program was loaded into EDSAC to carry out the following simple mathematical task:

FOR X = 1 to N PRINT X^2

EDSAC, the first true stored-program computer, 1949 *(EA)*

This results in the sequence 1, 4, 9, 16, 25, 36, 49 and so on, representing a small step for computing but a giant leap forward for computers. This machine was subsequently developed further for the tearooms firm J Lyons and Company, and in that guise was known as the LEO which stood for 'Lyons Electronic Office'. It was commissioned in November 1951.

At last one could say with confidence that the dream of Babbage to create a machine that could calculate in response to instructions from a human operator had finally been achieved. However, equally, one could assume that he would have found the machine and its use of a strange technology of glowing valves a very puzzling phenomenon.

COMPUTERS IN AUSTRALIA — CSIRAC

Despite the geographical isolation of Australia, from the earliest days and subject to the delay in the information being communicated, involvement in new technology has always been substantial. In the period before the Second World War, with the historical connection to Britain still well established, many of Australia's brightest university graduates routinely went to England to undertake higher research. From this invariably came an influx of new ideas once the graduates returned home. Mechanised computation and later electronic computation were typical products of this flow of ideas to Australia.

However, apart from this relationship with Britain, Australia's involvement with the early development of machine-assisted computation could be seen as a manifestation of a national interest in gambling and, particularly, horseracing. In the early part of the century, the noted engineer George Julius, later Sir George, came to the conclusion that there was scope to mechanise the process of on-course betting. His machine, known as the 'totalisator', was patented in New Zealand in 1913 and in Western Australia in 1916 and, in a progressively upgraded form, has provided the basis for the operation of the totalisator and betting facility, TAB, throughout Australia ever since. In addition, it was a technology that was successfully exported at an early stage and widely used internationally.

In the development of the computer, however, it was the mechanism of academic relationships coupled with involvement in the Second World War that provided the impetus for Australian computer developments.

During the early part of the war, Dr Trevor Pearcey was working on the development of microwave radar in the United Kingdom and, in the process, became familiar with the control of pulses using vacuum tubes and their storage using mercury delay lines. As a prelude to later proposals for Australia's first computer, this was invaluable experience.

In 1946, on his return to Australia, having had an opportunity to see the Aiken machine as constructed by IBM in action in the United States, he was in a position to influence the development of electronic computation research. At that time, Dr EG Bowen, the chief of the Radio Physics Laboratory of the Council for Scientific and Industrial Research, CSIR, later the Commonwealth Scientific and Industrial Research Organisation (CSIRO), had been contemplating three main lines of future research, computers not being one of them. Pearcey managed to persuade Bowen that electronic computers were an appropriate area of research to pursue and this led directly to the creation of Australia's first machine known as Mark 1.

Dr Trevor Pearcey operates Australia's first computer made at the Radio Physics Branch of the Council for Scientific and Industrial Research (CSIR); the computer was later known as the CSIRAC Mark 1 *(DEA)*

The teleprinter output of the CSIRAC Mark 1 *(DEA)*

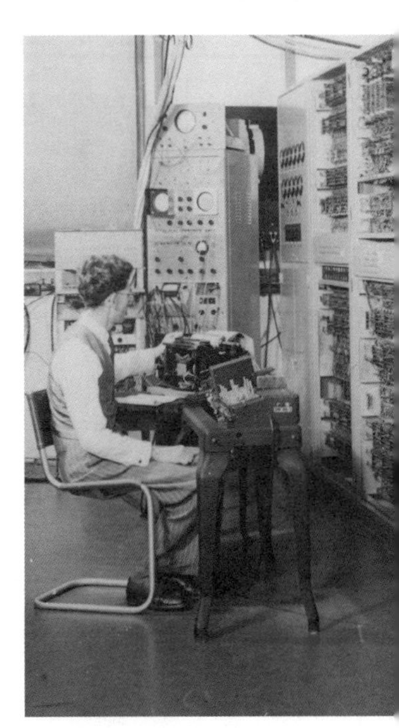

During the early part of 1947, Pearcey together with Maston Beard began work on the electronic computer project and by 1948 had prepared a report on how the machine should be made, based on a number of experiments. In 1948, reports of the work in Britain became available and the details of the machine created at Manchester University — Mark 1 — together with information about Pilot ACE at the National Physical Laboratory and EDSAC at the University of Cambridge, was also obtained. This information resulted from a visit by Pearcey to these institutions at that time.

Construction of the CSIR machine began and late in 1949 an initial test program was successfully run on the new machine. Full operation commenced in June 1951 and for the next 4 years the computer carried out computations on a number of significant projects in Sydney. During this time, a series of design improvements were made to enhance the performance and ease of use of the machine. With the construction of a computer at Sydney University based on the design of a machine created by the University of Illinois in the United States, the need for the Mark 1 came under review.

The Australian machine at Sydney University was dubbed SILLIAC, 'Sydney's Illinois Automatic Computer'. It was soon joined by a second

Left Maintenance of the CSIRAC Mark 1 *(DEA)*

Right Behind the CSIRAC Mark 1, showing the mercury delay-line memory in the horizonal tubes left of centre of the machine *(DEA)*

overseas-derived machine, the DEUCE from English Electric, purchased by what is now the University of New South Wales. This computer was given the name UTECOM, which stood for 'University of Technology's Electronic Computer'. At this time it also became apparent that no commercial organisation was prepared to take on the construction and sale of the CSIR-designed machine. With the competition provided by the two new machines in Sydney, it was decided to move Mark 1 to Melbourne.

The Mark 1 was therefore disassembled and moved to the University of Melbourne in 1955, renamed CSIRAC and continued to work there for the next 8 years. In 1964, CSIRAC was decommissioned and passed to the Museum of Victoria. Since then it has been preserved but displayed only twice in the intervening years, in 1980 and 1996.

COMMERCIAL APPLICATIONS

At Manchester University, Professor Williams based the design of a new computer on a quite different form of memory from that of the ENIAC in the United States, using cathode-ray tubes where the numbers were stored as fluorescent dots on the rear of the cathode-ray tube screen. The machine developed at Manchester was known as the MADM and commenced operations in April 1949. From this work came the Ferranti Mark I and this was put into commercial operation in February 1951.

Despite all of the work that had occurred in Britain during the war and after it, it is apparent that it was in the United States that commercial computers saw most intensive development. No doubt part of the reason for this was the same as had motivated the creation of business machines towards the end of the 19th century — lack of skilled operators. Whatever the basis, big companies and familiar names in the creation of business machines soon became interested and involved in the creation of this new electronic business machine.

In particular, Remington–Rand ultimately absorbed the business of Mauchly and Eckert, the Electronic Control Company, which was converted into its computer research division. Into this same division came another start-up computer company, Engineering Research Associates.

This new conglomeration at Remington–Rand produced another machine and another acronym. UNIVAC, 'Universal Automatic Computer', was developed to service a contract with the Census Bureau and, as with EDVAC in the United States and EDSAC in the United Kingdom, this machine made use of mercury delay lines for memory. In addition, it used magnetic tape for the first time to store data, thereby increasing the data transfer rate.

In the history of computing, UNIVAC established an enormous prestige in its involvement with the 1952 election. At a very early stage in the election of the new president and on the basis of a statistical program created for the task, UNIVAC correctly predicted a landslide victory for Eisenhower over his rival Stevenson. This remarkable prediction was made on the basis of only 6 per cent of the total vote, but produced results within quite small percentages of the actual ultimate figures.

The UNIVAC 1, designed by Mauchly and Eckert working with Remington–Rand, after its acceptance test of 1951 *(EA)*

IBM JOINS THE FRAY

At about this time and after a long period of seeming indifference to the new valve-based technology of office machines, IBM became interested in this new form of electronic computing. No doubt it had observed the success of its old rival, Remington–Rand, and had been stung by the publicity relating to the census results and the success of UNIVAC.

Its flirtation with electromechanical computers in the Harvard Mark I had no doubt also contributed to an initially high level of scepticism. The Harvard Mark I had turned out to be a quite expensive dinosaur, despite having effectively proved the validity of Babbage's model of computing. Part of the disenchantment of IBM with the Harvard Mark I was clearly associated with the behaviour of its principal developer, Howard H Aiken.

Thomas J Watson of International Business Machines *(IBM)*

Howard Goldstine, former associate of Mauchly, Eckert and Von Neumann, instructing at IBM Research Laboratories *(IBM)*

In describing the development of this machine, Aiken failed to write about the very significant part that IBM had played. This was totally to sour relationships with the company. Its chairman, Thomas J Watson, was said to have been furious at the slight and unforgiving thereafter.

IBM's first foray into electronic computing was somewhat tentative but successful nevertheless. It involved the application of electronic computation to an existing desktop electromechanical calculator known as a 'punch calculator' with the numerical designation 603. In both this machine and the subsequent 604, the inclusion of electronic computation provided an enormous improvement in performance. More to the point, it produced a machine that IBM was able to place in many offices in a highly profitable manner. Later, this same machine was further adapted for use by Northrop Aircraft Company to create a form of hybrid computer–calculator which used card programming. This machine was known as the CPC, or 'card-programmed calculator'.

As was happening with the burgeoning market in transistors, the advent of the Korean War developed pressure for the military application of computers. IBM was a particular beneficiary of this situation which allowed it to develop its first fully electronic computer based on Von Neumann architecture. Apart from the military demands, it appears that the success of UNIVAC in its involvement with census prediction also created intense pressure at IBM. Later, the situation was said to have generated a state of mind among the company officers that became known as 'panic mode', a description that has gone into the business vocabulary of the world.

IBM's first business computer, the 701, was developed in December 1951, but its first data-processing computer, the 702, was not ready until September 1953 and not finally delivered until 1955. This 4-year delay in matching the success of Remington–Rand might have been crippling in a lesser firm but, on the contrary, IBM responded to the situation and succeeded.

The mercury delay line form of memory developed for the earliest American machines was intrinsically slow. By comparison, the cathode-ray tube memory developed in Britain was much faster but somewhat unreliable in operation. IBM's response to this problem was to incorporate the newly developed magnetic core memory. Invented at the Massachusetts Institute of Technology as part of the Whirlwind computer project, it was incorporated into the IBM 700 series computers in 1953 and, in particular, the 704 and 705. Also importantly, in 1952 magnetic drum memory was applied to the 600 series computers developed by IBM and, by 1953, the 650 machine of IBM had become, in effect, the industry 'workhorse'. This was a reliable and useful machine and relatively far less expensive than its rivals. This resulted in the ultimate eclipse of UNIVAC and ensured the ascendancy of IBM despite its late entry into the field.

The commercial and technological race that developed now was as fierce as any before or since. By the end of the 1950s, of the larger organisations in the United States that had entered the field of electronic computing originally, only RCA, GE and Honeywell remained together with IBM and Sperry–Rand. This latter firm resulted from an amalgamation of Sperry Gyroscope and Remington–Rand. In addition to these larger organisations, three other smaller firms remained. These were Control Data Corporation (CDC), Burroughs and National Cash Register (NCR).

From 1960 to 1965, based on salesmanship, business acumen, leasing of machines as opposed to selling them and supported by the development of a comprehensive computer system in the 1401 machine, IBM came to dominate the field in the United States with perhaps 80 per cent of the available business. No doubt this was opportune for, in the same period, a dramatic decline in the use of punch-card business machines had occurred as a response to the introduction of electronic computers.

Part of the significant success of the 1401 machine, compared with its predecessor, the 650, was the introduction of transistors. Where the 650 had used a large number of large, hot, expensive and, ultimately, unreliable vacuum tubes, the 1401 now introduced smaller, cooler, cheaper and generally reliable transistors together with magnetic core memory. However, apart from these fundamental technical advantages, IBM also supplied the 1401 machine with an impressive high-speed peripheral printer which was able to print out 600 lines per minute. In addition, this computer system from IBM was associated with the development of the first major programming language known as RPG, 'report program generator'. This remained available and popular for many years.

During the 1960s, one of the significant reflections of IBM's organisational competence and research capability was its response to the perceived demand for compatible hardware. In a research report presented in December 1961, a new concept was developed under the acronym SPREAD, which stood for 'Systems, Programming, Review, Engineering and Development'.

This extremely bold document constituted a vastly expensive recipe for the development of IBM's future approach to computer systems for the full spectrum of possible users. The response to this report was undoubtedly strategically dangerous and potentially fatal in economic terms but led to an extraordinarily successful 'new product line' in the System 360.

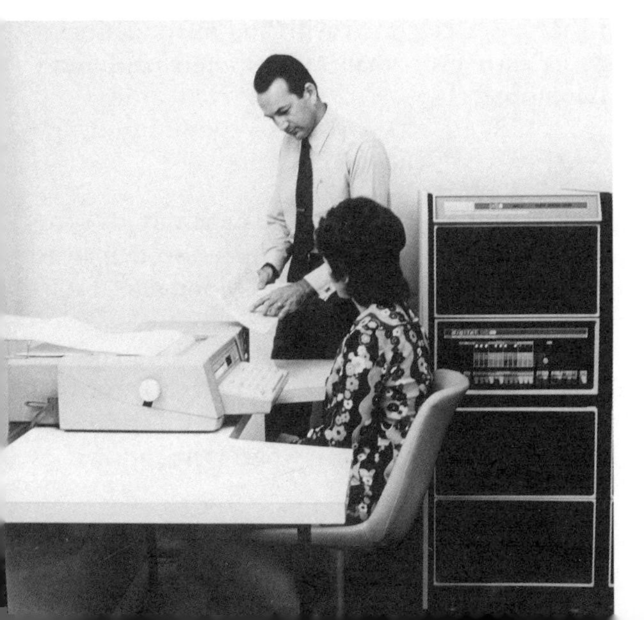

Left One of the range of small computers made by Digital Equipment Corporation (DEC), the PDP 8 minicomputer of 1965 *(EA)*

Right A British ICT computer of 1968, the 1902A, used transistor transistor logic (TTL) *(EA)*

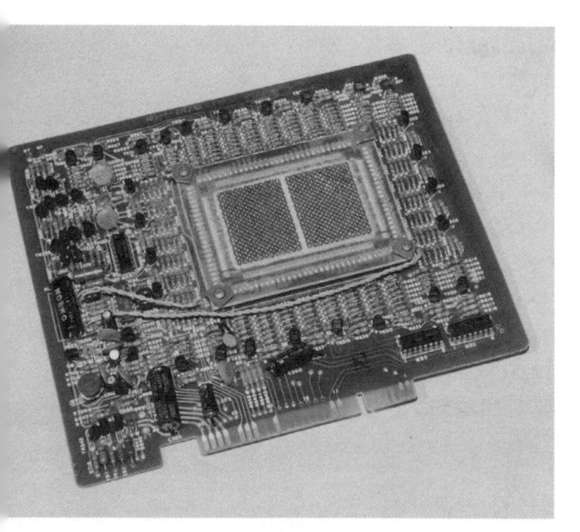

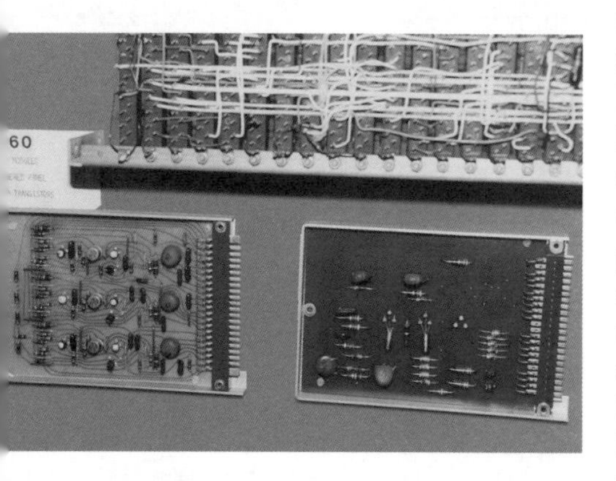

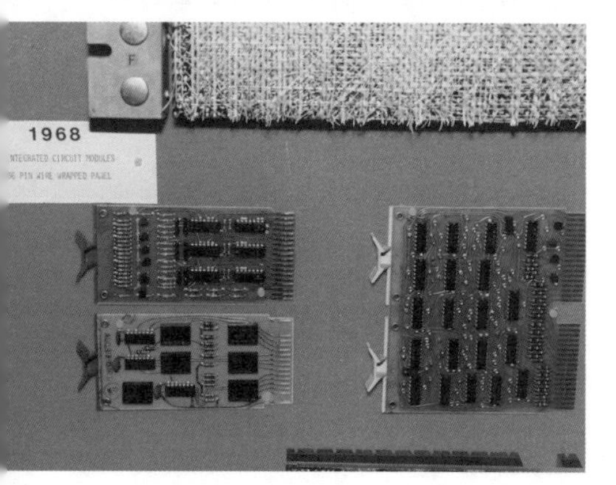

From the time of its launch in April 1964, this new series of IBM computers attracted orders for approximately 9000 machines — a demand that was not able to be met for some time. In spite of the fanfare of publicity with which the new machine had been presented to the public, in retrospect, the 360 series could not be seen as embodying the most advanced technology. Concurrently, Honeywell produced its 2000 series computer which was able to run IBM-generated programs but at a much faster rate than the comparable IBM machine. Further, in 1964, RCA produced its Spectra 70 machine which was also IBM 360-compatible but used integrated circuits with associated increases in reliability and reduction in running costs.

By the end of 1970, the number of rivals in the computing field had reduced yet again. By this stage, in the United States, in addition to the market leader, IBM, were Burroughs, UNIVAC, NCR, Control Data and Honeywell, known colloquially as the 'BUNCH'. In retaining its pre-eminent position in the field of computers, it can be seen that IBM's particular strength lay in its capacity to market its products rather than necessarily being at the leading edge in engineering and design. However, in the longer term, the failure of IBM to fully implement new technology and the decision to remain firmly wedded to the 360 system was to have destructive ramifications. It seems probable that IBM's long-term slide into unprofitability and the ultimate huge annual losses incurred in the early 1990s can be seen as having their genesis in this period when new, smaller machines were built and competed at a technically advanced level.

With the advent of the new machines, usually described as minicomputers, the computer began to be perceived as something akin to a conventional commodity or 'white good', no longer to be administered by a scientific hierarchy or 'priesthood' of computer technicians. Instead of being alienated from the end user, the new minicomputers were able to be operated without an intervening expert. Evidently this is a trend that has accelerated with the introduction of solid-state devices into ever-smaller computers and more recently, microcomputers.

Top Early toroidal core main memory with solid-state chips *(PRJ)*

Middle Germanium transistor system modules from 1960 *(PRJ)*

Bottom Integrated circuit system modules from 1968 *(PRJ)*

SOFTWARE DEVELOPMENT

Apart from changes to what is now described as the hardware of computers, another significant change now occurred which also helped to liberate users from computer experts. This was the development of software.

Early machines had been programmed in 'machine language', a highly efficient but frighteningly arcane method. Now an early form of simplified, computer-assisted program generation came with the development of 'assembly language'. Later again, what were known as 'compiler programs' were created to simplify the job of providing computers with instructions.

Undoubtedly, the most important of these compiler programs was FORTRAN, from 'formula translation', developed by the research department of IBM. In addition, another extremely important language, COBOL, standing for 'Common Business Orientated Language', was developed at the behest of the United States government as a means of achieving a degree of consistency in government departments.

Somewhat later again, another highly influential development was the creation of BASIC, standing for 'Beginner's All-purpose Symbolic Instruction Code'. This was created at Dartmouth College as a teaching aid by Professor John Kemeny and Thomas E Kurtz. This simple and 'user-friendly' language incorporated many of the structural and logical features of FORTRAN and was specifically designed to introduce people to computer programming. One of its most attractive features was that it was designed to be used in real time and was able to be edited or 'debugged' immediately so that program development was extremely rapid compared with its partial parent, FORTRAN. BASIC proved highly successful and was rapidly adopted by many academic institutions. Perhaps more importantly, despite its apparently humble beginnings as an aid to teaching, BASIC was adopted for use in most of the new generation of miniature computers that would appear in the next decade, microcomputers.

Integrated circuit logic board from 1971 *(PRJ)*

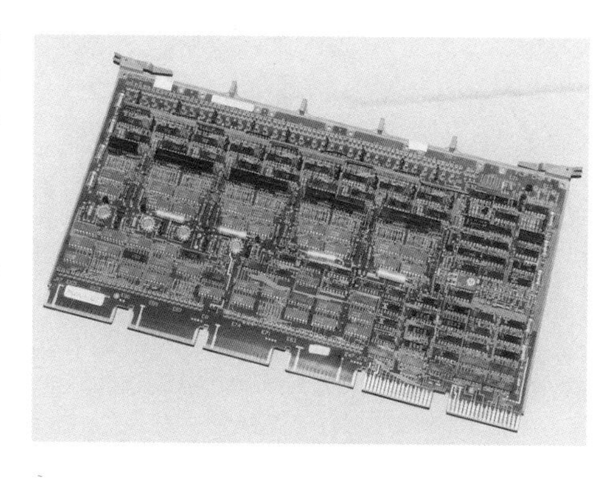

Integrated circuit logic board from 1980 *(PRJ)*

During the 1970s the creation of software was largely transferred from an association with the big computer companies to external contractors. This led to the development of a new industry which, initially, tended to be inherently chaotic and lacking in systematic methodology. This situation was responded to by the emergence of a new philosophical approach to the creation of computer programs, known as 'structured programming'. It also led in 1971 to the creation of a new language, PASCAL, which embodied all the principles of the new systematic structural approach. Another somewhat similar language, ADA, was sponsored by the United States government but has been generally far less popular.

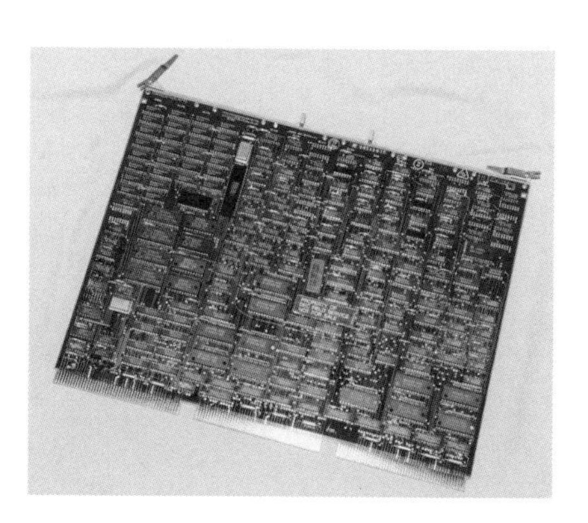

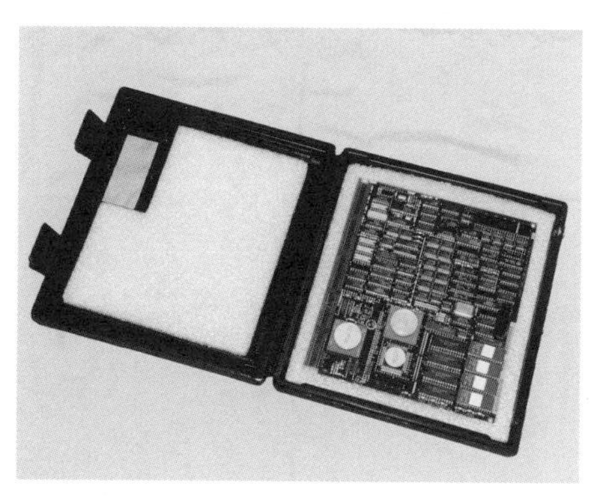

Far left High-density integrated circuit board with microprocessor from 1990 *(PRJ)*

Left Microprocessor board and supporting integrated circuits, 1990 *(PRJ)*

CHAPTER

15

1901 to 1995

Links in the Web

By 1901, when the first radio link had been established across the Atlantic between Europe and Canada and then to the Americas, a number of cable telegraphic links already existed across the Atlantic. Typical of these were the cables which had been laid by the Atlantic Cable Company, an amalgamation of British and American cable companies founded by two giants of the Victorian communications revolution. John Pender, a British industrialist, had been involved in the creation of the English and Irish Magnetic Telegraph Company and Cyrus Field had been responsible for setting up the New York, Newfoundland and London Telegraph Company. These two companies ultimately merged as the Atlantic Telegraph Company which laid the first successful cable between Valentia in the south of Ireland and Heart's Content in Newfoundland in 1866.

John Pender, as a company director, was later directly concerned with establishing the cable from London to the Far East which was laid on behalf of the Eastern Telegraph Company. Cyrus Field was also involved in this enterprise as an extraordinary director. This was the

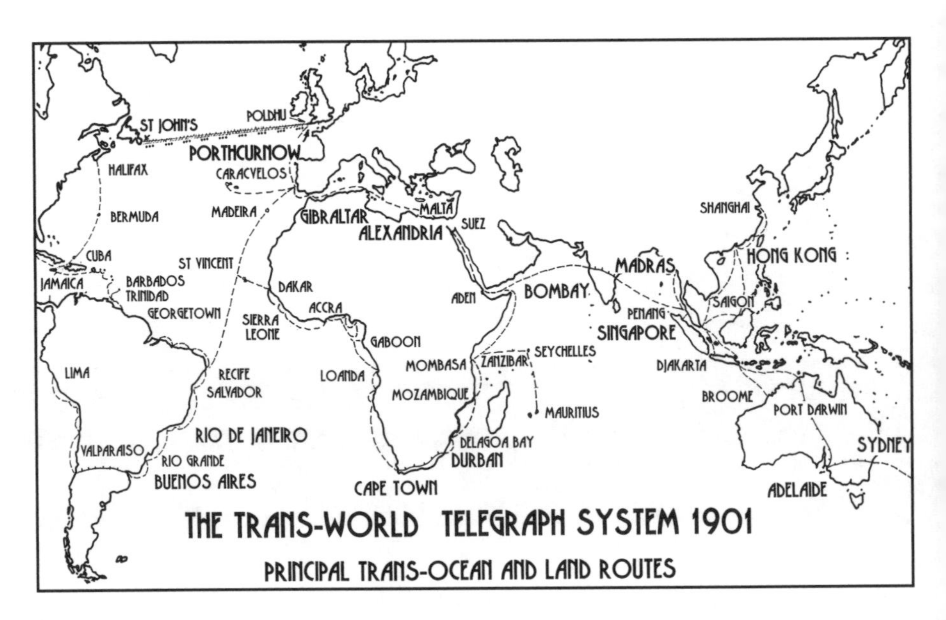

Map showing the main telegraph routes in 1901 *(PRJ)*

Telegraph 'repeaters' receiving a message in morse code and sending it to the next station *(PRJ)*

cable that was ultimately extended as far as Darwin in the north of Australia.

The link from Adelaide in South Australia to London, which had been opened on 21 October 1872, was to be the longest telegraphic connection in the world at that time, consisting of a series of undersea cables interconnected with land lines. It was soon supplemented by other links to places far distant from the commercial hub of Western Europe.

Between 1866 and 1900, fifteen cables were laid across the Atlantic to carry the burgeoning commercial traffic between Europe and the new world. Once laid across the cold, dark floor of the deep ocean, these cables proved to be robust and to have a remarkably long service life. Apart from the depredations of deep-sea trawlers which dragged up and broke the cables from time to time, they rested secure from damage. As time passed, however, and with demands for increasing message handling, the intrinsic limitations of these undersea cables became a severe restriction. By the 1920s, multiplex working saw data rates of 400 words per minute achieved, but this was at the extreme limit of a cable's capacity to pass data.

The length of undersea cables also produced significant restrictions in terms of their message-handling capacity. For this reason, during the latter part of the 19th century, a series of mid-oceanic islands developed enormous strategic importance. Such places, whose names had until the advent of cable been largely unknown or entirely exotic, became familiar to the public. Ascension Island, Guam, Midway Island and Cocos Keeling all provided the bases for intermediate cable stations. At these locations, for months at a time, employees of the cable telegraph services would receive the weak signals coming in, convert the morse code to text and then retransmit it by hand key. Thus the regenerated signals were sent on to their ultimate destination — a boring and repetitive task for the 'repeaters' but fundamental to the success of the early telegraph system.

Unfortunately, in some places, the educational level and enthusiasm of such operators was low and, to add to the problem, their command of English could also be quite limited. In some parts of the Victorian cable network, garbling of messages was very common and much resented by the originators and recipients when significant sums had been paid for the messages to be carried.

An Ericsson wall telephone of 1895 *(PRJ)*

WIRELESS COMPETITION

Between 1901 and 1927, the undersea telegraph cable network continued to expand, providing telegraphic services of increasing sophistication. During the latter part of this period, the early morse telegraph system started to be displaced by the use of the electromagnetic teletype, and the 'telex' ultimately became a common feature of business communications.

Although the trans-Atlantic wireless telegraph of Marconi, established in the far west of Ireland in 1907, had begun to compete successfully with the trans-Atlantic cable, it was not until the opening up of the Beam Wireless telegraph service in the 1920s that cable was at last to meet its most serious rival. Over a period of just a few years, revenues of the cable companies were dramatically reduced, so much so that strategic amalgamation was considered essential by the British government as a means of ensuring the survival of the cable companies.

Apart from the reduced cost of the wireless telegraph service compared with cable telegraph, radio communication was also able to provide an intercontinental telephone service for the first time. This began in 1927 and also helped to draw business away from the cable and wired telegraphic network.

During this period, despite a degree of optimistic posturing by the directors of the cable companies, in the United Kingdom pressure was now applied to achieve fresh technical development. As an early response to the competition with wireless, such developments as the 'regenerator', designed to replace the human operator, were directed at arresting the 50 per cent reduction in cable traffic that occurred once short-wave beam wireless was established. During a relatively short period of time, the old manual repeater stations were converted to automatic 'regenerator' operation. This had a significant impact on both the accuracy and speed of data transmission. In addition, the reduction in manpower to operate the service made it possible to reduce the cost of sending messages and improved the profitability of the cable system as well. At a technical level, it was discovered that 'loading' of the undersea cables with suitable inductors could allow a significant increase in data rate, and this also contributed to the improvement in the cable telegraphic service.

Despite the economic and operational upheaval that the introduction of short-wave telegraphy produced, there can be little doubt that it helped to develop the climate of competition which has so benefited the public in later years. Since the 1920s, the availability of telephone and data services has gone from being a highly expensive luxury to an everyday commodity that people expect to have access to at reasonable cost.

Left Candle-stick telephones from 1910 to 1920 *(PRJ)*

Right The evolving wall-mounted telephone from 1910 to 1920 *(PRJ)*

UNDERSEA CABLE TELEPHONY

Despite the enormous success of the telegraphic cable system, for many years telephony was not feasible over any significant distance using an underwater connection. However, after the best part of a century of technical development in undersea cable techniques, a joint American, Canadian and British effort led to the creation of the first trans-Atlantic telephone cable known as TAT-1. The joint contract between the British Post Office, the Canadian Overseas Telecommunications Commission and the American Telegraph and Telephone Corporation was signed in 1953 and the operation went ahead with far less drama than had attended the laying of the first trans-Atlantic telegraph cable in 1866.

The American section across the Atlantic involved two cables, one for each direction of the telephone signal, whereas the Canadian section consisted of a single two-way cable. Interestingly, this system made use of thermionic valves, despite the availability of recently invented transistors. The reason for this was that in 1953 the transistor was a mere 6 years old and involved a still largely untested technology over the long term. In designing repeaters for the trans-Atlantic cable which were essential to amplify the rapidly fading voice signal, total reliability over a 20-year period was considered essential. In this respect the technology of valves was far better known in 1953 and therefore they were

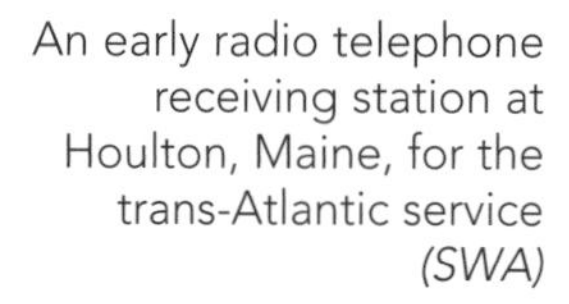

An early radio telephone receiving station at Houlton, Maine, for the trans-Atlantic service *(SWA)*

incorporated into what was intended to be perhaps the most reliable system of telecommunications in the world at that time.

Valves used in the deepsea repeaters were type 175HQ with three valves for every repeater. Plate voltages were held at 50 volts and heaters operated at 18 volts. TAT-1 was able to accommodate 36 simultaneous conversations on radio frequency channels between 12 kilohertz and 174 kilohertz. TAT-1 was obsolete before it was laid.

SATELLITE INTERLUDE

In a book entitled *Ways to Space Flight* published in 1928, Herman Oberth had made suggestions concerning radio communication via satellite. Although unaware of this prediction, the noted science fiction writer Arthur C Clarke promoted the same notion of satellite communications in an article contained in the British magazine *Wireless World*, in February 1945.

In 1957 satellites appeared in space above Earth for the first time, much to the consternation of the American people. Sputnik was carried into space in October of that year by a ballistic missile of the Union of Soviet Socialist Republics (USSR) and created something akin to panic among the American military. Although of no strategic or military value, and only capable of generating a marker signal in the high-frequency short-wave band, Sputnik generated intense pressure in the United States to create a competitor. However, before this could occur, it was soon followed by Sputnik II, a much larger satellite which created an even greater furore in the US military establishment.

In July 1962, the United States at last produced the necessary competitor with Telstar I designed at the Bell Laboratories of the American Telegraph and Telephone Corporation (AT&T) and this was placed in orbit. This satellite was powered by solar panels and had a capability to communicate over two super-high-frequency channels at 6.39 gigahertz and 4.17 gigahertz for the 'up link' and 'down link' respectively.

As any telephone user in the early 1960s will remember, the irritating counterpoint of this dramatic new technology was the half-second delay between the completion of a sentence and its reception at the far end. Compared with the undersea cable without this problem, this delay was not only quite perceptible but disconcerting, if not frustrating and irritating. Conversationalists, with some difficulty, became accustomed to a 'polite pause' mode of communication. Until this new rhetorical device was applied, interruptions and mangled conversations were an inevitable part of using the

Top The control room in New York for the Bell Radiophone service *(SWA)*

Bottom Radiophone operator at the New York exchange on the trans-Atlantic service *(25Y)*

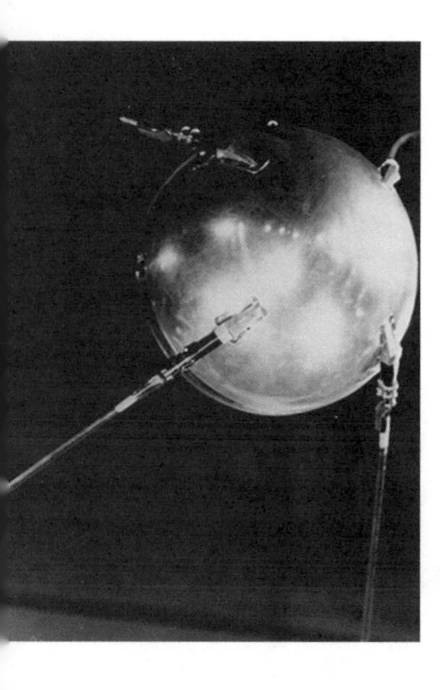

Sputnik *(NASA)*

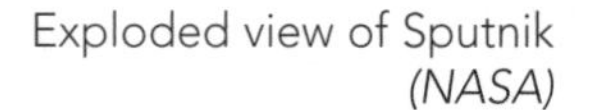

Exploded view of Sputnik *(NASA)*

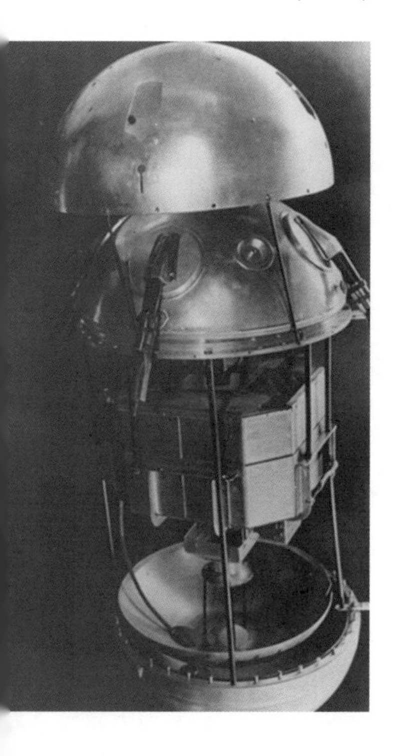

satellite communication path. The emergence of the improved undersea cable links some years later was a huge relief for the many users who had become accustomed to suffering poor telephone communications during this satellite period.

A fundamental problem with Telstar 1 was that it operated in a low orbit and its velocity was such that coverage of a given area lasted for only 20 minutes or so before the satellite disappeared to another part of the world. Evidently a far more satisfactory approach compared with the elliptical orbit of 1530–9000 kilometres (950–5600 miles) was a geostationary orbit. In this arrangement, the satellites were established in space at a height so that they would orbit at the same speed as the rotation of the earth. Because of this, their relative position would not change. Equally, being stationary relative to the Earth's surface, the footprint over which they would operate at ground level would also be constant.

The first of these new geostationary satellites, SYNCOM 1, failed to achieve orbit in February 1963. However, its successors, SYNCOM 2 launched in July 1963 and SYNCOM 3 in August 1964, were successful. In the United States it was realised that this new technology required a special-purpose organisation to administer it. This led to the establishment in 1962 of an operating and administrative authority in the United States known as the Communications Satellite Corporation or COMSAT. It was set up to have sole control of the establishment and administration of satellites flown from America, a situation which was to create a considerable degree of anxiety and annoyance in the Bell Corporation.

SYNCOM 3 was notable for providing a relay of television signals from the Tokyo Olympics in 1964, although this was confined to a single black-and-white channel.

The technical successor to the SYNCOM range of satellite, INTELSAT 1, which was known as 'Early Bird' initially, was put into orbit in April 1965. Over the 20 years between 1965 and 1985, the INTELSAT satellites, commencing with number 1 and going up to number 6, saw an increase in telephone channel capacity from 240 channels to 35 000 channels. In addition, where INTELSAT 1 could accommodate only a single black-and-white television channel when the voice channels were inactive, INTELSAT 6 was able to handle two colour television channels simultaneously with the voice channels.

These satellites were to have a huge impact on the expectation of the general public for world news. Evidently the transfer of television images from continent to continent was unaffected by the half-second delay which applied inexorably to telephone conversations.

There can be little doubt that world television, which has been made possible by such communication satellites, has had a major impact on the public. Since the initiation of a world-spanning television service based on satellites, the public has been confronted with daily happenings from all corners of the globe. Disasters, wars, sporting events and sometimes even peace conferences flicker across our daily TV news, leaving us almost overwhelmed but unable to ignore the torrent of information. Truly, to quote the words of Arthur C Clarke, what has been achieved is 'A future when, whether they like it or not, for better or worse, all men on earth are neighbours'.

INTELSAT V — an artist's impression *(TEL)*

CABLE REVIVAL

The successful introduction of wireless telegraphy for trans-Atlantic messages in 1907 had created ripples on the tranquil surface of telegraphic cable business. In the 1920s, with the advent of short-wave communications both for telegraphy and telephony, these ripples became a 'tsunami' wave.

In the United Kingdom, such a situation was seen as strategically undesirable and from it came the enforced amalgamation of Marconi's wireless telegraphy system with British cable interests to form a new conglomerate, the Cable and Wireless Company. The financial stability of the cable companies was restored but competition could not be stifled forever.

Forty years later, wireless communication with satellites to provide intercontinental data and voice communication services was, for a while, to be just as devastating to the cable networks as had been the earlier impact of the beam short-wave wireless. However, technology was at hand to assist in a resurgence of the cable systems. The principal factor in this new era was the transistor. Applied to the trans-Atlantic telephone cable of 1970, TAT-5, this new device provided a great improvement in the capacity of the cable. However, beyond the transistor, another fundamentally different technology was in the process of development — the 'fibre optic' cable.

In 1975, AT&T had produced a fibre-optic cable capable of carrying 50 000 telephone calls on land with repeaters set at a separation of 11 kilometres (7 miles) and over the next 8 years there were major improvements made to this figure. By 1983 it was apparent that a serious rival to the older copper-cored undersea cable was now available.

Telephone evolution, 1920 to 1960 *(PRJ)*

When this new technology was applied to the newest trans-Atlantic cable at that time, TAT-8, it allowed 40 000 simultaneous telephone conversations to be held. By comparison, its immediate predecessor, TAT-7, which had used a copper core, had been able to accommodate a mere 4200 simultaneous conversations. Only a couple of years later, in 1990 there was a further improvement with TAT-9 which was able to accommodate double the number of calls possible on TAT-8.

Today, having leap-frogged over the satellite revolution in data and voice communications which began over 20 years earlier, once again the primacy of cables is being threatened by new satellite systems. A proposal to create a world-encompassing network of satellites to provide communication with users of handheld telephones on the ground seems likely to create a fresh revolution in trans-world communication. Best known of these is a system that originally proposed 77 satellites and for that reason was named Iridium, after the element with the atomic number 77.

In one of his predictions there will be a small rebuff to that otherwise impeccable prophet of the future, Sir Arthur C Clarke. In a paper entitled 'Space Communications and the Global Family' presented in 1986, he said:

> So let us start with our old science fiction friend, the wrist watch telephone. Frankly I don't believe in it. I'm not going to stand like an idiot holding my arm in front of my face. The telephone of the future will be a waist belt box — just like the Walkman and its successors — with a very light ear-piece and microphone, working through an optical or electromagnetic link so that one does not get continually entangled in tiny wires.

A modern fibre-optic undersea cable is brought ashore at Bondi Beach, Australia, in 1982, linking across the Pacific Ocean to Canada *(ALC)*

Although there are certainly portable telephones available on the preferred pattern of Arthur C Clarke, today the majority of portable telephone users have modules that fit in the hand to be held in front of their faces. As for the existing terrestrial cellular network phones, the newer models designed to work with satellites such as the Iridium network all appear to be handheld. It seems the familiar pose of the telephone user, with hand close to mouth, is likely to perist for quite a while yet. When the first bionic telephone implant is produced, then it may be an entirely different situation and Sir Arthur may be proved correct.

Cable-laying vessel of Cable and Wireless, lying at Circular Quay, Sydney, in 1996 *(PRJ)*

COLD WAR STRATEGY

Shortly after the cessation of the Second World War, in the immortal words of Winston Churchill an 'Iron Curtain' fell across Europe as former allies became implacable foes. For more than 40 years from the end of the war in 1945, the Union of Soviet Socialist Republics (USSR) was to remain the enemy of the United States and its allies in the western part of Europe, although in the latter part of this period a gradual thaw in relations set in. During this period, the North Atlantic Treaty Organisation (NATO) was established as a bulwark against communism but this did not mean that more distant America was any less concerned with the military might of Russia.

American sensitivity and suspicions of Russia and its dictator, Joseph Stalin, were aggravated by the explosion of a Russian atom bomb in August 1949. The subversive activities of British university graduates and intellectuals for a time also soured the relationship between the United States and the United Kingdom, at least in regard to the exchange of 'intelligence'. Apart from this, however, American fears of airborne attack involving nuclear weapons led to a mad scramble for a defensive strategy. One response was a committee that was set up to deal with air defence systems and engineering with the acronym ADSEC. Under the chairmanship of Professor GE Valley, it inevitably became known as the 'Valley committee'.

Another important step was the establishment of the Lincoln Laboratory at Massachusetts Institute of Technology (MIT) in 1951. It was here that a scientist, JCR Licklider, was employed to set up a Human Engineering Group and where his interest soon turned to computers and their capability for powerful interaction with human cognition. Licklider left MIT in 1958 to join Bolt, Beranek and Newman, a consulting firm, with the promise of access to computational resources to allow his research interests to develop further. We will meet him again shortly.

By 1963 a variety of steps, derived in part from the conclusions of the 'Valley committee', had led to the creation of a defensive strategy referred to as the 'semi-automatic ground environment'. This system, usually referred to by its acronym SAGE, had 23 distributed centres at which were located computers to hold data and provide communications capability. These machines were provided by IBM, a contract which helped to confirm this organisation's position as the pre-eminent computer manufacturer during this period.

Despite all the work involved in setting up SAGE, it had been devised to deal with a perceived threat from conventional jet-engined bombers. By the time SAGE was implemented, this threat no longer existed due to the introduction of intercontinental ballistic missiles in the 1960s. Despite its obsolescence, however, SAGE was undoubtedly important in supporting the development of the new computer industry. Further, the need for lightweight and compact components certainly accelerated the development of printed circuits, solid-state devices, core memory and new systems of data storage.

At a more mundane level, the connected network of communications between these distributed centres was later applied to the problem

A drum of fibre-optic cable ready to wind into the cable-laying ship's hold *(ALC)*

Fibre-optic cable in the cable-laying ship's hold *(ALC)*

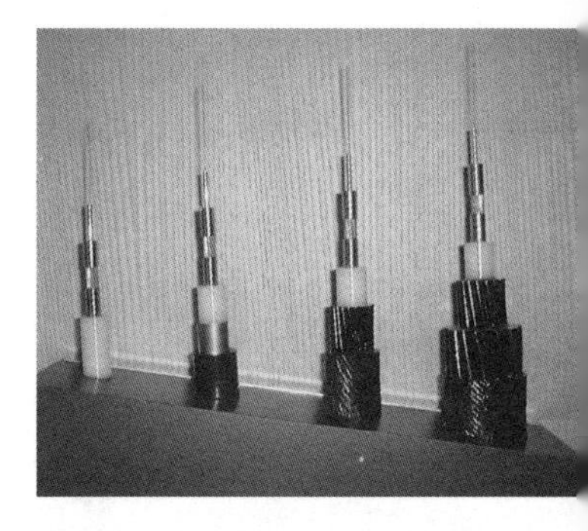

Various levels of armouring of fibre-optic cable from 'no armour' in deep ocean locations to 'heavy duty' armouring used to protect against damage at the inshore end of the cable *(ALC)*

of booking and ticketing as required by the airline industry, with networks set up using dedicated leased telephone lines. Used initially by American Airlines, this communications technology soon spread to Pan American and Eastern Airlines, allowing real-time reservation and ticketing with access to an enormous centralised database.

Left Typical antenna array in cellular telephone network *(PRJ)*

Right Typical terrestrial-based antenna array for satellite signal reception, Optus satellite link ground station at Belrose, north of Sydney *(PRJ)*

ARPA IN THE UNITED STATES

If the explosion of the Russian atom bomb in 1949 could be seen as the genesis of the 'reds under the bed' panic of the McCarthy era, then the flight of Sputnik in October 1957 was one of the major factors underlying America's decision to create a strategic defensive system. This would, at a much later time, grow into the massively expensive technology known as the Strategic Defence Initiative, involving lasers with offensive capability. However, in the early days, the ambitions of the United States were largely confined to a defensive capability, although the Strategic Air Command of nuclear bomb-laden aircraft certainly had a major offensive potential even though ostensibly to be used only in a retaliatory role.

One of the key areas of an effective defensive system was soon perceived to be a communications network that was not intrinsically vulnerable to nuclear attack. In order to cope with the perceived threat from the Soviets, in 1957, at the instigation of President Eisenhower, the Department of Defence located at the Pentagon in Washington created a new department known as the Advanced Research Projects Agency (ARPA). Within this department, a subordinate organisation known as IPTO, the Information Processing Techniques Office, was set up.

The first director of IPTO was the scientist we met earlier, Joseph Licklider. He had been trained in psychoacoustics before developing an interest and expertise in computers, and came to the position at ARPA from the consulting firm Bolt, Beranek and Newman. Interestingly, given the direction of subsequent events, Licklider also came to his new job with a remarkable appreciation of the potential power of the

Typical antenna array in cellular telephone network for Telstra in Australia *(PRJ)*

computer. In this regard he had developed some perceptive notions of the ultimate symbiosis of man with the computer and the desirability of achieving greater use of computers by 'time-sharing' methods. In addition, his prior involvement with Bolt, Beranek and Newman could be seen as the beginning of a long and fruitful association of this firm with ARPA and, ultimately, in the creation of the Internet.

As a firm, Bolt, Beranek and Newman was a somewhat unusual base for computer-related activities in this early period. The firm had been set up by the partners as a consultancy in building acoustics, Bolt being an architect and physicist and Beranek an electrical engineer, and Newman was later recruited with qualifications in architecture and physics. Despite this construction industry background, computing science was to become a major source of income. After a disastrous problem of litigation over failed acoustic design work on a concert hall, computing-related projects became one of its principal consulting activities.

Once installed at IPTO, Licklider established a personal network of 'top' computer experts from a variety of universities and research organisations. This personal network was no doubt the motivation for Licklider to develop the idea of a network of interconnected computers, able to share the developing ideas and research of professional colleagues.

Above ABC TV main antenna at Gore Hill, Sydney, and satellite link dish antennae for receipt of television and data from satellites *(PRJ)*

Commercial satellite receiving dishes in the industrial area at Gore Hill, Sydney *(PRJ)*

After a relatively short period, Licklider went back to an academic post at MIT and was followed in the director's chair at ARPA by Ivan Sutherland and not long after, in 1966, by Robert Taylor. By this stage the primary emphasis of ARPA to respond to the evident technical prowess of the Soviets in launching successful space and ballistic rockets had abated somewhat. Although the integrity of vulnerable communications links across the United States was still a concern, for Taylor the need to provide a network to connect the scientific community was more fundamental. If this led to 'spin-off' benefits to military research, then well and good, but a network of computers would allow more efficient use of scarce computer resources. Perhaps even more important, given the funding responsibility of ARPA for much of the research work at universities and elsewhere, it would help to avoid duplication of scientific effort.

With these considerations in mind, Taylor now proposed a network development project. His timing and presentation must have been appropriate because it took just 20 minutes to persuade the Director of ARPA, Charles Herzfeld, and the money was allocated.

OTHER CONTRIBUTORS — RAND AND NPL

While ARPA represented the womb in which a new communications network was about to be created, certain elements were required from elsewhere to achieve a fully viable system. One of these elements was an idea that came from Paul Baran who was doing research at the RAND Corporation, a defence-related organisation owned by the US government. This project had also been concerned with the vulnerability of communications links across the continent. The AT&T cables were seen as intrinsically highly vulnerable to attack, and Baran had conceived of a network of nodes and wired links as a more robust arrangement. The intrinsic redundancy of the linkages in the 'distributed network' that Baran suggested in a research paper of 1965 would ensure that, if any particular link was cut, several other paths were available for data to travel along. In addition, by breaking up the data into 'message blocks', alternative paths could be followed, depending on availability. By this means, the integrity of messages would be assured as they travelled to their intended destinations, irrespective of the route that the component parts of the message had taken to reach them.

The Baran model of a distributed network depended on the notion of a series of unattended and self-contained switching nodes through which the data would be routed to its destination. Unfortunately, due to the intransigence of AT&T, this project was suspended in 1965 but came to the attention of Taylor at IPTO just in time to influence the ultimate form of the new network in the use of packet switching.

While Baran's work was being undertaken at RAND in the United States, in the United Kingdom a similar project had been conceived of by Donald Davies of the National Physics Laboratory (NPL). Davies had earlier been part of the team established after the Second World War to exploit the technology of electronic computers. This had been the group in which Alan Turing had worked and at which the Pilot ACE computer had been developed.

Despite the very different situation from which the work proceeded — the British Isles' proximity to Europe and, by definition, the Soviet Union — a remarkably similar framework for communications had been arrived at by Davies. As with Baran's network, Davies had suggested an element that he called a 'data packet', one line of text long and travelling at the same transmission rate. However, compared with the response of AT&T, Davies had received not only cooperation but enthusiasm from British Telecom. It seemed that this organisation was keen to see its infrastructure of cables used more efficiently, and digital methods and packet switching were perceived as the future of communications. In this context, Davies is credited with developing the notion of 'packet switching' as an activity distinct from the operations of particular 'host' computers.

ARPANET

A project director was necessary to make the ARPA network a reality and for this purpose Taylor approached a computer academic, Larry Roberts, who was working at the Lincoln Laboratory at MIT. After some time and some quite Machiavellian manoeuvres, Taylor achieved his goal and Roberts joined ARPA at the tender age of 29.

As the idea of a network of linked computers developed under the supervision of Robert Taylor and the control of Larry Roberts, the method of switching and guiding the data packets was considered and refined. This soon led to the idea of a special-purpose computer to carry out this task, physically separate from the computers needing to access the network. These machines were called interface message processors (IMP).

In July 1968 a specification of the IMP was sent to 140 potential contractors with the result that the biggest organisations refused to tender for the work. The contract was ultimately awarded to Bolt, Beranek and Newman, who had based their proposal on a Honeywell DDP 516. This machine was one of the new minicomputers using integrated circuits and its triumphant incorporation in the initial version of the Internet, ARPANET as it was to be called, probably also marks the start of IBM's slide into economic disaster with its first serious failure to grasp the significance of smaller, more 'user-friendly' computers.

At Bolt, Beranek and Newman, the responsibility for implementing the contract that had been awarded to the firm by ARPA was put into the hands of Frank Heart who had been hired in 1966. He had learned about computers from visiting professor Gordon Welchman, who had been intimately involved in the development of the Colossus computer at Bletchley Park. Subsequently, Heart had worked on the Whirlwind computer, which later had led to the SAGE project, described earlier. He had then moved to MIT to the newly created Lincoln Laboratory. Given a capacity for assembling talented and productive project teams, it seems that Heart was ideally suited to creating both the machines and the software needed to make the IMP and ARPANET a reality.

Part of Heart's responsibility was to ensure that the network and the passage of data were intrinsically robust and capable of adapting to removal of linkages for whatever reason. In part this was achieved by the use of hardware to undertake 'checksum' integrity checks and to slow the transfer of data to a minimum. In the interests of reliability of the checking process, a 24-bit checksum was used. In addition, in order to carry out maintenance in the event of hardware or software failure, the concept of 'remote diagnostics and system checking' was also developed. This allowed the consultants and contractors, Bolt, Beranek and Newman, to carry out a supervisory and maintenance function from their offices rather than having field maintenance crews travelling around over enormous distances to rectify faults. In the same context, to cope with failures of the power supply, the 'watchdog timer' was introduced to allow automatic restart of computers where such a situation had occurred.

Given its military role as the switching element of a defensive network resistant to nuclear attack, appropriately the DDP 516 used by Bolt, Beranek and Newman was built like a battleship, weighing in at just on a tonne and painted dark grey. Costing US$80 000 in 1967 dollars, this 'minicomputer' was about the size of a large refrigerator. Despite this gargantuan size (by modern standards), it had a mere 12 kilobytes of main memory, although this was in the form of non-volatile 'magnetic core' memory. The machine had a data capacity of 6000 words and this allowed it to be programmed in 'assembly' language, the forbidding predecessor of modern computer languages although a slight improvement over machine code. Ultimately, although the Honeywell machine had its own internal editor software, the assembly code was produced on a PDP-1 owned by Bolt, Beranek and Newman and was typed in with a Type 33 teletype terminal. Later, assembler code so created was fed into the Honeywell using punched paper tape.

The IMPs constructed by the Bolt, Beranek and Newman team were intended to be connected to the AT&T cable network via conventional 600-ohm twisted pair going from the machine to the local exchange. Data travelling at 50 kilobits per second flowed to and from the IMPs and the local telephone exchange. From the IMPs the data then flowed to and from the associated host computers. This led to the need to separately specify the form of protocol for moving the data between the IMPs and the host computers. This task was carried out externally to Bolt, Beranek and Newman by a committee of academics from the various intended user institutions and was based on a detailed specification produced by Robert Kahn of the consultant team. This led to the production of the first network control protocol (NCP) in 1970 and in turn to the first file transfer protocol (FTP) in 1972. Included in these protocols was the principle of 'adaptive routing' that would allow a data packet to move through the network by alternative paths, depending on availability.

Once initiated, the ARPANET quickly escalated and, from four connected computers at major universities in 1969, by 1972 the number of connected machines had jumped to 24. This increase in connections has followed an exponential growth path ever since.

The file transfer protocol, FTP, which was established in 1972 together with Telnet, allowed major differences of hardware in host computers to be accommodated. By the addition of some supplementary routines, FTP also became the basis for the first boom in 'e-mail'. Developed by Ray Tomlinson, the programs known as SNDMSG and READMAIL became immensely popular. Tomlinson is also notable for having introduced the @ symbol to act as a separator in addressing data. Within only a year of the introduction of FTP, 75 per cent of ARPANET traffic involved e-mail, a rate of growth that continues unabated to the present time.

INTERNET

Initially, ARPANET was exclusively land-based but it was soon realised that a radio network for mobile operations should be accommodated and, later, the idea of international linking via commercial satellite was to be catered for. It was realised that the protocols to deal with this form of traffic would require more elaborate specification and, in order to develop the necessary software, Vinton Cerf of Stanford University was employed. Under his supervision and guidance, the transmission control protocol and internet protocol were developed and are now conventionally referred to as TCP/IP. This work appeared in a paper defining the new approach in 1974 but took some time to achieve universal application.

Ultimately, the whole of the ARPANET and associated radio and satellite networks were converted to TCP/IP in January of 1983 and this marks the beginnings of the Internet as it now operates with its trans-world characteristics. At this same time, another extremely important development was the concept of the 'domain name'. This label allowed any particular packet to travel to its destination via directions from a number of intermediate nodes and without the ultimate address being required to be known at such intermediate points.

NFS NET

The ARPANET had been instituted with a hard-wired 'backbone' of cable links which allowed a data transmission rate of 56 kilobits per second. By 1984 it was apparent that this was inadequate to cater for the burgeoning traffic on the net. Beyond that, ARPANET was also a very expensive system to institute for the smaller and less well endowed academic establishments. However, despite these problems, there was no doubt that it was well on the way to becoming indispensable for the interchange of ideas between universities, as the use of e-mail continued to expand ever more rapidly. With the Internet connection to overseas centres, the pressure for greater data capacity once again became painfully apparent.

In 1984 the National Foundation of Science (NFS) introduced a new high-speed 'backbone' allowing a data transfer rate of 1.5 megabits per second and this was operational by 1988. Shortly after this, the number of computers connected to the new backbone was 170 but within another 5 years this had grown to 4500 machines. Once again, data flow demanded an upgrading and at this stage the NFS Net was converted to a 45-megabit per second 'backbone' capacity. Apart from the data flow capability of this new system, it was also attractive financially, costing about a fifth of the cost of setting up an ARPANET site, which was US$100 000 in late 1970s dollars.

By 1995, the problem of running a network had become such as to convince the US government that it should remove itself from the communications business altogether. Accordingly, direct access to the NSF Net was terminated in 1995 and four commercial, private sector contractors were appointed to administer connections. With this commercialisation of the US component of the Internet, advertising became possible and this, in turn, led to a major invasion by commercial interests in the last few years.

THE INTERNET AND THE WORLD WIDE WEB

Based on the new, high-speed NFS Net, Internet traffic continued to expand at an incredible rate. A particular factor in sustaining the rapid growth and in cementing the use of TCP/IP was the introduction of the SUN Microsystems range of workstations. These machines came equipped with Berkley UNIX and TCP/IP and created an enormous demand. This was further expanded as Ethernet, developed by Bob Metcalfe at Xerox PARC in 1973, became commercially available. This allowed complete local area networks (LAN) to be connected to NSF Net directly and from there to the older ARPANET.

Only a few elements were now lacking to distinguish what was available then from what is available now. One of these elements was a more elegant method of sending e-mail than the previous attachment to the FTP. In the Simple Mail Transfer Protocol (SMTP) this was provided.

However, the counterpoint to this new method of control was the need for a more specific method of addressing and this was achieved by the use of the new 'domain name system' (DNS). This embodied a hierarchical or branching arrangement so that the now familiar address endings constituted the limbs of a vast tree of addresses. The familiar *edu*, *com*, *gov*, *net*, *org* and *int* will be expanded to respond to the massive growth of the Internet in only the last few years.

So superior was the new system in terms of access and speed that, progressively, the older ARPANET 'backbone' was rendered redundant and finally in 1990 it was shut down. However, such an apparently serious event occurred with scarcely a ripple as, by this time, alternative links were available via the NFS Net. ARPANET had served it purpose and quietly passed into history as the battleship grey IMPs were progressively turned off and the slow links severed.

In 1990 an element was added to the Internet that was to radically chage its capabilities and convert it to the World Wide Web as it is now known. This was a capacity to transfer graphical and multimedia information across the Internet as digital packets of data. To the text-based information exchange of the Internet has now been added audio and video data transfer with the promise of television and high-fidelity stereo sound distribution in the future, limited only by the capacity of public and private links.

As a tribute to the growing international nature of the Internet by this time, the necessary work was carried out in Switzerland at the CERN physics laboratory in Geneva. The work of a team led by an unassuming and modest researcher, Tim Berners-Less, this new capability had undoubtedly propelled the Internet to world prominence in just a few years. This was soon complemented by the development of the first successful Web browser, Mosaic, later to be better known as Netscape, and, at last, the public had access to the full resources of the Internet. Microcomputers had been the other essential component that made all this possible, as the power of universal access to the data and information resources of the World Wide Web was unleashed to an eager public.

CHAPTER

16

1927 to the present

Communications on the move

For many years after 1927, the only way to send a voice across the oceans of the world and between the continents was by radio. Such a connection evidently fuelled a demand for personal contact that could not be met by the undersea cables of the period but, ultimately, with changes in technology, this was catered for in 1957 as earlier described.

Apart from the reliance on radio for long-distance intercontinental communications, voice contact between persons on the move away from the fixed links of wire and cable also had to be provided for through the medium of radio. The development of the radio telephone became a valuable feature of shipboard life, particularly for more affluent passengers, as the use of the short waves increased. In the remoteness of the inland of major continents and away from the reach of the conventional telephone network, radio also became the basis of voice contacts.

In Australia in the 'Outback', the system introduced by Flynn and Traeger was developed from something reliant on morse code communications to one in which voice communications were available. With

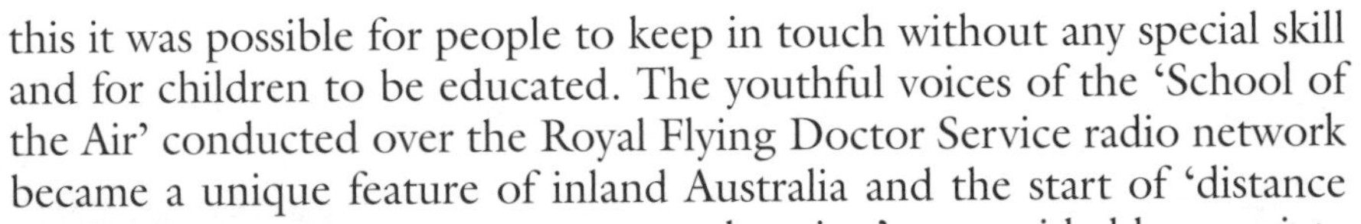

Portable radio communications used by the Vienna Fire Brigade in the 1930s *(WIT)*

this it was possible for people to keep in touch without any special skill and for children to be educated. The youthful voices of the 'School of the Air' conducted over the Royal Flying Doctor Service radio network became a unique feature of inland Australia and the start of 'distance education' as provided by a variety of educational establishments since that time.

One other technical change that has occurred since the Second World War has been the introduction of single sideband (SSB) operations at high frequencies. Although generally confined to specialised communications and particularly to the operations of radio amateurs, this system has allowed a significant increase in available electromagnetic spectrum space for users. With this system, a voice link can be comfortably accommodated in 3.5 kilohertz. By comparison, using conventional double-sideband, amplitude-modulated signals, the radio broadcasters on both the medium waves and at higher frequencies occupy 9 or 10 kilohertz.

No doubt the reluctance to embrace SSB by the broadcasters is a function of the greater complexity of SSB radios but also relates to the problem of providing a reasonable level of fidelity for music in particular. Where the requirement for communication is purely to achieve good voice contact, SSB is more than adequate. However, providing for music is a different problem and unlikely to be solved in the short term. Beyond this, as satellite broadcasting of both voice and television becomes increasingly common, it is probable that the use of the short-wave bands will decline somewhat. In addition, the introduction of digital broadcasting in the ultra high frequency (UHF) bands will undoubtedly have an impact on the use of both the medium wave bands and the short waves.

Despite the susceptibility of the short waves to severe fading and other forms of distortion and interference, for the less developed parts of the world where cost of the receivers is a serious consideration, this mode of broadcasting remains extremely attractive. For this reason, it is anticipated that the short waves are unlikely to be abandoned for a long time.

Apart from the continued use of the high-frequency (HF) bands by commercial and international interests, it is interesting that there has been a major renewal of interest in this part of the radio frequency spectrum by the military forces in developed nations. This is very much a reflection of concern with the vulnerability of satellites to attack, particularly in the case of the geosynchronous variety, which sit at a fixed location in space.

MILITARY AND CIVILIAN COMMUNICATIONS

No doubt fostered by the needs of the military and particularly through advances during the Second World War, the concept of portable radio communications developed over the period from 1927 to the 1970s. In the war years, the large, heavy, clumsy backpacks discussed in an earlier chapter were replaced with progressively refined, lightweight, compact apparatus. Even before the advent of transistors in the 1950s, miniature valves had allowed the designers to create apparatus that even by today's standards was of relatively modest proportions.

Typical of this newer, more compact communications equipment from the Second World War was the handy talky which, in one box not much bigger than a contemporary telephone handset, provided communications on the battlefield over quite respectable ranges. Apart from compelling an interest in such new compact technology, the war years taught some very bitter lessons concerning the optimum mode of operation of radio associated with terrain. With the realisation of the fundamental restrictions that applied to high-frequency transmission operating in 'ground wave' mode, the military moved to progressively higher and higher frequencies for short-range communications as required on the battlefield. Frequencies between 40 and 80 megahertz became common and the efficiency of the new short-antenna systems helped to justify the cost of such a major undertaking.

Radio communications on board the *Flying Scotsman*, United Kingdom, in 1937 *(WIT)*

However, two matters were to prompt other major changes in the military use of radio communications. Firstly, the use of 'jamming' as a means of disrupting command and control procedures forced a review of methods. Secondly, the interception of messages became a matter requiring new methods of encryption. In more recent times, the introduction of radio communications that can hop from frequency to frequency at a very rapid rate, made possible by the availability of solid-state technology and microcomputers, has made combating both these problems somewhat easier.

One significant impact of military radio technology was the availability of large amounts of high-grade equipment at the cessation of hostilities in 1945. The surplus market was for a while flooded with apparatus that had cost thousands of dollars to provide to the military services. Surplus to the requirements of the newly reduced fighting forces, this apparatus was eagerly received by the radio amateur community. Unfortunately for the radio amateurs, this was a situation that did not last for very long. As the threat of communism developed in the late 1940s and the Korean War broke out, procurement of new and even more compact apparatus followed.

The availability of the war surplus radio helped supply an army of radio amateurs who had been trained as signallers during the war years. Their interest in the medium of radio and knowledge of portable operations and use of compact mobile apparatus all fed into a latent public desire for access to portable communications.

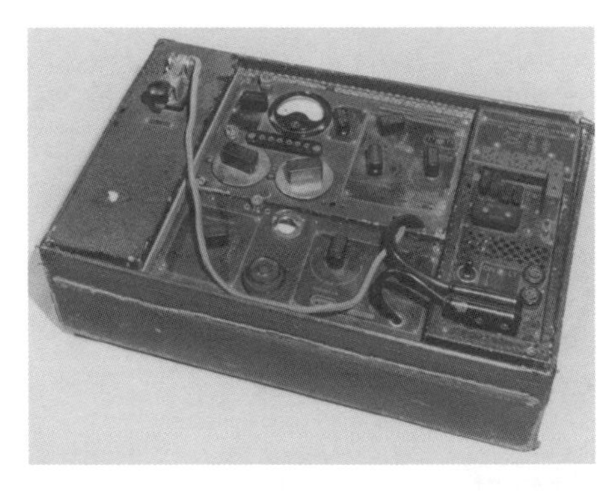

The B2 suitcase clandestine radio of 1941 *(PRJ)*

CLANDESTINE COMMUNICATIONS

Apart from radio communications equipment developed for use on the battlefield, another interesting area of technical effort during the years of the Second World War went into producing radio transmitters and receivers for clandestine purposes. As a reflection of the early success of the German army in defeating the French, and the need of the British Intelligence Services to communicate with the many resistance groups that sprang up to harass the invaders, some special apparatus was necessary to meet the demanding situation this presented.

In the last years before the Second World War, apparatus intended for 'undercover' operations emanating from Britain had been heavy, clumsy and a terrifying liability when considered in the light of a vigilant and violent opponent in the German occupation troops. Typical of this early pre-war period was the Mark XV set made in Britain on the basis of radio amateur techniques of that time. Weighing a formidable 20 kilograms (45 pounds), this set was arranged in two large wooden boxes and would have been extremely conspicuous to carry. Using two tubes, a 6F6 crystal-controlled oscillator stage driving a 6L6 power amplifier tube, the transmitter could produce a signal that could be heard in the United Kingdom from most parts of Europe, depending on the time of day and the frequency used. A single crystal was used to control the frequency of operation and this was changed to set the particular part of the spectrum to be used.

Special Operations Executive (SOE) clandestine transmitter and receiver, Type 21 Mark II *(PRJ)*

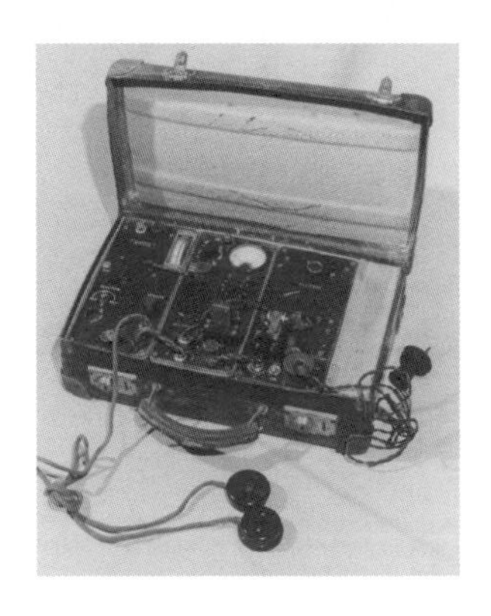

The fundamental deficiencies of this set were soon responded to and replacement sets came from the skilled hands of Polish technicians whom the British Intelligence Service established in a small workshop at Letchworth. Two of the Polish transceivers, combined transmitters and receivers in a common case, were masterpieces of the art of miniaturisation for that time. The BP 3 and the AP 4 represented a significant improvement over the earlier Mark XV, being small enough to be carried around in small suitcases. Relatively light and compact even by contemporary standards, these sets allowed agents in France and even further afield to maintain contact with the British Intelligence Service.

Early in the war, Churchill ordered that a new organisation be created to 'set Europe ablaze'. Initially, this new aggressive force, the Special Operations Executive (SOE), relied for its communication

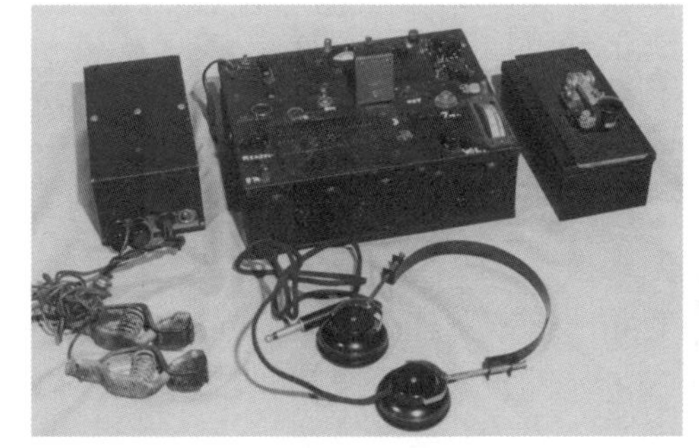

Right Steam-powered electricity generator for use with the Type A Mark III of 1944 *(PRJ)*

Far right The Type A Mark III suitcase clandestine radio transceiver of 1944 with vibrator power supply (from 6-volt car battery) and spares box *(PRJ)*

service on apparatus supplied by the Secret Intelligence Service (SIS). However, a considerable level of interservice rivalry led to SOE demanding that it arrange its own communications facilities and initially this was to be provided by such sources as the Polish technicians. However, the demands for light weight and portability led to the creation of British sets amalgamating the best of Polish and American technology and the production of some outstanding transceivers.

Produced quite late in the war and used in the Far East and in Burma, the Type A Mark III set is a good example of this compact, lightweight apparatus. Compared with pre-war radios, the Type A Mark III weighed just 2.2 kilograms (5 pounds) and was able to produce a very respectable 5 watts of power. It, too, was crystal-controlled but had a receiver that could cover a range of frequencies from 3.2 megahertz to 8 megahertz. This allowed transmission and reception frequencies to be split, making the job of the German direction-finding teams that much harder. In addition, the Type A Mark III was able to be run from a 6-volt power supply via a 'vibrator' power pack. Having a portable power supply allowed transmission to continue even when the mains supply was cut off. Turning off the mains was a standard technique of the Gestapo for localising the source of a British agent's signal, but having a constant source of electricity available neatly foiled the technique.

In addition to the 'vibrator' supply, the Type A Mark III was able to be run from a variety of different generators. These ranged from thermoelectric generators to wind generation and, far and away the most exotic, an alternator generator driven by a miniature steam engine. This could be run on the fire produced with pieces of wood and a steady water supply and therefore was particularly suitable for remote locations such as the jungles of Burma.

The Type A Mark III could be fitted into a small fibre suitcase of the type used by many schoolchildren. For dropping by parachute, the radio set and its vibrator power supply were fitted into two steel boxes with internal felt strips to cushion the shock of landing. A 6-volt battery was, at that time, a standard item of automotive apparatus and therefore readily available in France, even with a German army of occupation.

CLANDESTINE OPERATIONS

Even in its innocuous fibre case, an agent's transceiver was a dreadful liability and its discovery guaranteed that its owner would be put up against a wall and shot. Many brave radio operators suffered precisely this fate at the hands of the German occupying forces. By the beginning of the war, methods of radio direction-finding were very well developed and the German forces used highly efficient and speedy methods of detection of clandestine transmissions. By comparison, the whole process of transmitting coded messages was in the early days a long, tedious and exacting exercise, likely to put the operator in a most hazardous position.

These problems were met by separating transmission and reception times and constantly moving the apparatus to new locations. Further, after the loss of many operators, they too were separated from their radios while in transit.

In the postwar period, technical developments to make the interception of clandestine traffic more difficult included the introduction of 'burst transmissions'. These involved the pre-recording of morse code messages on magnetic tape and their subsequent transmission at high speed to make the task of detection much harder.

Just as important to the improvement in clandestine operations was the invention of the transistor because of its impact on the weight and size of radios. But even more important was the removal of the need for a 'heater' in the valve and the substantial current that it had drawn. Again, this made the provision of power very much simpler than in the valve era, even with the most efficient heaters that drew a fraction of the current of earlier types.

As for the military, the erection of efficient antennas for clandestine operations presented particular problems. Not only could the antenna usually not be erected externally or high up and clear of blanketing objects, but usually low power was an inevitable attribute of the apparatus. Conversely, the use of continuous wave morse code transmissions made it possible to achieve trans-continental ranges with only a few watts of power. Apart from this, compared with the use of voice communications, encoded letters sent in morse code represented a far more secure and compact method of maintaining contact with controllers in Britain.

LATER DEVELOPMENTS IN CLANDESTINE RADIO

Apart from transmitter–receiver combinations, now usually referred to as transceivers, there was a major need for access to news in Europe after the occupation by the German army. A large number of home-made radios were constructed by enterprising technical people on the Continent, but obtaining suitable parts was always problematical and the results could be less than useful where simple regenerative circuits had to be relied on. Apart from anything else, the capacity of the regenerative straight set to radiate a signal if improperly operated could have disastrous results where the occupying forces were listening for anything of this sort.

The answer to this need was a well-built superheterodyne radio and this was produced in the latter part of the war for use by the partisans in France and the Maquis. In 1944, as a prelude to the invasion of France on D-day, 6 June, many messages were passed over the conventional overseas service of the BBC, embedded in routine announcements. For the reception of this type of traffic, a new compact radio was developed. Known as the MCR 1, this radio used newly available miniature valves, 1T4, 1S4 and 3V4, and came complete with a miniature mains power supply. It was usually supplied in a metal box used for biscuits and was therefore generally known as the 'biscuit radio'. The MCR 1 provided an admirable response to the needs of the French Resistance forces, since it was quite small and, with plug-in coil boxes, able to cover the whole of the broadcast band and also most of the short waves. In the prelude to D-day and in the weeks that followed, immense damage was done to supply systems of the retreating German army by the efforts of determined French patriots supplied and directed from Britain using the services of the BBC and relayed by the new small receivers.

Although transistors undoubtedly revolutionised the form of radio communications apparatus after 1947, there was a limit to miniaturisation set by other components. Tuning capacitors and coils, essential to operation on the high frequencies at that time, could not be made substantially more compact. For this reason, the gain with introduction of transistors was, as much as anything, in the reduction of the power demanded compared with valve sets.

An interesting example of this early transistorised era can be seen in the clandestine transceiver of the American firm Delco, produced in the early 1960s for the use of the CIA and subsequently sold to Australia for use in jungle environments such as Malaysia and Vietnam. Through the inclusion of the single, multipart battery within the casing of the radio, a relatively compact package was created, with the considerable advantage that it was almost completely watertight and could be immersed in water for a considerable period or buried in wet ground — all very useful attributes for apparatus intended for clandestine use and concealment from the watchful eyes of the enemy.

Top The 'biscuit tin' receiver, the MCR 1, for support of the French Resistance movement, using the new miniature battery-powered valves *(PRJ)*

Middle A secretly built transceiver made by the Danish resistance, located at the Resistance Fighters Museum in Copenhagen *(PRJ)*

Bottom A fully transistorised clandestine transceiver made by Delco in the United States from the early 1960s *(PRJ)*

PORTABLE RADIO

In the period immediately after the Second World War, the availability of a new generation of miniature all-glass valves allowed radio manufacturers to create small and compact portable radios for the consumer market. During this period, a number of radios of about the same size as a hardcover book were produced and the influence of military and clandestine apparatus of the war years was quite clear.

Examples of such miniaturised valve radios came from a variety of sources and the photographs in this chapter give a good impression of what was available. The one major problem of these valve radios remained the need for a high-current filament to be heated up so that the electron emission in the vacuum could occur. The large proportion of the space in these early compact valve portables was taken up with two batteries, one to supply the high tension for the anode connection and the other to light up the heater or cathode. Batteries were both expensive and short-lived in these radios and their use was very much confined to those with deep pockets and a predilection for expensive toys.

The discovery of transistors in 1947 fundamentally changed the situation and, as earlier described, the Regency Mark 1 created a demand for what would soon become a completely pervasive new device, the portable transistor radio. As a particular product of the electronic growth of Japan in the postwar period, the 'trannie' was to become the

Above left An HMV valve portable from about 1947 *(PRJ)*

Above middle The Marconiphone valve portable from 1948, Model P20B *(PRJ)*

Above right Transistor portable from 1960, the Little Nipper *(PRJ)*

preferred accoutrement of any young 'swinger', only to be replaced years later by the ubiquitous 'Walkman'. Significantly, both the original Japanese transistor radio and the 'Walkman' came from the same source, the Sony Corporation, which has continued to produce state-of-the-art radio and electronic equipment.

By the end of the 1960s, the pervasive nature of the transistor radio had seen a huge decline in the cost of the device. At this stage, more sophisticated radio apparatus began to appear, designed in the days of valves but now rendered practical by the reduced cost of transistors as economies of scale took hold. In the 1970s the availability of the Wadley loop design allowed the creation of equipment with a completely unprecedented level of performance and accuracy of frequency selection. At this time, once again, Japanese design was applied, in particular to apparatus designed for the radio amateur. However, for the general public, the epitome of this design approach was to be seen in the Wadley loop circuit as applied in a South African designed Barlow Wadley portable radio.

Unfortunately, the Barlow Wadley portable was significantly less easy to use than contemporary transistor portables, particularly on the ordinary broadcast band, and it was far less popular than it deserved to be. However, as a short-wave radio, its performance was, and remains, formidable, even compared with the best of today's synthesised receivers. For those in the know, it is a radio well worth obtaining.

The next major change to consumer portable radios could be seen as associated with the introduction of the integrated circuit. This device with its multiple transistor stages allowed frequency synthesis to be provided quite cheaply as a substitute for the Wadley loop method of frequency determination. In the 1980s, again the formidable Sony Corporation developed a number of remarkable short-wave portable radios, initially in the form of the ICF 2001 and more recently in the far smaller ICF 7600D. Despite the high performance of these two radios, the early solid-state devices were rather 'hungry' in relation to power supplies. Even using alkaline batteries, these radios would not allow such a power source to last for very long and a mains power supply was, in reality, a necessity for sustained usage.

In recent years, new radios from Sony, Sangean and Grundig have incorporated the same form of frequency synthesis as pioneered in the earlier consumer radios by Sony but with low-voltage, low-current-

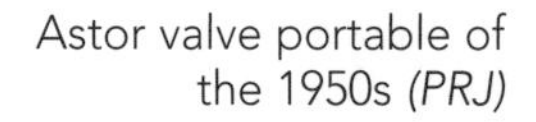

Astor valve portable of the 1950s *(PRJ)*

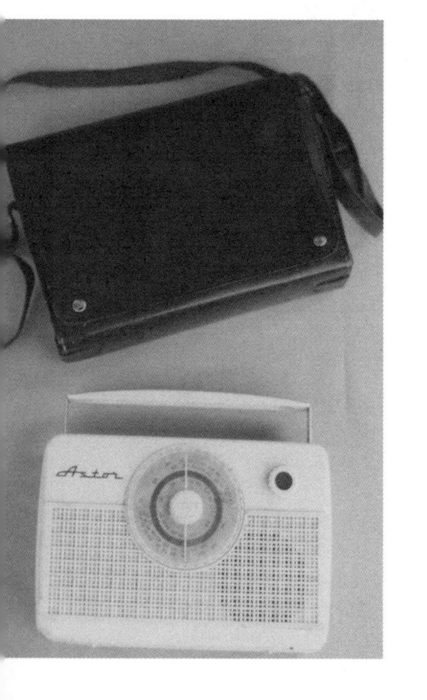

drain devices. In addition, the emergence of surface mount, ultra-small supplementary components has allowed even smaller radios to be produced with enhanced features. However, a feature that is usually absent in all the more recent radios is a capacity to resolve single-sideband transmissions. No doubt this is because of the limited level of interest exhibited by the general public and the degree of skill required to resolve SSB in any case.

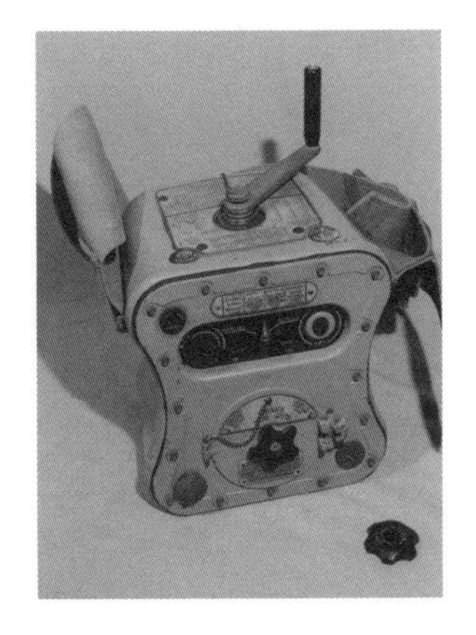

The 'Gibson Girl' emergency transmitter of 1944 made in the United States *(PRJ)*

EMERGENCY PORTABLE RADIO

During the years of the Second World War, many pilots who had been forced to parachute into the sea or leave a sinking aircraft were rescued because of the availability of a special form of portable transmitter. Originally produced for use by the German airforce, the Luftwaffe, the portable emergency transmitter was copied and improved for both the Royal Airforce and the United States Airforce.

The US version of this device, known as the 'Gibson Girl', was made in large numbers and used for a number of years after 1945. The name was associated with the wasp-waisted form of this radio, related to the shape of fashionably dressed young ladies of the Victorian era and the fashion designer, Dana Gibson. Perhaps unsurprisingly, the British and German devices had a far less flamboyant shape and therefore did not attract comparable affectionate 'nicknames'.

The particular attraction of these emergency devices was that they did not need any skill to operate. All that was required was the intelligence to inflate a hydrogen balloon with the generator provided. This device produced hydrogen by decomposition of sea water and involved a tin can filled with chemicals which was attached to an aluminium tube. This, in turn, was coupled to a rubber balloon to which the antenna was then attached. All these items came in a relatively compact emergency kit which included the radio transmitter. Turning a crank handle would produce a cry for help in morse code which was automatically produced within the 'Gibson Girl'.

The 'Gibson Girl' was progressively replaced over 20 years by more modern emergency radio transmitters. Initially these were restricted to operations on the marine emergency frequency of 500 kilohertz but, in subsequent versions, this was expanded to include a range of frequencies in the short-wave band, and therefore such radios were more

Below left The Barlow Wadley XCR 30, full-range high-frequency portable communications receiver using the 'Wadley loop' drift-cancelling circuit *(PRJ)*

Below middle The Sony 2001 digital display, key-pad-operated transistor radio able to resolve CW and SSB on the HF bands *(PRJ)*

Below right The Sony 7600D digital display and SSB-capable transistor radio with full HF coverage and also FM broadcast radio *(PRJ)*

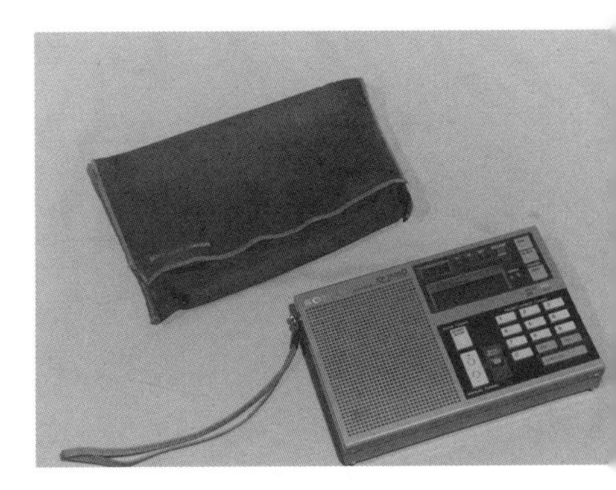

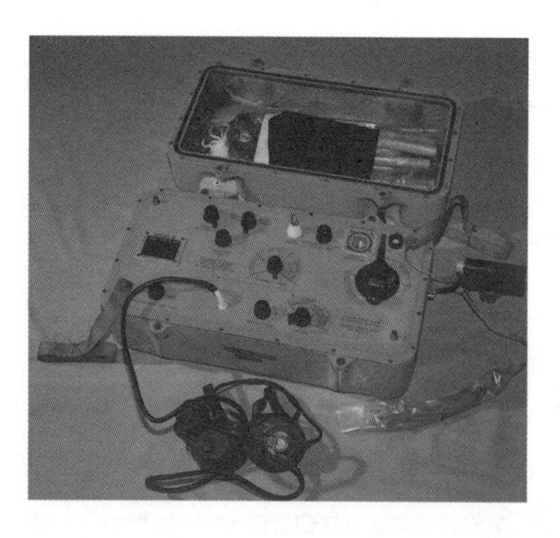

Right Successor to the 'Gibson Girl' — automatic morse code sender sends SOS on high-frequency band with a hand-powered generator *(PRJ)*

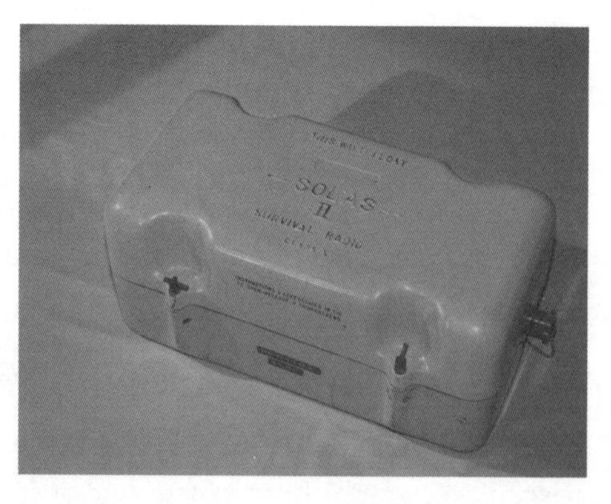

Far right The 'Solas' (safety of life at sea) emergency transmitter/receiver sealed up for storage — in this condition, the equipment will float *(PRJ)*

suitable for low power and long range. This later generation of emergency radio was somewhat larger than the 'Gibson Girl', and featured two crank handles and a plastic rather than aluminium casing. This device also allowed voice contact to be made, once the rescuers were close enough to receive the signal.

In more recent times, much of the radio apparatus of long and short waves has been completely superseded by the introduction of the IMARSAT system of marine communications through access to satellite. This is generally comparable with the INTELSAT international system for transmitting and receiving television and telecommunications. Even more recently, this has been supplemented by access to the global positioning system (GPS), a 'spin-off' from the American strategic system of ballistic missile guidance used to such devastating effect in the Gulf War.

PERSONAL COMMUNICATIONS

Demand by various commercial organisations such as taxi companies and couriers saw the introduction of VHF mobile radio (initially the exclusive domain of civil authorities such as the police, fire brigade and ambulance) in vehicles in the postwar period. At first in the low VHF range at above 40 megahertz, high demand and crowding of the allocation saw a shift of new services to above 150 megahertz and in recent times to the low UHF range at above 400 megahertz.

The early use of single-channel 'simplex' operation, in which only one end of the radio link could transmit at a time, was soon replaced by 'half duplex', with the use of different transmit and receive frequencies. In this latter mode, the use of a base station to control the operations of a number of mobile stations became common.

Sony 'Walkman' personal tape recorder and FM stereo/AM radio *(PRJ)*

In the 1960s, radio amateurs began to explore the use of VHF coupled with small handheld radio transceivers, and for a while this was matched only by military and quasi-military operations such as the police and other civilian services. In the following decade, the availability and demand for citizen-band (CB) radio communications, both personal and mobile in vehicles, developed in the United States but soon swept around the Western world, providing a freedom of access to personal communications that could scarcely have been imagined in 1945.

What was peculiar in this explosive development of the CB band in the 1970s was the use of a high-frequency allocation at 27 megahertz, given the military experience earlier described. This, coupled with a requirement for low-power operation, was no doubt intended to contain the CB operation to purely a local service. Not surprisingly, regulations to control power level and antenna efficiency were of limited utility and many CBers managed to operate over thousands of miles using 'skip' operation. Many of these same CBers finished up with radio amateur licences at a later stage, so perhaps the infringement of operating requirements had a positive aspect.

In more recent times, a move to provide the CB fraternity with an allocation up in the UHF range at above 440 megahertz has seen an interesting social differentiation of users occur in Australia and perhaps elsewhere. In the country and the 'Outback', it is this higher frequency that has become the 'backbone' of much personal communication where a conventional telephone is not practical. The farmer on his tractor out in the wheatfield now routinely talks to the 'Missus' with a UHF CB. By comparison, in the cities, the HF CB seems to have been taken over by a very motley and variable collection of communicators, some sane and sensible and some quite the converse. One group of users that remains firmly wedded to the HF CB at channel 8 in the Australian system are the long-haul truck drivers who can be heard discussing everything from the occurrence of the next police trap to the state of the weather on any highway on which large articulated trucks are found. In addition, four-wheel-drive addicts also find 27 megaherz a useful and cheap way to keep their convoys together during inland trekking.

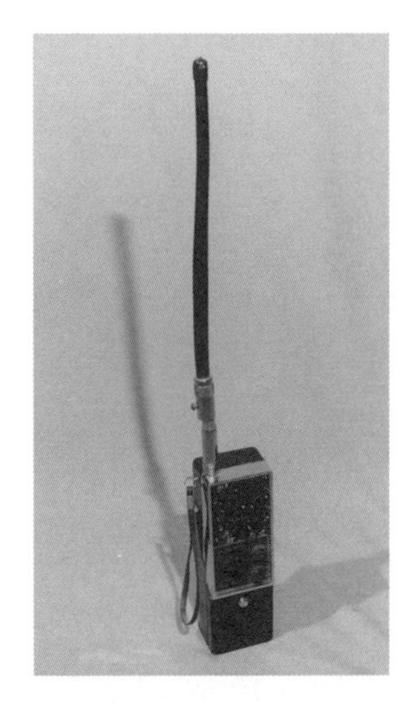

Top Typical handheld citizen-band (CB) transceiver supplied by Dick Smith Pty Ltd in Australia — 27 megahertz AM with three crystal-locked channels, 1978 *(PRJ)*

Bottom ICOM radio amateur 2-metre band VHF FM handheld transceiver of 1980, Type IC 2A *(PRJ)*

As demand has developed from more and more users for access to the radio spectrum for business and other purposes, new and more efficient systems have been introduced. In particular, the introduction of 'trunked' systems, in which a group of users share a number of channels that are made available on a demand basis, has allowed a far greater efficiency of spectrum use. Provided that users are prepared to accept minor delays in accessing the channels, an enormous increase in the number of users able to access one another at any one time can be achieved.

For example, with five alternative channels in operation and at a 5 per cent level of service, 580 mobile users can be accommodated with a mean waiting time of 3.3 seconds. This can be compared with a 'simplex' arrangement in which two users are able to talk on a single channel at any given time. Also very significant for future operations has been the development of a cellular pattern of base stations to handle distributed mobile operators.

CORDLESS TELEPHONE

A parallel development to the use of personal radio transceivers was the introduction of cordless telephones. These devices had a small base transmitter and receiver connected to the conventional wired telephone system and a handset that could communicate with the static base — hence the image of Hollywood starlets floating on 'Li-Los' in kidney-shaped swimming pools, sipping champagne while chattering to distant admirers in a time before the 'cellular' telephone became a technical reality.

Cordless phones usually operated at 1.7 megahertz and 49 megahertz to allow full duplex operation involving natural conversational speech. An inexpensive system and immensely popular, for some time it caused the conventional telephone companies considerable anxiety. However, more importantly, the cordless telephone showed the desirability of a fully portable personal telephone system involving a greater level of security. Justifiable concerns were held regarding security when it was possible to talk to overseas locations via a neighbour's base and hence have his account billed for the exercise.

PORTABLE TELEPHONE

During the late 1970s, 'analog' telephone systems were introduced in various parts of the developed world, based on a cellular configuration of base stations to communicate with personal handsets. Initially, these telephones tended to be relatively cumbersome, contained in a transportable box with the handpiece set in a cradle. However, compared with later and more compact telephones, such analog devices had relatively efficient antennas and a substantial degree of reserve power, making contact in difficult operating conditions remarkably reliable.

Progressively, in the last few years, a digital cellular network based on the European global system of mobile communications (GSM) has been introduced in Australia, with a substantial loss of satisfaction to many former users of analog telephones. Part of this problem appears to do with the penetration of the digital cellular network and its coverage away from the main urban areas. In a country the scale of Australia, it is not surprising that the cellular system tends to be concentrated around the urban centres and along the main road and rail systems. This inevitably leaves large areas inadequately covered by base sites and creates an associated level of irritation by users who previously were able to 'get in' with the old analog sets and now suffer from 'drop outs'.

'Drop outs' remain one of the frustrations of using personal portable telephones but with their minuscule power output this is the price that has to be paid for mobility. There is also concern that because of the proximity of concentrated electromagnetic radiation to the skull of a user there may be a connection between brain tumours and this new manifestation of the modern age. At present, research is somewhat inconclusive in relation to this issue, although some South Australian research tends to suggest that there may indeed be a cause-and-effect relationship.

The cellular portable telephone system constitutes an extension of the conventional wired system with a series of switching centres controlling the mobile network. Through these switching centres, the handling of transfer of moving handsets from one cell to the next, now known as 'hand off', makes continuity of speech comparable with the traditional static telephone. Only occasionally can the spill from adjacent channels and users be heard faintly, and again this is a small problem, given the advantages of instant access to the conventional system.

As users of the personal telephone move around the system, the switching centres interrogate all the mobiles and sense their relative signal strengths. As movement occurs, so instructions are transmitted to change to communications with a new cell, the whole exercise being under the control of a computer system. Without the speed of computerised control, no doubt the complexity of the telephone network could not be made to operate in a fashion that so nearly imitates the reliability of the fixed telephone system.

One side effect of this interrogation of portable telephones in the network, not commonly appreciated, is that the position of any individual can be tracked in the control computer and records kept. Truly, 'big brother is watching' if necessary. Beyond that, until full digital telephones are in use coupled with full encryption of signals, interception of conversations will continue. As has occurred, much to the embarrassment of certain notable individuals, the analog service is able to be intercepted by the keen radio hacker using a 'scanner' type of radio receiver.

HAMS AND HACKERS

Amateur radio transmitter and receiver of 1925 *(PRJ)*

Running through the history of radio communications and the development of the computer is a significant but somewhat underrated character — the technical or interested non-professional 'tinkerer'. Indeed, it has been suggested that Marconi himself could reasonably be seen as the ultimate electronic tinkerer who was able to turn his passion into an industry.

In the field of radio communications, the electronic tinkerer soon became known as a radio amateur or, less respectably, a 'ham'. This latter term has lost the status it originally enjoyed since the appearance of a notorious and unflattering although affectionate portrayal by the British comedian Tony Hancock. As with all really funny and penetrating comedy sketches, Hancock's portrayal was amusing because of the strong element of truth that underlay what he presented.

In more recent times, the stories of Stephen Wozniak and Steven Jobs of Apple fame and of Paul Allen and Bill Gates of Microsoft fame show the impact that computer dabblers have had and the respectable commercial positions that they have achieved from relatively humble beginnings. These days, microcomputer tinkerers are usually referred to as 'hackers', although the term is now more associated with software tinkering than construction of microcomputer hardware.

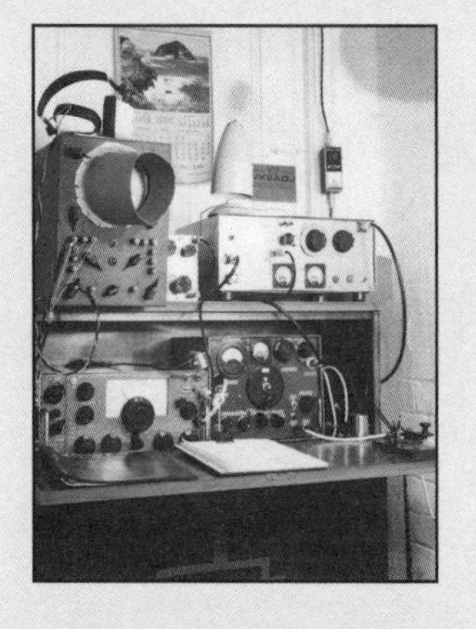

Above The author's first 'homebrew' SSB transmitter and receiver, 1975 — transmitter with transistors and valve output plus valved linear amplifier; receiver all valves with product detector *(PRJ)*

RADIO AMATEURS

Given the length of time that has elapsed since the birth of radio, in many respects it is remarkable that radio amateur activity still exists at all. This is because, with all new technologies, the level of initial involvement and support of enthusiasts seems inevitably to be related to the novelty of the technology. Thus the support of enthusiasts for steam engines, motor cars and comparable modern systems has declined over time as the matter of interest has become commonplace.

The earliest radio amateur organisation appears to have been the Junior Wireless Club of New York City, established in January 1909. This was soon

Right The author, VK2AQJ *(far left)*, with the stalwarts of the amateur broadcast service in New South Wales, VK2KFU, VK2ZTM, VK2EFY and VK2AEJ *(PRJ)*

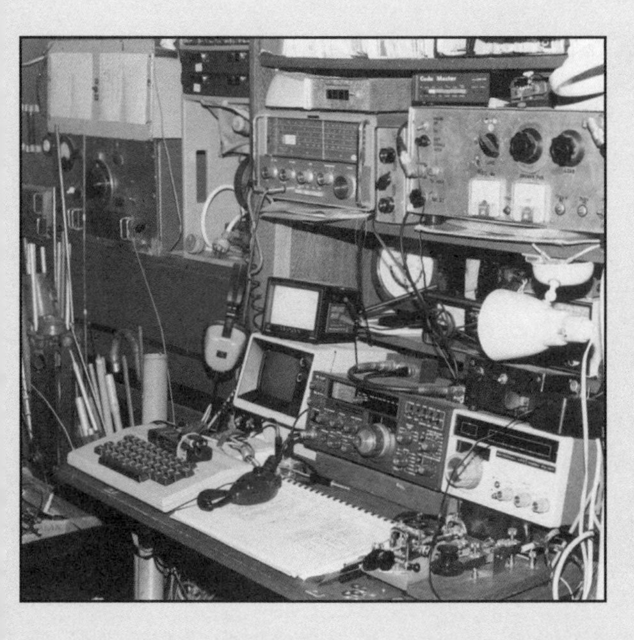

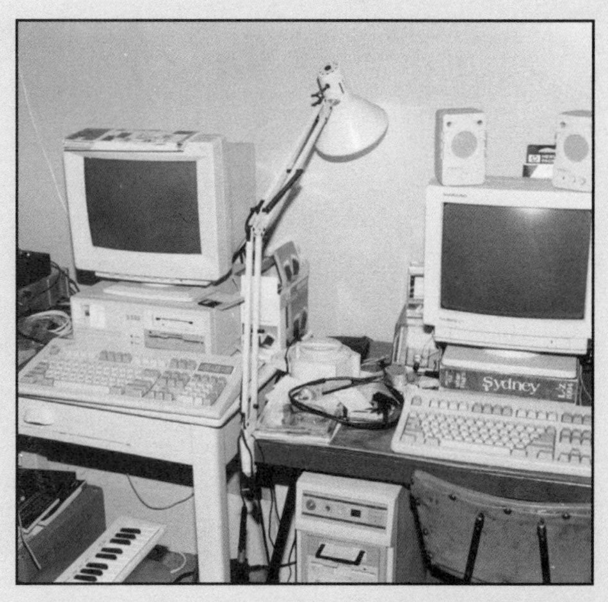

Left The author's all solid-state amateur radio station, VK2AQJ, in 1995 *(PRJ)*

Right The author's computer packet radio station in 1996 plus the workstation using a 486 DX4 of 1995 *(PRJ)*

followed by the institution of the Wireless Association of America, a relatively short-lived organisation. On the other side of the Pacific Ocean, in March 1910, the Wireless Institute of New South Wales was formed with Frank Leverrier, QC, as its first president and with the call sign XEN. The secretary of this new organisation was Walter H Hannam, a participant in Douglas Mawson's expedition to the Antarctic. His call sign at this time was XQI. The Wireless Institute of New South Wales was the genesis of the Wireless Institute of Australia which still remains active today.

On 5 July 1913, the London Wireless Club was formed and became the basis of the current organisation, the Radio Society of Great Britain. In May 1914, as the clouds of war were rolling over Europe, Hiram Percy Maxim and Raymond Tuska set up the American Radio Relay League, which has gone from strength to strength to the present day.

What propelled radio amateurs into a position of almost notoriety was access to the short waves and the application of valve technology which so radically changed the face of communications in the 1920s. The first one-way contact was achieved by an American radio amateur, Paul Godley, who went to Scotland to achieve success. This was followed in November 1923 by the first two-way contact across the Atlantic between the French station of Leon Deloy (call sign 8AB) and KB Warner (1MO) with the traffic manager of the American Radio Relay League and John L Reinartz (1XAM).

Later, the first British station to achieve such a two-way contact was JA Partridge (g2KF) with KB Warner (u1MO), with Leon Deloy (f8AB) also on frequency. This contact was achieved at a frequency of between 108 and 118 metres.

The New Zealand station of Frank D Bell (z4AA), located at Palmerston West, made contact with WB Mangner (u6BPC) in San Pedro, California, in September 1924. This feat was somewhat overshadowed by Frank Bell in a contact with a British station, g2SZ, the School at Mill Hill in London on 29 October 1924 — about as far as a signal can travel without going back towards its source. It was another 7 months before an Australian amateur, Charles D Maclurcan (a2CM), located at Strathfield, managed to make such a contact. This was with EJ Simmonds (g2OD) of Gerrards Cross in Buckinghamshire, north of London.

The home of amateur radio broadcasting in Dural, New South Wales, Australia *(PRJ)*

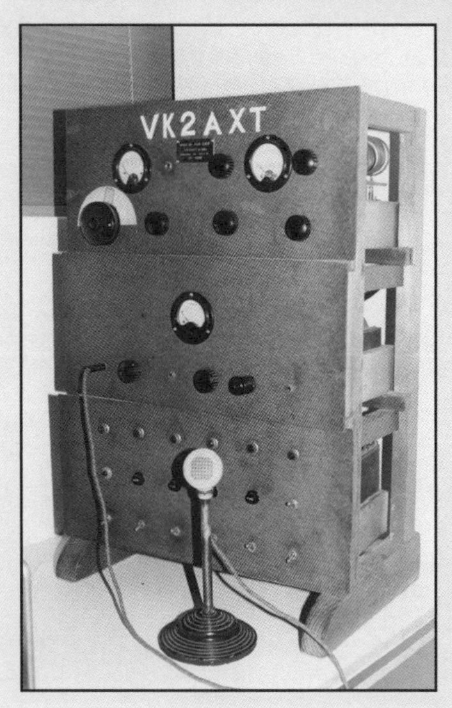

An AM transmitter and receiver constructed by VK2AXT in 1955 *(PRJ)*

The short waves had made long-distance communications a possibility, so that radio amateur traffic now developed into a commonplace activity to mirror the development of commercial long-distance telegraphy. However, access to international communication meant that now the selection of prefixes related to particular countries could no longer be provided on an unregulated basis. Following an international conference, the old lowercase informal prefixes were dropped and a range of new uppercase prefixes was adopted that persist to the present day. In some instances the parentage of the newly allocated call sign prefixes was not hard to detect. Great Britain, for example, was allocated the 'G' prefix and New Zealand received 'ZL'. However, at this time Australia lost its 'a' and received 'VK' instead, and the United States was allocated 'W', which bore little relationship to the earlier 'u'.

Over a period of 100 years, radio has changed from being an extraordinary novel and interesting technical activity to one that is totally pervasive and taken for granted by the general public. In the face of this destructive force of public familiarity, the important factor has been the emergence of new and dramatic developments, all still falling under the general heading of radio communications.

Thus the early interest of individuals in morse code or continuous wave transmission has been expanded into voice communication using amplitude modulation and, later, single sideband. In addition, frequency modulation has been embraced together with moves from the high frequencies up to VHF and UHF since the end of the Second World War. The introduction of data transmission systems such as radio teletype and more recently 'amtor' and 'packet radio' have mirrored developments in the industrial and commercial domain. Beyond this, radio amateurs have moved into television and radio broadcasting in a somewhat limited fashion, determined by the licensing regulations under which amateurs operate.

Radio amateurs have involved themselves in satellite technology although, admittedly, this has been possible only by 'riding on the coat tails' of the commercial organisations which have provided space in their launch vehicles.

Left Amateur television station studio at Gladesville (New South Wales) with main antenna at Gore Hill *(PRJ)*

Right Radio amateurs operate the commemorative station at Cabot Tower, St John's, Newfoundland, in 1996 *(PRJ)*

In many respects, involvement in satellite activity exemplifies the ultimate problem of radio amateurs, 100 years after the commencement of their area of technical interest. Because of the extent of financial resources and the sophistication of the products in most areas of radio technology, those with an interest have had to become reconciled to using commercially developed and produced equipment. It has become simply unfeasible to compete at the private level with the apparatus that the manufacturers are able to develop or build, or achieve a level of performance remotely capable of matching what is on commercial offer. Hence, what is commonly known as the 'black box' syndrome has arisen and this has done much damage to enthusiasm and repelled a generation of young people who might, in other times, have become involved.

COMPUTER HACKERS

At the risk of offending a large group of 'dyed-in-the-wool' traditional radio amateurs, it seems fairly clear that the inheritors of the mantle of electronic experimentation are now generally to be found sitting in front of microcomputers. Since the mid-1970s, when the microcomputer started to become a viable and useful machine, it has rapidly passed through the 'interest group cycle' of novelty followed by reduced technical interest and ultimately public acceptance due to its availability. Again, as with the radio communications cycle, what has kept interest in microcomputers alive are the burgeoning applications and new sophistication of the hardware. This has become most obvious in recent years with availability of the Internet which provides all the novelty and amusement that could have been anticipated in the early days of radio, say after 1920.

Although a good deal of computer construction is still possible and upgrading of hardware is undertaken by many people, it is probably in the area of software that the major hobby developments have occurred. In this context, the development of the 'shareware' concept, with purchase of a licence suggested after trying out the goods, has helped to justify the efforts of individuals. Without such financial support, a whole area of new and interesting applications for various purposes would not have occurred.

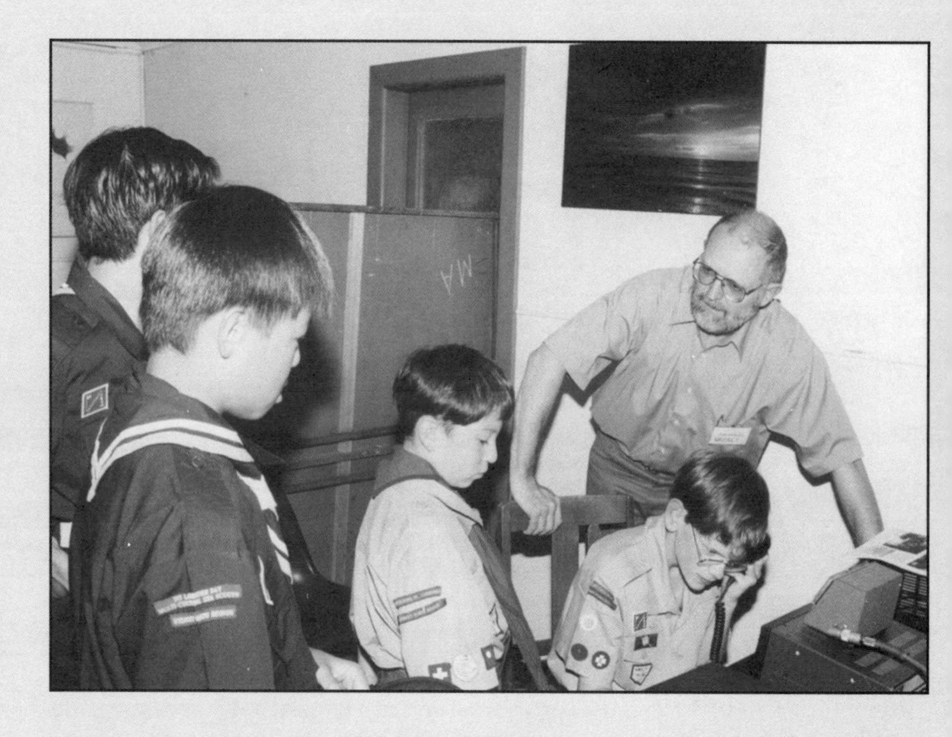

Running the 'Jamboree of the Air' (a Scout Association event) in 1998 — VK2AEJ supervises a party of Cubs and Scouts at the HF rig *(PRJ)*

TECHNOLOGICAL 'TINKERERS'

In terms of the interest of individuals, new technologies go through a predictable cycle of novelty and fascination, acceptance, pervasiveness and total public acceptance. At the start of the process, new users of the technology learn about what can be done through a process of self-help and mutual teaching. They may also be involved in developing new applications of the technology which are absorbed into the mainstream of development.

Ultimately, as the new technology reaches maturity, the manufacturing of it passes into the domain of major industrial organisations. From then on, changes tend to be related to detail and sophistication rather than to basic principle, and result from high-level and expensive industrial research. At this time, the enthusiastic 'tinkerer' generally moves on to a new area of interest.

The face of young Australia — Scouts and Guides operate an Internet communications station at 'Jamboree on the Internet' in 1998 *(PRJ)*

Looking back, it is apparent that the enthusiastic technological 'tinkerer' has had a highly beneficial impact on the development of the communications revolution that has ultimately resulted in the Internet and the World Wide Web. Given that, it seems imperative that whatever can be done to provide the basis of involvement of young people in technological developments can only be beneficial in the longer term.

In the arguments that are starting to gain some credibility with central government, based on 'economic rationalism' and exemplified by the sale of radio frequency spectrum, it is essential that the longer term advantages of maintaining space for those who like to experiment should continue. Equally, the benefits of exposing children to computers at an early age seem quite unarguable. Those who succeed in the future will do so on the basis of brain power and a mature understanding of the power and application of computer technology.

CHAPTER
17

1975 to the present

Microcomputer revolution

Shortly after the discovery of wireless communication in the late 1890s, a new breed of electronic dabbler and experimenter came into being, anxious to exploit the marvellous new medium of radio. Later known as radio amateurs, these inveterate experimenters treated the advancing fields of electronic and radio science as a sort of adult playground but involving a very serious form of entertainment — the exploration of the unknown.

As the technology changed so, too, did the horizons of the radio amateurs and, with the advance from radio to television and then to other modes of communication, there too were the experimenters, looking to exploit the novel and exciting. Even a cursory glance at the history of radio telecommunications and electronics indicates the extent to which inspired amateurs have contributed to the development of the art.

With the advent of computers, a new generation of electronic enthusiasts grew up and started to explore the new technology. In this they shared with the previous generation the same motivation of technical curiosity that had applied during the early development of radio. Now many had been exposed to the power of the mainframe computer. Perhaps there was an opportunity for these enthusiasts to have access to such capabilities on a restricted basis and with a much smaller machine.

Prior to 1975, although there had been a number of attempts to produce a miniature type of computer, what provided the impetus for the amateurs to be involved in small computers was the availability of a new low-cost machine in kit form.

Of the earlier attempts to reduce the size of computers, perhaps most ironic was a portable model produced by IBM in 1973 which the engineers had called SCAMP. Inertia and lack of foresight saw this project abandoned, to be revived nearly 10 years later. However, at this time, IBM was effectively forced to respond to the developments that had been initiated by the amateur fraternity rather than determining itself how things should be.

THE ALTAIR 8800

In 1975, Micro Instrumentation and Telemetry Systems (MITS) was a financially 'challenged', small-scale electronics company, located in New Mexico and seriously in need of a new, 'breakthrough' product. The head of the firm, Ed Roberts, decided that there was scope to produce a small-scale, economical hobby computer. This machine was aimed squarely at all the bright young men who longed to get access to the real thing but lacked the resources or availability of access rights. Roberts approached Intel and the subsequent negotiation allowed him to purchase the new 8080 microprocessor for $75 compared with the conventional $360 that Intel was charging for this device on the open market.

The machine that Roberts designed at that time can be seen now as remarkably primitive and lacking in utility. A review of the inadequacy of its specification is likely to amuse even the most ignorant computer user of the present day. With an 8-bit Intel 8080 central processing unit and running at a clock speed of 2 megahertz, the machine had a memory capacity of 256 bytes (a quarter of a page of text). Data entry and program control were via toggle switches on the front panel and output was to 36 light-emitting diodes which displayed answers in binary code. The machine had no capacity to store data but what no doubt excited the imagination of a huge number of amateurs was its design. Incorporating a 16-slot 'backplane', it could be seen as allowing the hobby user to add boards to the basic machine and thereby enhance its repertoire of functions.

The machine that was described in *Popular Electronics* in January 1975 and named the 'Altair 8800' was an instant and huge success. As a result, MITS was inundated with orders for the new device. It cost the substantial sum of US$397 in 1975 dollars and involved quite a lot of spending money for the average electronic tinkerer of that period. It also required assembly and this, in itself, depended on the skill of the enthusiast if a working machine was to be achieved. Despite this, thousands were ordered, to the extent that production capacity of MITS was severely overstretched, and this was a problem that ultimately led to the failure of Roberts' business.

MICRO-SOFT

Even in the capable hands of an enthusiastic army of electronic amateurs, the Altair remained a relatively limited device but its repercussions were profound. In order to make it more useful, a couple of young enthusiasts developed a compact version of an already popular academic training language to operate on it. This version of BASIC was intended to be loaded onto the Altair using a punched paper tape and, despite the practical problems of such a system, it did allow the machine to be used for real work, if of rather limited scope. The young enthusiasts were Paul Allen and Bill Gates III. Their firm, Micro-Soft, was to grow over the years into the firm with which almost every microcomputer user has a 'love–hate' relationship. Sometime later the hyphen was dropped and the firm became Microsoft.

COMPUTER PROGRAM FOR MICROCOMPUTERS (CP/M)

Another software development from this period was the emergence of CP/M or the computer program for microcomputers, designed by Gary Kildall of Digital Research. This disk operating system was specifically designed to allow the 8080 CPU of Intel to be used in a microcomputer environment until replaced by an operating system known initially as Q-DOS. For some years this operating system was the industry standard. Many of its features were later transferred into the disk operating system that would become standard in the following decade, MS-DOS.

THE APPLE

Apart from such new ventures of a commercial type, what also occurred in this period was the sudden growth of computer user groups. To one of these groups, located in the San Francisco Bay area, came two young men, Stephen Wozniak and Steven Jobs. Wozniak, or 'Woz' as he was and remains known, was a radio amateur and an inspired tinkerer, whereas Jobs was always the aspiring business tycoon. Their construction and business enterprise, which created a new microcomputer they called the 'Apple', was the starting point of the next serious revolution in electronics — access by the general public to computer resources.

This machine was soon followed by a new and enhanced design which was designated the Apple II and had a keyboard for data entry with BASIC contained in read-only memory (ROM). Coupled to a colour monitor, this machine produced a great step forward in personal computing performance and allowed serious work and analysis to be undertaken.

For a while, the Apple II was the favoured machine in the classroom environment. This capability was soon emphasised by the appearance of the first of the 'killer aps', business applications for microcomputers. This was Visicalc, the first of the 'spreadsheet' accounting systems which quickly became a runaway success.

Released to the market by Dan Bricklin and Bob Frankson in December 1979, Visicalc (Visible Calculator) took up a mere 25 kilobytes of memory but was a formidable piece of software that made it possible for anyone to produce a chart of accounts with the accuracy of a trained accountant. For a while after 1980, the Apple II and Visicalc were the supreme example of the new microcomputer art. This was not to last for very long.

The Tandy Corporation TRS-80 personal microcomputer of 1978 *(PRJ)*

MICRO CHALLENGER

Concurrently with the development of the Apple series of computers, other challengers were appearing. Radio Shack, the electronic hobby arm of the Tandy Corporation, produced the TRS-80, known somewhat contemptuously in Australia as the 'Trash 80'. This machine, costing $399, appeared sometime after 1977 and was comparable in power to the Apple as was the machine made by Commodore known as the PET. However, by comparison with the Apple, these contenders lacked the attraction and capability of the Wozniak–Jobs device.

Another machine from this period that was to start the author on his journey towards computational 'nerdery' was developed in Ohio and appropriately called the Challenger Superboard. In a technical sense, this little-known machine developed by Ohio Scientific should have been more of a challenge to the main players than it ever managed to be in fact. This was because it incorporated all the features that made the Apple II such a success, including the use of the Mostek 6502 central processing unit. It had Microsoft BASIC in 8 kilobytes of ROM, 8 kilobytes of random access memory (RAM) and a built-in keyboard, and could be coupled to a television set for use as a monitor. It was also able to be enhanced with add-on amateur-designed boards. In the less-than-skilful hands of the author and supported by members of CHAOS, the Challenger computer user group in Australia, this machine was enhanced over 18 months to be a reasonably respectable small business machine with 48 kilobytes of RAM and dual 8-inch disk drives with a storage capacity of 1.4 megabytes each — comparable to the capacity of a modern diskette but physically nearly ten times bigger.

Right The Ohio Scientific 'Challenger 1P' microcomputer of 1979, incorporating the 'superboard' *(PRJ)*

Far right Inside the 'Challenger 1P' showing the 'superboard' which could be purchased separately, 1979 *(PRJ)*

Far left The Osborne 01 'Browncase' of 1982 *(PRJ)*

Left The Osborne 01 'Greycase' of later in 1982 *(PRJ)*

This machine was able to do serious work and, with the assistance of a primitive and rather unfriendly word-processor package, it allowed the author to start typing up his reports at home, rather than relying on an increasingly reluctant spouse or office staff in their spare time.

Unfortunately, the software was bug-prone and the odd page or more of text tended to go off into cyberspace, which did not engender much confidence for tackling bigger reports. In addition, the disk operating system was incompatible with CP/M and this is probably what ultimately killed off this otherwise useful microcomputer.

The Challenger was ultimately put into the cupboard and a second-hand Osborne II 'Greycase' was substituted using CP/M and supplied with a Z 80 CPU. The path to future Intel products was now established. With this replacement machine came bundled Wordstar 2.2, SuperCalc and DBase. These software products were probably worth far more in the long run than the machine itself because they were stable and reliable and a good basis for setting the author on a path to expanded usage of the power of the microcomputer.

PORTABLE MICRO

The Osborne was ostensibly a portable machine but the reality was that it weighed in at over 20 kilograms (9 pounds) and it took a considerable degree of determination and strength to carry it around. Most owners conceded that it was more a 'luggable' machine than a truly portable device. It went to meetings of the Ausborne User Group where other enthusiasts busily swapped software — 'review' was the conventional term applied to this process.

Even for that time, the Osborne had some serious shortcomings. It had a minuscule built-in screen display which could show only 40 characters at a time — to see the right-hand side of the text, one had to shift across using control keys. An optical magnification system was built by the author using a flat Fresnel screen and this helped considerably. Ultimately, when a proper green screen monitor was obtained and driven from an add-on video card, the full line of text was able to be displayed — a huge improvement.

Apart from the visibility problem, the Osborne had two 5¼-inch floppy disk drives with the pathetic capacity of 90 kilobytes — a tremendous step backwards after the capacity of the 8-inch drives on the old Ohio Challenger. Disk-doubler hardware was soon installed and at 180 kilobytes the machine was more or less usable, particularly as it had access to Wordstar — not a very friendly program compared with modern software, but for 1984 it was 'magic'. Designed by Seymour Rubinstein of MicroPro in 1979, it originally cost $450 and, over a period of 5 years, nearly a million copies were sold. It was to be used by a whole generation of writers and is still lurking around, a tribute to successive efforts to modernise it to suit the new graphic user interfaces.

Unfortunately for the Osborne, its birth coincided with the decision of IBM to investigate this new upstart technology of the microcomputer. Following an internal memorandum from a senior manager, WC Lowe, an approach was made to purchase the Apple Company from Wozniak and Jobs but was rejected. As a result, IBM decided to make its own machine and the project was given to Don Estridge to develop. Unfortunately, he was killed somewhat later in an air crash, but not before the new IBM product, its first personal computer, had been developed.

Top The CP/M-based Bondwell microcomputer of 1984 *(PRJ)*

Bottom The personal computer from IBM, the IBM PC of 1982 *(PRJ)*

THE IBM PERSONAL COMPUTER (PC)

The personal computer (PC) used an 8080 16-bit CPU with 64 kilobytes of random access memory and cost the formidable sum of $2880 when the machine was launched in September 1981. The appearance of the IBM PC was assisted by a subtle advertising campaign featuring Charlie Chaplin of silent film fame and the new product was soon a sensational success. Before long it had reduced the former market dominance of Apple to a mere 10 per cent.

The appearance of the first IBM PC went a long way to killing off all the earlier machines with their limited memory and CP/M (or worse) non-standard operating systems. Apple managed to stave off the attack and, in that respect, was one of the few exceptions to IBM-induced company mortality. In addition, despite the enormous popularity of CP/M at that time and its potential to be reworked to suit the new IBM machine, the chance was lost. When the executives of IBM came to visit Digital Research to discuss the use of CP/M in the new PC, Gary Kildall, the designer and owner of CP/M, was away on the wings of his aircraft and unavailable. The opportunity that IBM had offered was lost because it was the youthful Bill Gates who waited in the wings of the theatre of opportunity. Digital Research's most famous product, CP/M, was denied a new existence driving the IBM PC. Instead, the ultimate opportunist, Bill Gates, was invited to include an operating system with the Microsoft BASIC that was being considered by IBM.

Gates's reaction was to rush off and buy a suitable DOS from a small Seattle firm, Seattle Computer

Products, for $30 000. Known by that firm as Q-DOS (Quick and Dirty DOS) it was appropriately renamed Microsoft DOS and the deal with IBM was consummated. Further, in negotiating this deal, Gates demonstrated his acumen for business in that the software was licensed to IBM as had been the case with MITS, rather than selling it outright. At an early stage, this astute step would begin the process by which Gates became the world's richest businessman by the end of the next decade.

The Compaq personal computer, a clone of the IBM PC XT of 1994 *(PRJ)*

In retrospect, for the user of MS DOS and later Windows, this must be seen as a great tragedy. CP/M, through its development as the operating system known as MPM, was capable of proper multi-user, multi-tasking and time-sharing, which Microsoft has still to completely achieve in Windows 95, many years later. Windows NT or its later manifestation, Windows 2000, appears to be a different story.

Despite the instant success of the new IBM PC, its design and the manner in which the architecture of the new machine was made accessible to the general public was to prove the basis of its downfall. Contrary to the ingrained policies and secrecy of IBM relating to the design of hardware, anyone could see how to design add-on hardware for the PC and a roaring market in enhancements developed almost immediately. Far more serious was the arrival of the 'clone' machines, some licensed by IBM but far more not. Of the former class of machines came the author's next acquisition.

The Osborne did very well for a couple of years but, in the end, in the face of newer more demanding software, a change had to be contemplated. In the same context, the Ausborne Users Group also was going through a period of transition as the IBM compatibles became an inevitable requirement for continued usefulness. Of the early clones of the PC but, in many respects, better built were the machines made by Compaq and the next machine to grace the author's work table was a Compaq Portable. Again the word 'portable' had to be viewed with a degree of optimism because this machine was probably 25 per cent heavier than the already heavy Osborne and needed the muscles of Hercules if transportation over any distance was seriously contemplated. However, this machine came with a 5-megabyte hard drive and one 360-kilobyte floppy drive and was really a pretty useful machine. Its major defect was a small display screen, although the resolution was similar to the Hercules standard and to the later VGA. It handled Wordstar Version 5 very well, apart from other useful software that was by then available.

Later, as the technology of disk drives improved, a single 720-kilobyte diskette drive together with a single 360-kilobyte drive were substituted for the old 360-kilobyte drive. Because of the improvements in the hardware, the two new drives were able to be fitted where the old single drive had lived previously.

The Apple Macintosh of 1984 with the graphic user interface (GUI) and mouse *(PRJ)*

MACINTOSH AND THE GUI

Despite the initial dominance of IBM and its new line of machines using Intel CPUs, Apple Computers was about to unleash a new system of software. This software, together with the new microcomputer that had been developed, would produce what is known as a 'king hit'.

During the 1970s, Xerox Corporation at its Palo Alto Research Centre (PARC) in Silicon Valley had worked on methods to make computers more accessible and easier to use. One of the most important results of this work, though difficult to implement at reasonable cost, was what would later be known as a graphic user interface or GUI. This approach had, in turn, built on work carried out at Stanford Research Institute in 1965 funded by ARPA under the directorship of Joseph Licklider. At that time, the researcher Douglas Engelbart had envisaged the device that most modern microcomputer users now either love or hate, the 'mouse', as the data input device has come to be known.

Steve Jobs of Apple was aware of this work and, in 1983, looking for a way to combat the emergence of the IBM PC, thought about the possibility of using a GUI on a new machine. The first attempt was embodied in a machine called the LISA but it was to sink almost immediately, being too slow, too expensive and quite cumbersome. However, the general idea of the GUI was now developed by a specially recruited 'gang' of piratical, bright young men and the results of their labour was the Macintosh, an outstanding success. Indeed, so successful was

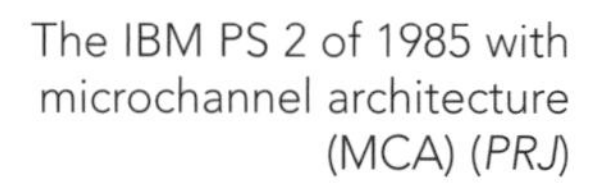

The IBM PS 2 of 1985 with microchannel architecture (MCA) (*PRJ*)

the new system of working using a mouse to point and graphical symbols called, somewhat inappropriately, 'icons', Microsoft was forced to try to produce an equivalent product.

The Microsoft product that later emerged was called Windows, the initial version appearing in 1985 after a massive program of work. Reputed to have involved 80 programmer years of development time, the product was initially a massive failure. This was mainly a reflection of the level of power and speed of IBM-compatible machines of that period. The Intel 80286 was simply not fast enough or powerful enough to cope with the demands of a graphical display and not until the emergence of the 80486, with a built-in maths co-processor running at 33 megahertz, could the requirement for graphics be fully answered.

An Olivetti laptop portable of 1994 with a 486 running at 25 megahertz, a 120-megabyte hard drive, 4 megabytes of RAM and a black-and-white VGA screen *(PRJ)*

Windows Version 2 came out in 1988 and Version 3 in 1990. By this time IBM clones were more or less capable of accommodating the demands of the software, and the product 'took off'. However, because Windows was really a sort of 'front end' to Microsoft DOS, it was still far from an adequate substitute for the Apple Mac OS.

Microsoft has spent the last decade or so trying to match the Apple software and started to get near that goal in 1995 with the release of Windows 95. In addition to emulating the general consistency of the Apple OS, Windows 95 manages to handle 'multi-tasking' reasonably well most of the time, but not always. As sorrowful users know, when Windows 95 crashes, it can be even more spectacular than when the same thing happened with its predecessor, Windows 3.1.

As a tribute to its fundamentally brilliant design, the Apple Macintosh operating system has remained relatively stable and still handles the addition of new hardware to the basic computer in a fashion that IBM clone users have to admire. By comparison, on the IBM clone machines, 'plug and play' software capability usually copes with new hardware, but not always.

A Tandy Corporation TRS 80 for commercial use with two 5¼ -inch floppy disk drives *(PRJ)*

MAINFRAME DINOSAURS

Given its part in creating the present microcomputer environment, it is quite remarkable that, in the process, its progenitor IBM was very nearly successful in carrying out the extremely painful process known in Japan as 'sepuku' or, more often and erroneously, 'harakiri'. This act of ritual self-disembowelment is the final recourse of the failed warrior.

By opening up the architecture of its first PC, IBM certainly managed to stabilise and standardise the architecture and operations of microcomputers but it also managed to create the seeds of a marketing disaster. The production of 'clones' of the IBM machine moved offshore to Japan, Singapore, Taiwan and Korea and, in the main, these new manufacturers paid no licensing fees to IBM. Painfully, too, many of these new machines offered more power and at far lower price than the American product and sold worldwide. However, not only were the IBM microcomputers slaughtered in the rush but also the mainframe industry in which IBM had been pre-eminent became itself a target.

The new microcomputers had access to increasingly powerful CPUs and the price of random access memory, in solid-state form, quickly tumbled so that extremely powerful machines could be placed on every worker's desk. At last anyone could have a complete machine that was significantly more powerful and flexible than the mainframe of only 10 years previously. It was a case of computing power to the people, without the need for a sort of computer priesthood to control the day-to-day operations of the machines.

The classic mainframe computer became a dinosaur and it took an economic and marketing disaster of epic proportions for the computer industry to realise what had happened. A significant period of decline has only recently started to turn around and the large-scale centralised machine to again be profitable. Among developments that have assisted this change is the advent of massive parallel processing which may yet prove to be the salvation of the mainframe. For all the miniaturisation that has been achieved in hardware, ultimately the demand for more and more computing power is likely to lead back to larger machines.

In 1981, the local government authority that the author worked for at that time purchased a machine to handle all of the accounting processes and ostensibly to provide a local area network (LAN) for the transfer of data files produced in departments. This was to be accomplished by linking Phillips CP/M-based word processors to the new computer. The mainframe machine had a main memory of 1 megabyte with 10 megabytes of memory in several hard drives the size of washing machines and cost the best part of A$100 000. The machine did the job that it was supposed to do, but the LAN was never made to work. A recent telephone call revealed that the software provided in 1981 for the municipal accounts and budget continues to be used although the original hardware has been replaced by new equipment.

Compared with what was available in the microcomputer field only 10 years later, it can now be seen that what was purchased in 1981 was a very expensive investment. Also it is apparent that the hardware became obsolete with extraordinary rapidity. But then who could have guessed how quickly the microcomputer revolution would develop, once the machines were available and once the demand for them was revealed?

As a final small irony in this story of the rise of the microcomputer, in 1995 IBM, which had been instrumental in launching Bill Gates to unprecedented wealth with its purchase of rights to use the operations system known as MS DOS, agreed to purchase the Apple OS for use in the IBM Power PC laptop. What greater endorsement could the designers at Apple have received! An associated matter is that, as for the development of radio in the 1920s, all the hard work of the amateurs that created the revolution was gradually absorbed into the operations of a few big organisations detached from the beginnings of the microcomputer revolution.

Today, the major organisations that remain in the software area are Microsoft, Novell, Lotus and Corel. In contrast, the hardware situation is very different with Asia still producing the majority of machines and parts in an increasingly competitive market where prices continue to fall as power continues to increase. Three chip manufacturers dominate the market in CPU manufacture at present with Intel head and shoulders above the rest, which includes AMD and Cyrix.

CHAPTER

18

Beyond 2000

Micro-comms

From the bridging of the Atlantic in 1901 by the first tenuous radio signal, the creation of the Internet and the World Wide Web has followed a complex and multistranded path. Whatever the twists and turns in the route, however, it can be seen that a driving motivation has been to respond to the demand for more rapid and useful communication.

In the achievement of a world system of communication, a dream of humanity is well on the way to realisation — instant communication between any person on the planet and any other person and the immediate availability of information, however complex. This is very much what early science fiction writers had in mind when describing the impact of 'telepathy' on society, the work of AE Van Vogt many years ago being a good example. Certainly, for earlier ages, what is now available to the world community in the World Wide Web would be seen as something akin to 'black magic'.

The ability of the World Wide Web and the Internet to provide information and convey messages both written and voice-based around

the planet has grown as rapidly as anything that has happened in the communications field in a century of development. In that, it is apparent that the Internet is now meeting a latent demand that would have amazed earlier generations of communications engineers and computer scientists.

It is also apparent that the capacity of the Internet to convey information will soon extend well beyond what is currently seen as its capability limit. Already, radio broadcasting that has been confined to the short-wave and medium-wave bands for most of the 20th century is available on the Internet. All that is required is a software 'add-on' to the basic Web browser, called 'Real Audio', and the BBC, Voice of America and the Australian Broadcasting Corporation can be heard without any of the usual background interference of static and other overlapping stations.

Parallel with the conventional telephone service, it has been possible for both amateur and commercial organisations to use digitisation of voice messages to achieve a very economical alternative to the conventional wired system. No doubt, at this very

An Alcatel digital cellular telephone with a personal digital assistant (PDA) built-in; text entry is via a stylus with character recognition *(ALC)*

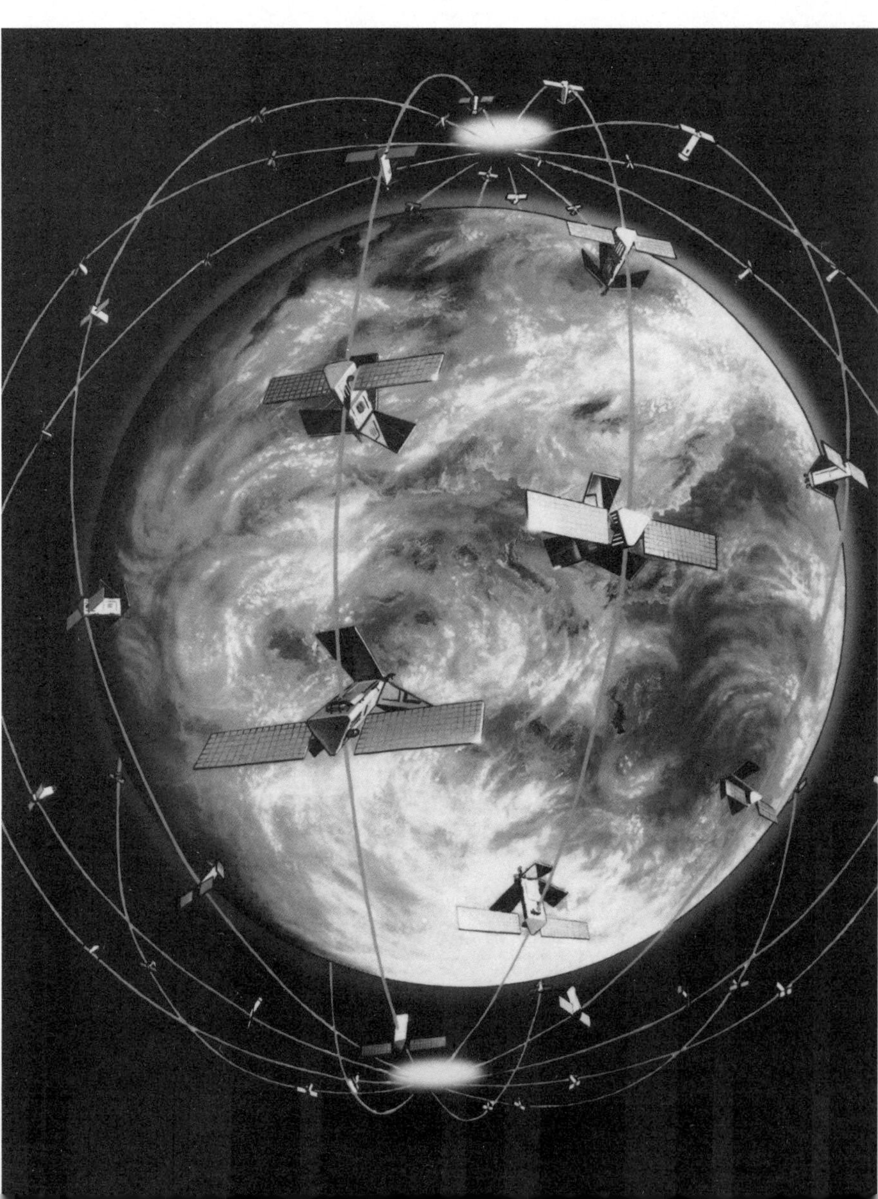

The Iridium low earth-orbiting satellite system (LEOS) *(IRID)*

moment, this development is causing a degree of anxiety among the conventional telephone service providers, whose service charges and revenue will undoubtedly be affected by the new development.

Even the allure of television, a service that went so close to killing conventional radio broadcasting when it first appeared, is starting to suffer in a serious manner from the impact of Internet competition. In a recent survey carried out in Australia, it was clear that a transfer of community interest to the Internet had seen a major decline in patronage of conventional 'free-to-air' television broadcasting. Inevitably this has had an impact on the support of television advertising and no doubt accounts for the rapid proliferation of advertising on the Internet and the spread of the infernal 'cookie' as a means of determining the extent of penetration of such activities.

Beyond the issue of advertising and its baleful impact on the usefulness of the Internet is the extent to which, in only the last few years, the Internet has undergone rapid commercialisation and use for business and trading purposes. With a connection to the World Wide Web it is now possible for the most remote and inaccessible business to sell its products via the Internet and, in recent times, examples have occurred in such unlikely places as in the isolation of Australia's Outback. One example of this is the enterprising maker of leather whips who has found a world demand for a product that otherwise would be so obscure that even local sales would be restricted to the declining numbers of bullock drivers in the far north.

Australia, as a particularly distant and isolated part of the world, separated from the commercial system of the northern hemisphere, has benefited in particular from the expanded connection that is available in the Internet. Now with e-mail flashing across the surface of the earth and through space at the speed of light, no place is too distant or too inaccessible. With the availability of cellular telephones and satellite communications, anyone can remain in contact however apparently remote the location from which a call is made.

In many homes, the availability of a microcomputer and a modem has made it possible to gain access to the most extraordinary range of resources at the cost of a local phone call and the time charge of a commercial Internet service provider (ISP). A world of library resources and databases of every kind provide the basis of information that only a few years ago would have involved months of research. Now all is available with the assistance of a Web browser like Netscape or Explorer. The only proviso is that, at present, what is available is largely unverifiable and one must rely on the institution or individual whence the information comes as to its accuracy. Sources can easily reflect the biases of their presenters and this can lead to error rather than truth.

No doubt, in the future, verification of material will become a matter that is handled by some sort of Internet world supervisory body. In that regard, completely ungoverned operations on the Internet seem

An Iridium satellite in position with solar cells deployed *(IRID)*

A handheld digital telephone by Kyocera for use with the Iridium satellite system — it can be used with the local cellular network when not in its Iridium docking shoe *(IRID)*

inevitably to require the control of the users on some democratic basis. The emergence of 'netiquette' is probably the first clear reflection of the need for some agreed basis of operation of a system which is intrinsically so capable of being chaotic. Concern by private users and government at the unbridled dissemination of pornographic material is also symptomatic of the capacity of the Internet to be a channel for socially unacceptable behaviour. Across this issue runs the different sensibilities of various ethnic and religious groups.

In very recent times, the wonder of modern communication and the capacity of the global positioning system (GPS) allowed a British yachtsman in extraordinary peril to be snatched from the freezing waters south of Macquarie Island at the edge of the Antarctic. Only a generation ago, Tony Bullimore would undoubtedly have foundered with his overturned vessel, his fate completely unknown. In fact he was pulled from the sea by the crew of an Australian Air–Sea Rescue aircraft, having survived above the freezing water inside the hull of his inverted yacht.

Given the extraordinary growth of the Internet in the last few years, it is immensely tempting to speculate on where this new development in world communications and trade is likely to take us in the future. Although there have been many highly relevant speculations on the future, what is more than obvious is that frequently it is something quite unexpected that creates the greatest ultimate change. Radio waves represent perhaps as good an example as any.

In the Victorian era, the notion of energy being able to move through a vacuum and to carry intelligence would have seemed quite bizarre. In the space of a mere 5 years, the discoveries of the pioneers Lodge, Branly and particularly Marconi created the basis of a completely new technology that became the foundation stone of many of the developments discussed in earlier chapters.

While the electronic and scientific experts were in the process of developing the computer using conventional valves, others were at work finding a substitute for the valve in a new solid-state device, the transistor. These two lines of work later converged to give the world the desktop computing revolution that is now upon us. If one or other of those two seminal steps had not occurred, the world would be almost unrecognisable today.

The uneasy peace that followed the Second World War saw the development of weapons of destruction at a level that previous generations would have found simply inconceivable. Out of the fears of the 'Cold War' with the Soviet Union came the beginnings of the wired network, ARPANET, on which rests the miracle of the Internet.

All that can be guaranteed of the future is that it is unlikely to be anything like anyone expects. Providing that humanity does not sweep itself away by a world war or the inadvertent release of some malignant new artificial disease or, for that matter, disappears as a result of the spread of old diseases, the future looks extremely interesting and exciting. The characteristics of humanity are likely to remain as a stable underpinning to whatever technical developments are likely to arise, however unexpected. However, in that recognition lies the core of a concern for the future health of the planet as a whole.

As revealed by the images conveyed from its distant satellite the Moon, the Earth swims serenely in a sea of darkness, infinitely lonely and fragile. From space the sprawling multitudes of humans are invisible and all that can be seen are the swirling clouds, the blue seas and occasionally the land masses — a small 'spacecraft' indeed to carry all the life that at present one can be certain exists in the universe. Maintaining the vessel that we are all committed to in a habitable state will clearly tax the ingenuity of future generations of scientists. Perhaps, in a major way, the extraordinary power of the Internet to provide communications will help to advance the process of maintaining the only home that humanity can look to for the foreseeable future.

Reconciling the racial and religious attitudes of humanity so that world harmony is achieved and the future success of the race guaranteed is something in which the Internet may well play a fundamental part. Future generations will look back at the period at the end of the 20th century and judge the efforts of humanity so far. It is hoped that what will be seen is something that has contributed to a more desirable future for our descendants.

By then, fundamental discoveries in cosmology and the science of matter may have provided a completely unexpected means of communication and travel. Harnessing 'singularities' and the tantalising 'black hole' may involve new directions as profoundly unexpected as would today's science and technology be for the 16th century Renaissance man par excellence, Leonardo da Vinci.

His dreams and speculations have become our commonplace experience but, even in his dreams, the full scope of our science would be baffling in the extreme. Even to look ahead 50 years, we are in very much the same situation, but what can be anticipated is the altogether unexpected and, in the end, that is the only certainty.

Postscript

It was bitterly cold and a vicious wind was roaring out of the North Atlantic, battering at the eaves and walls of the old building. In the sky above, a tiny patch of material could be seen diving and soaring in almost uncontrollable gyrations as a small group of men hauled on a rope attached to it. Almost 150 metres in the air, the 2-metre kite was a barely visible part of the plan to contact far-away Poldhu through the copper wire hanging from it and connected to apparatus in the unused Fever Hospital below.

Inside, shivering with the cold, sat a young man, formally attired in a tweed shooting jacket and, somewhat incongruously, a high-crowned tweed hat. Opposite him stood attentively a dark and silent figure, his assistant. For a long time the young man had sat silent and intent with a peculiar-shaped earphone receiver pressed to his ear as he listened to the sound of the natural universe, to all the clicks and bangs and the static of the ether, unsullied by human technology.

Suddenly, faintly, through the noise of the natural ether came the anticipated sound. The listener stiffened as he heard three, short crackling spurts of noise. Unmistakable now against the background, the three short dots of noise came again. The young man smiled faintly and, passing the headphone to his assistant, said enquiringly, 'What do you hear, Kemp?'

A slight pause and Kemp, screwing up his face with concentration and then grinning broadly, said, 'It's Poldhu, Mr Marconi — you did it!'

And so he had, but more than he could possibly have imagined.

STRANDS IN THE WORLD WIDE WEB

1822	Babbage's paper on the difference engine no. 1 published
1844	Morse telegraphic link between Washington and Baltimore in the United States established
1851	International adoption of the continental code based on the morse code
1865	James Clerk Maxwell demonstrates the theoretical existence of electromagnetic waves
1866	The *Great Eastern* lays the first trans-Atlantic cable between Valentia in Ireland and Heart's Content in Newfoundland
1871	Death of Babbage
1872	Telegraph cable between Darwin and Adelaide in Australia established
1874	Guglielmo Marconi is born in Bologna, Italy
1876	The telephone is patented by Alexander Graham Bell
1879	The principle of the coherer is discovered by DE Hughes
1884	Paul Nipkow invents the scanning disk, the basis of early television
1887	Heinrich Hertz demonstrates the existence of electromagnetic waves
1888	Thomas Edison discovers the 'Edison effect' in light bulbs
1890	JA Fleming investigates the 'Edison effect'
	First US census using Hollerith equipment
1892	Edouard Branly and Oliver Lodge independently invent the coherer
1894	Marconi begins his experiments with wireless at the family villa near Bologna
1896	Marconi moves to England and receives the first patent for wireless telegraphy
1897	The Marconi Wireless Telegraph and Signal Company is created
1898	Oliver Lodge patents the tuned circuit for radio signals
	Commercial installation of wireless in ships
1899	Installation of wireless in the vessels of the Royal Navy
	First military use of radio in the Boer War
1901	Marconi receives signal from Poldhu at St John's, Newfoundland (12 December)
	Reginald Fessenden describes the detection of continuous wave signals by heterodyne mixing
1902	Introduction of the magnetic detector (Marconi)
1904	JA Fleming patents the diode, two-element valve (tube)
	Continuous wave transmission demonstrated by Fessenden using an alternator
	Introduction of the multiple tuner (Marconi)
1905	Trans-Tasman Sea wireless demonstration, Australia, by the Marconi Company
1906	Lee de Forest patents the triode, three-element tube (valve)
	Fessenden's Brant Rock telephony experiments
1907	Clifden (Ireland) to Glace Bay (Nova Scotia) trans-Atlantic message service commenced
1909	The SS *Republic* sinks following collision
	Cyril F Elwell introduces the arc transmitter to the United States
1910	Wireless Institute of Australia formed in New South Wales, Australia
	Dr Crippen apprehended on the SS *Montrose*
	C Maclurcan transmits music from his station 2CM at Sydney, Australia
1911	AA Campbell Swinton describes an all-electric television system
	Douglas Mawson's expedition to the Antarctic commences
	Thomas J Watson joins NCR
	Ernest Fisk arrives in Australia as resident engineer for the Marconi Company

1912	Sinking of the RMS *Titanic*
	De Forest invents the audion tube amplifier
	Radio station POS (later VIS) opened in Sydney, Australia
1913	Edwin H Armstrong invents the regenerative detector circuit using positive feedback
	Amalgamated Wireless (Australasia) Ltd formed with Ernest Fisk as technical manager
1914	First World War starts
1915	The Gallipoli campaign
1916	Marconi commences experiments with the 'short waves'
1917	Zimmermann telegram deciphered and passed to the US government; United States enters the war
1918	Transmission from Caernarvon received in Sydney, Australia, on the long waves
	Armstrong patents the 'superhet'
	End of the First World War (November)
1919	Radio Corporation of America, RCA, created
	Alexanderson alternator installed at Caernarvon
1920	Radio broadcasting experiments initiated by Marconi Company employees
	Dame Nellie Melba broadcasts from Chelmsford, England
	Regular broadcasting from Pittsburg radio station, KDKA, commenced
1921	First direct contact from the United States to Britain by radio amateurs (Paul Godley on 210 metres)
1922	British Broadcasting Company created (12 November)
	First BBC broadcast station opened, London-based 2LO
	First radio telegraph on the short waves from the United Kingdom to Switzerland
1923	Vladimir Zworykin proposes and patents an all-electronic television system
	Broadcast station 2SB commences service in Sydney, Australia (23 November)
	First two-way contact between the United States and the United Kingdom by radio amateurs
1924	Direct signal to Sydney, Australia, via the short waves from Poldhu (30 May)
	Broadcast station 3LO commences service in Melbourne, Australia (13 October)
	First two-way contact between the United Kingdom and Canada
	IBM created in the United States
1925	Grimeton radio station commences operation
	Reverend John Flynn tests the first inland radio in Outback Australia
1926	Short-wave beam radio-telephone service to Canada opened from the United Kingdom
	John Logie Baird opens experimental television service using the Nipkow disk with call sign 2TV
	Rugby long-wave station opened by the British Post Office operating on 16 kilohertz
1927	Short-wave beam service from the United Kingdom to Australia, South Africa and India commenced
	Alfred Ewing presents paper on his 'War Work'
1928	Philo Farnsworth demonstrates an all-electronic televison system using a camera called the 'image dissector' and an image display based on the cathode ray tube
	Rugby radio station commences operation on the short-wave band
	Flight of the *Southern Cross* from the United States to Australia by Charles Kingsford Smith and Charles Ulm
	The differential analyser of Vannevar Bush installed at Massachusetts Institute of Technology
	Australian national broadcasting service commenced
1929	Creation of Cable and Wireless Ltd, from merger of Marconi Company with other cable communications companies
	Introduction of pedal wireless system in the Australian outback

1929	BBC experimental television service commenced using the Baird system
	Zworykin develops the kinescope, an image display device based on the cathode ray tube
1930	Lilienfeld patents a form of primitive field-effect transistor
1931	Zworykin develops the iconoscope as an answer to the Farnsworth 'image dissector'
1934	Baird adopts the Farnsworth 'image dissector' camera as part of the system offered to the BBC, using a scan rate of 240 lines
	Watson Watt and Rowe visit the acoustic station at Hythe
	CEM Joad resolution at Oxford University — 'Under no circumstances will this house fight for King and Country'
1935	Patent dispute over primacy of electronic television in the United States is ruled in favour of Farnsworth
	First tests of radio location using short-wave reflections in the United Kingdom (the birth of British radar)
	Conrad Zuse sets down the concept of a binary computer, the Z1
1936	Dual-system national television service commenced by BBC from Alexandra Palace with the Baird and EMI systems used on an alternating basis
	Paper 'On Computable Numbers' by Alan Turing
	German troops retake the Rhineland
1937	BBC adopts the EMI system of televison with a 405-line scanning system using the 'Emitron' camera
	Death of Marquese Marconi
	Japan invades mainland China
1938	RCA produces the image iconoscope, incorporating the best elements of the 'image dissector' and the iconoscope (almost identical to the British 'emitron')
	Paper on the use of Boolean algebra in designing counting and switching networks by Claude Shannon
	German troops move into Austria and the 'Anschluss' is proclaimed
1939	NBC commences regular television broadcasting
	Farnsworth Television and Radio incorporated, broadcasting from Fort Wayne, Indiana
	Second World War begins
	RCA loses royalty battle with Farnsworth and is forced to pay licensing fees on his patents
	William Shockley joins the Bell Laboratories to study solid-state phenomena
1940	Commercial televison broadcasting in the United States commences from WBNT
1941	Colour television demonstrated by NBC in the United States
	FCC adopts the NTSC standard of 525-line, 30 frames per second, black-and-white TV
1943	The Harvard Mark 1 by IBM put into operation
1944	D-day landings announced, 6 June
	The 'Colossus' begins work at Bletchley Park
	The 'Market Garden' operation and the Arnhem debacle, September
1945	End of the Second World War
	Arthur C Clarke predicts the introduction of the communications satellite
1947	The point-contact transistor is invented by Shockley, Brattain and Bardeen
1949	Farnsworth Television and Radio Company sold to International Telegraph and Telephone
1951	RCA demonstrates a new electronic colour television system compatible with existing black-and-white television
	UNIVAC passes its acceptance test
1952	GW Dummer of Royal Radar Establishment, United Kingdom, proposes a monolithic block of components without connecting wires

1953	NTSC (based on RCA system) colour television adopted in the United States
	The trans-Atlantic telephone cable TAT-1 laid by a US–UK–Canadian consortium
1954	Texas Instruments manufacture the Regency TR 1 pocket transistor radio
1955	BBC opens an FM broadcasting station at Northam in Kent
	Shockley Semiconductor Corporation founded by William Shockley
1956	Black-and-white television service on 625 lines commenced in Australia; coverage of the Olympic Games in Melbourne
	Bardeen, Brattain and Shockley awarded the Nobel prize for creation of the transistor
1957	Russia launches Sputnik
	ARPA established in the United States and Joseph Licklider becomes first director of IPTO
	Eight engineers leave Shockley Semiconductors to form Fairchild Semiconductors
	Programming language FORTRAN made available
1958	Jack Kilby invents part of the solid-state integrated circuit
1959	Robert Noyce invents the other part of the solid-state integrated circuit
	Integrated circuit announced by Texas Instruments
	Fairchild files a patent for the planar process for making transistors
	FORTRAN II made available
1960	COBOL programming language released
1961	Fairchild Semiconductors releases the first commercial integrated circuit
1962	United States launches Telstar Mark I
	Television signals sent from America to Europe via satellite
1963	D Engelbart patents the first 'mouse' as a pointing device
1964	G Moore propounds the notion that integrated circuits will double in complexity every year from then on which is later known as 'Moore's Law'
	BASIC programming language developed by Professor J Kemeny and TE Kurtz at Dartmouth College, United States
1965	The first INTELSAT flown from the United States
	Digital Equipment Corporation (DEC) market the PDP 8 minicomputer
1966	Robert Taylor initiates the ARPANET project
1968	Intel formed by Robert Noyce and Gordon Moore
1969	First four IMPs installed at universities in the United States — ARPANET initiated
	The design of the first microprocessor, the 4004, created at Intel by Marcian Hoff (Ted)
1970	Xerox opens the Palo Alto Research Centre, PARC
	G Hyatt files a patent for a microprocessor he describes as involving 'single integrated circuit computer architecture'
1971	Texas Instruments produce the 'Pocketronic' electronic calculator
	Fairchild produces the first central processing unit, the 4004
1972	File transfer protocol introduced (FTP)
	Intel produces the 8008 microprocessor, accessing 16 kilobytes of memory and using an 8-bit configuration
1973	Gary Kildall writes CP/M, Control Program for Microcomputers
1974	Winterbotham reveals the 'Ultra Secret'
	TCP/IP protocol established by Vinton Cerf
	Intel produces the 8080 microprocessor running at 2 megahertz and accessing 64 kilobytes of memory (6000 transistors)
	UNIX programming language is released by Bell Labs
1975	The Altair 8800 advertised in the United States via *Popular Electronics* at US$439
	Fibre-optic cable developed

1975	CP/M applied to the 8080 microprocessor
	Bill Gates and Paul Allen found Micro-Soft (later Microsoft) and license their BASIC to MITS for use on the Altair
1976	Steven Jobs and Stephen Wozniak found Apple Computer Company on 1 April
	CP/M patented by Digital Research and created by company founder, Gary Kildall
	The 6502 microprocessor produced by MOS Technology and designed by C Peddle
1977	The Apple I microcomputer developed and sold
	The Commodore Pet 2001 produced using a 6502 chip
	The Tandy TRS-80 microcomputer produced
1978	Intel produces the 8086 microprocessor running at 4.7 megahertz (29 000 transistors)
1979	Visicalc introduced for use on the Apple microcomputer, designed by D Bricklin and R Frankson
	Micropro releases Wordstar wordprocessing program written by S Rubenstein
1981	The IBM personal computer launched using an 8088 running at 4.7 megahertz with 64 kilobytes of RAM
	The Osborne 01 computer introduced using a Z 80 microprocessor loaded with CP/M, BASIC, Wordstar and SuperCalc
1982	Intel produces the 80286 running at 6 megahertz (134 000 transistors and speed of 0.9 MIPS)
	Microsoft provide MS-DOS for the IBM PC
	Compaq Computer introduces the Compaq Portable PC using an 8088 running at 4.77 megahertz
1983	Apple LISA with GUI (graphical user interface) introduced
	Lotus 123 produced for MS-DOS, designed by J Sachs and M Kapor
	Osborne Computer Corporation files for bankruptcy
	Microsoft announces Windows development
1984	NFS Net established
	Apple produces the Macintosh using a Motorola 68000 CPU with 32 bits and running at 8 megahertz with the GUI
	Microsoft releases MS DOS 3.1 for the PC
	2400-baud modems start to appear
1985	Windows Version 1 introduced 2 years after first announcement
	Intel produces the 80386DX with 275 000 transistors running at 13 megahertz
1986	Compaq produces the Deskpro 386 using the 80386 running at 16 megahertz
	IBM produces the PC XT using a 80286 with a capacity to address 640 kilobytes of RAM with

1986	1.2-megabyte floppy drive and 20-megabyte hard drive
1987	Commodore produces the AMIGA 500 using the Motorola 68000 CPU with 512 kilobytes of RAM
	IBM produces the PS/2 series including the Model 30 (8086 at 8 megahertz), the Model 50 and 60 (80286 at 10 megahertz) and the 80 (80386 at 20 megahertz), all using the Micro Channel Architecture (MCA)
	IBM introduces the Video Graphic Adapter (VGA) with 16 colours at 640 by 480 or 256 colours at 320 by 200
1988	Microsoft produces DOS 4, able to address disk partitions above 32 megabytes
	Ashton-Tate produce DBase 4 for MS DOS
1989	Intel produce the 80486 running at 25 megahertz and integrating the 80386 CPU with the 80387 maths co-processor (1.2 million transistors running at 20 MIPS)
	Intel produces the 80486 running at 33 megahertz (27 MIPS)
	WordPerfect 5.1 released
1990	Microsoft produces Windows Version 3
	G Hyatt awarded patent for 'single chip integrated circuit computer architecture' after a 20-year wait
	ARPANET shut down
1991	Microsoft produces MS-DOS 5 with Q Basic to replace GW Basic
1992	NEC produces a double-speed CD-ROM drive
	Microsoft produces Windows 3.1 and sells 1 million upgrade copies in 7 weeks
	Microsoft produces Windows for Workgroups 3.1 allowing network operation
	Intel introduces the Pentium with 32-bit registers, a 64-bit data bus and able to address 4 gigabytes of address space (3.1 million transistors running at 112 MIPS for the 66-megahertz version)
1994	Intel produces the 100-megahertz Pentium running at 166 MIPS
	Mosaic Communications produces the Netscape Navigator Version 1 browser for the Internet
1995	Windows 95 introduced
	The Internet becomes publicly accessible as the World Wide Web
1997	Digital portable telephones begin to make analog phones obsolete
1998	Integration of portable telephone and personal digital assistants make mobile Internet access possible
1999	The Iridium satellite network commences
2001	Anniversary of the first trans-Atlantic radio signal

References

Admiralty Handbook of Wireless Telegraphy, vols 1 and 2, HMSO, London, 1938.
Aitken HGJ, *Syntony and Spark*, Wiley, Sydney, 1976.
Aitken HGJ, *The Continuous Wave*, Princeton University Press, NJ, 1985.
The American Institute of Electrical Engineers Annual Dinner and Mr Marconi, *The Electrical World and Engineer*, New York, 18 January 1902.
American Radio Relay League, *From Spark to Space,* ARRL, Novington, USA, 1989.
Australian Corps of Signals, *Signals,* Halstead Press, Sydney, 1944.
Austin BA, Wireless in the Boer War, in *100 Years of Radio*, Proceedings of the IEE conference, September 1995, IEE, London, p. 44.
Baker WG, *Magnetism and Electricity,* Blackie, London, *c.*1900.
Baker WJ, *The History of the Marconi Company*, Methuen, London, 1970.
Barthols, BR, *WIA Book,* Wireless Institute of Australia, Mulgrave, 1985.
Barty-King H, *Girdle Round the Earth,* Heineman, London, 1979.
Bateman J, *The History of the Telephone in NSW,* Telephone Collectors Society of NSW, Sydney, 1980.
Bean CE, *Anzac to Amiens,* Penguin, Melbourne, 1946.
Beaumont J, *Australia's War — 1914/1918,* Allen and Unwin, Sydney, 1995.
Beesley P, *Very Special Intelligence,* Sphere, London, 1977.
Belrose JS, Fessenden and Marconi: Their Differing Technologies and Trans-Atlantic Experiments During the First Decade of this Century, in *100 Years of Radio,* Proceedings of the IEE conference, September 1995, IEE, London, p. 32.
Bennett JM et al., *Computing in Australia,* Australian Computer Society Ltd, Hale & Iremonger, Sydney, 1994.
Blake GG, *History of Radio Telegraphy and Telephony,* Radio Press, London, 1926.
Blaney G, *The Tyranny of Distance,* Sun, Melbourne, 1966
Bleakley J, *The Eavesdroppers,* AGPS, Canberra, 1991.
Bowyer C, *Fighter Command,* Sphere, Glasgow, 1980.
Braun E & Macdonald S, *Revolution in Miniature,* Cambridge University Press, NY, 1978.
Bridges TC & Tiltman HH, *Master Minds of Science,* George G Harrap, London, 1937.
Bridgman L, *Aircraft of the British Empire,* Sampson Low Marston, London, 1937.

British Air Forces — Special Publication, *Illustrated London News and Sketch*, 1946.
British Broadcasting Corporation, *Twenty Five Years of British Broadcasting 1922–1947*, BBC, London, 1947.
Brown AC, *Bodyguard of Lies*, WH Allen, London, 1976.
Bryant A, *Turn of the Tide*, Fontana, London, 1965.
Buchan J, *The Battle of the Somme — First Phase*, Thomas Nelson, London, 1916.
Bull K, Samuel Morse, *Amateur Radio Action*, vol. 11/10, p. 45.
Bull K, The Father of Modern Radio, *Amateur Radio Action*, vol. 10/4, p. 21.
Bull K, Edison, Bell and the Carbon Microphone, *Amateur Radio Action*, vol. 8/5, p. 43.
Burton N, Radio Genesis Australis, *The Mariner*, March/April 1968, p. 206.
Butler R, *Breaking the Ice*, Albatross, Sydney, 1988.
Calcaterra J, The Boston Television Party, *Radio News*, May 1931, p. 986.
Calvocoressi A, *Top Secret Ultra*, Sphere, Glasgow, 1979.
Campbell-Kelly M & Aspray W, *Computer*, Basic Books, New York, 1996.
Cameron AR, The Story of the Overland Telegraph Line, Lecture to the South Australian Postal Institute, 11 October 1932 (unpublished).
Chambers EW, Unpublished correspondence, Marconi Archives at Great Badow — Letter dated 4 May 1972 and paper entitled Saga of Australian Wireless 1888 to 1920.
Chapman EH, *Wireless Today*, Oxford University Press, London, 1941.
Churchill WS, *The World Crisis 1911 to 1918*, Four Square, London, 1923.
Churchill WS, *The Second World War*, vols 1 to 6, Cassell, London, 1948.
Claricoats J, *World at Their Fingertips*, Radio Society of Great Britain, Potters Bar, UK, 1967.
Clark A, *Barbarossa*, Penguin, Harmondsworth, UK, 1965.
Clarke AC, *Profiles of the Future*, Pan, London, 1962.
Clarke AC, *How the World Was One*, Victor Gollancz, London, 1992.
Cockaday LM, *Wireless Telegraphy for All*, Herbert Jenkins, London, 1923.
Cocking WT, *TV Equipment*, Wireless World and Iliffe joint publication, London, 1940.
Collier R, *The Sands of Dunkirk*, Fontana, London, 1961.
Collins AF, *Radio Amateur's Handbook*, Thomas Y Crowell, New York, 1922.
Constable A, *Early Wireless*, Midas, Tunbridge Wells, UK, 1980.
Cross J, *Red Jungle*, Robert Hale, London, 1958.
Crowther JG & Whiddington R, *Science at War*, HMSO, London, 1947.
Davis P, *Charles Kingsford Smith: The World's Greatest Aviator*, Summit Books, Sydney, 1977.
Deacon R, *A History of the British Secret Service*, Granada, London, 1969.
Deighton L, *Blitzkrieg*, Triad Grafton, London, 1979.
DeLacy B, Who Did Invent the Superheterodyne, *HRSA Radio Waves*, April 1995, p. 25.
DeSoto CB, *200 Metres and Down*, Amateur Radio Relay League, West Hertford, USA, 1936.
DeVries L, *The Book of Telecommunications*, John Murray, 1962.
Dinsdale A, De-bunking Television, *Radio News*, January 1931, p. 593.
Dowsett HM & Walker LEQ, *Handbook of Technical Instruction for Wireless Telegraphists*, 7th ed., Iliffe, London, 1943.
Dowsett HM & Walker LEQ, *Handbook of Technical Instruction for Marine Radio Officers*, 9th ed., Iliffe, London, 1950.
Duncan R & Drew CE, *Radio Telegraphy and Telephony*, John Wiley, New York, 1929.
Eaton JP & Haas CA, *Titanic — Destination Disaster*, Patrick Stephens, Wellingborough, UK, 1987.
Eisenhower DD, *Crusade in Europe*, William Heineman, Melbourne, 1948.
Enever T, *Britain's Best Kept Secret*, Alan Sutton, Stroud, UK, 1994.
Edwards R, *Panzer*, Arms and Armour, London, 1993.
Ernst Alexanderson ..., *Electronics Australia Yearbook 1975/76*, p. 14.
Erskine-Murray J, *Wireless Telegraphy*, Crosby Lockwood, London, 1914.
Eunson M, The Early Telegraph Systems, *Amateur Radio Action*, vol. 9/7, p. 18.
Eunson M, Telegraphy, Morse Code and All That, *Amateur Radio Action*, vol. 9/9, p. 27.
Evans C, *The Making of the Micro*, Victor Gollancz, London, 1981.
Evans C, *The Mighty Micro*, Victor Gollancz, London, 1979.
Evans JV, New Satellites for Personal Communications, *Scientific American*, April 1998, p. 60.
Fairley P, *Television Behind the Scenes*, Severn House, London, 1976.
Fjermedal G, *The Tomorrow Makers*, Tempus, Washington, 1986.

Fleming JA, *Radio Telegraphy and Radio Telephony,* Longmans Green, London, 1916.
Fleming JA, *The Thermionic Valve,* Wireless Press, Sydney, 1919.
Fleming JA, *Principles of Electric Wave Telegraphy and Telephony,* Longmans Green, London, 1919.
Foot MRD, *SOE — 1940–1946,* Mandarin, London, 1984.
Fox B & Webb J, Colossal Adventures, *New Scientist,* 10 May 1997, p. 39.
Freed L, *The History of Computers,* Ziff-Davis, Emeryville, USA, 1995.
Fuchs V & Hillary E, *The Crossing of Antarctica,* Cassell, London, 1958.
Garlinski J, *Intercept,* Dent, Melbourne, 1979.
Garratt GRM, *The Early History of Radio — From Faraday to Morse,* Institute of Electrical Engineers UK in association with the Science Museum, London, 1994.
Geddes K & Bussey G, *Television — The First Fifty Years,* Philips Electronics, Croydon, UK, 1986.
Geeves P, Australia's Radio Pioneers (4 parts), *Electronics Australia,* May 1974, p. 26; June 1974, p. 30; July 1974, p. 34; August 1974, p. 50.
Geeves P, Marconi and Australia, *AWA Technical Review,* vol. 15, no. 4, 1974, p. 131.
Gernsback S, *1927 Radio Encyclopedia,* Vintage Books, Palos Verdes, USA, 1974.
Giles JM, George Augustine Taylor — Some Chapters in the Life, Supplement to *Building: Lighting: Engineering,* 24 November 1957.
Gill EW, *War, Wireless and Wangles,* Basil Blackwell, Oxford, 1934.
Goldstein W, *The History of Television,* Portland House, Godalming, UK, 1991.
Hafner K & Lyon M, *Where Lizards Stay Up Late,* Simon and Schuster, New York, 1996.
Halloran AH, Scanning Without a Disc, *Radio News,* May 1931, p. 998.
Hammond J, The Father of FM, *73 Magazine,* February 1982, p. 50.
Hancock HE, *Wireless at Sea,* Marconi International Marine Communications Co Ltd, Chelmsford, UK, 1950.
Harclerode P, *Para — Fifty Years of the Parachute Regiment,* Orion, London, 1992.
Harclerode P, *Arnhem — A Tragedy of Errors,* Arms and Armour, London, 1994.
Harcourt E, *Taming the Tyrant,* Allen and Unwin, London, 1987.
Harfield A, *Pigeon to Packhorse,* Picton, Chippenham, UK, 1989.
Harman N, *Dunkirk — The Necessary Myth,* Jove, New York, 1980.
Hawkhead JC, *Handbook of Technical Instruction for Wireless Telegraphists,* Marconi Press Agency, London, 1913.
Hawkhead JC & Dowsett HM, *Handbook of Technical Instruction for Wireless Telegraphists,* Wireless Press, Sydney, 1915.
Heinlein RA, *The Door Into Summer,* Panther, London, 1960.
Hill J, *Radio, Radio,* Sunrise Press, Bampton, UK, 1993.
Hill J, *The Cat's Whisker,* Oresko, London, 1978.
Hills A, Terrestrial Wireless Networks, *Scientific American,* April 1998, p. 74.
Hinsley FH & Stripp A (eds), *Code Breakers,* Oxford University Press, Oxford, 1994.
Hodges A, *The Alan Turing Home Page,* from the Internet: www.turing.org.uk/turing/
Horrocks B, *A Full Life,* Fontana, Glasgow, 1960.
Hough R & Richards D, *The Battle of Britain,* John Curtis, London, 1989.
Hoy HC, *40 OB or How the War Was Won,* Hutchinson, London, 1934.
Hoyt EP, *Japan's War,* Arrow, London, 1986.
Hughes DR & Hendricks D, Spread Spectrum Radio, *Scientific American,* April 1998, p. 82.
Hurley F, *Shackleton's Argonauts,* McGraw Hill, Sydney, 1948.
Hutchinson G, *Baird,* The Author, Hastings, UK, 1985.
IBM, *IBM — The Thomas J Watson Research Centre,* IBM Research Communications Department.
Ingersoll R, *Top Secret,* Partridge, London, 1946.
Jack Hillary and Associates, *Surf's Up — Alternative Futures for Full Service Networks in Australia,* AGPS, Canberra, 1995.
Jahnke DA & Fay KA (eds), *From Spark to Space,* American Radio Relay League, Novington, UK, 1989.
James RR, *Chindit,* Sphere, London, 1981.
Jensen PR, *In Marconi's Footsteps 1894 to 1920: Early Radio,* Kangaroo Press, Sydney, 1994.
Johnson B, *The Secret War,* Arrow, London, 1978.
Jolly WP, *Marconi — A Biography,* Constable, London, 1972.
Jones BE, *Small Electric Apparatus,* Cassell, London, 1913.
Jones G, Babbage: Architect of Modern Computing, *New Scientist,* 29 June 1991.
Jones RV, *Most Secret War,* Hamish Hamilton, London, 1978.

Kahn D, *The Codebreakers,* Sphere, London, 1973.
Kahn D, *Hitler's Spies,* Arrow Books, Sydney, 1978.
Kates J & Smith N, Thomas Edison — Radio Prophet, *Amateur Radio Action*, vol. 4/9, p. 18.
Kingsford Smith C, *The Southern Cross Story,* Seal, Sydney, 1995.
Knight L, The Coming of the Superhet (Parts 1 and 2), *Radio Bygones*, Christmas 1990, p. 10; February/March 1991, p. 28.
Kozacuk W, *Enigma,* Arms and Armour, London, 1984.
Ladd J & Melton K, *Clandestine Warfare,* Guild, London, 1988.
Lankshear P, *Discovering Vintage Radio,* Federal Publishing, Sydney, 1992.
Leasor J, *Green Beach,* Corgi, London, 1975.
Leggatt DP, 80 Years of British Television, International Conference on History of Television, November 1986, London IEE Publication, No. 271, p. 8.
Leutz CR & Gable RB, *Short Waves,* CR Leutz, 1930.
Lewin R, *Ultra Goes to War,* Grafton, London, 1988.
Liddell Hart BH, *History of the First World War,* Pan, London, 1973.
Liddell Hart BH, *History of the Second World War,* Pan, London, 1973.
Lodge OJ, *Signalling Across Space Without Wires,* London, 1899.
Lorain P, *Secret Warfare,* Orbis, London, 1983.
Lord W, *Lonely Vigil,* Pocket Books, New York, 1977.
Lord W, *A Night to Remember,* Longman Green, London, 1956.
Lubsynski HG, Some Early Developments of Television Camera Tubes ..., International Conference on History of Television, November 1986, London IEE Publication, No. 271, p. 60.
McKay F, *Traeger, the Pedal Radio Man,* Boolarong Press, Brisbane, 1995.
MacKeand JCB & Cross MA, Wide-band High Frequency Signals from Poldhu in *100 Years of Radio*, Proceedings of the IEE Conference, September 1995, IEE, London, p. 26.
Mackenzie C, *Gallipoli Memories,* Panther, London, 1965.
MacKinnon C, Harold Alden Wheeler and the Hazeltine Company, *HRSA Newsletter*, July 1993, p. 17.
Macksey KJ, *Panzer Division — The Mailed Fist,* Macdonald, London, 1968.
McMahon M, *A Flick of the Switch,* Vintage Radio, Palos Verdes, USA, 1975.
McMahon M, *Radio Collectors Guide 1921–1932,* Vintage Radio, Palos Verdes, USA, 1973.
McMahon M, *Vintage Radio,* Vintage Radio, Palos Verdes, USA, 1973.
Marceil WS, The First Radio Broadcast, *Radio Bygones*, Christmas 1992, p. 4.
Marconi G, *Wireless Telegraphy,* Royal Institution, London, 1900.
Marconi G, *The Progress of Electric Space Telegraphy,* Royal Institution, 1902.
Marcus G, *The Maiden Voyage,* Allen and Unwin, London, 1969.
Martin J, *Telecommunications and the Computer,* Prentice Hall, Englewood Cliffs, NJ, 1990.
Masefield J, *Gallipoli,* Heinemann, London, 1917.
Masterman JC, *The Double Cross System,* Granada, London, 1979.
Mawson D, *Home of the Blizzard,* Heinemann, London, 1915.
Melton HK, *CIA Special Weapons and Equipment,* Sterling, New York, 1993.
Miller CE, Radio Goes to War, *Radio Bygones*, August/September 1989, p. 3.
Millward JD, A Brief History of Telecine, International Conference on History of Television, November 1986, London IEE Publication, No. 271, p. 86.
Miles WGH, *Admiralty Handbook of Wireless Telegraphy,* HMSO, London, 1925.
Ministry of Information, *By Air to Battle,* HMSO, London, 1945.
Montgomery BL, *The Memoirs — Montgomery,* World Publishing Co., New York, 1958.
Moorhead A, *Gallipoli,* Wordsworth, Ware, UK, 1997.
Moyle A, *Clear Across Australia,* Nelson, Melbourne, 1984.
Murray J, *Calling the World,* Focus Books, Sydney, 1995.
Muscio W, *Australian Radio,* Kangaroo Press, Sydney, 1984.
Nalder RFH, *The Royal Corps of Signals,* Royal Signals Institution, London, 1958.
Occleshaw M, *Armour against Fate,* Columbus, London, 1989.
O'Dell TH, Marconi's Magnetic Detector, *Electronic and Wireless World,* August 1993, p. 666.
Parker W, From Altair to AT, *Australian PC World*, August 1985, p. 29.
Pelton JN, Telecommunications for the 21st Century, *Scientific American*, April 1998, p. 68.

Peverett AM, Some Early Radio Receivers, *Wireless World*, November 1972, p. 510.
Perret B, *A History of the Blitzkrieg*, Jove, New York, 1989.
Phillips VJ, *Early Radio Wave Detectors*, Peregrinus with Science Museum, Stevenage, UK, 1980.
Pickworth G, The Spark that Gave Radio to the World, *Electronics and Wireless World*, November 1993, p. 937.
Pickworth G, Marconi's 200 kW Trans-Atlantic Transmitter, *Electronics and Wireless World*, January 1994, p. 28.
Pocock RF & Garret GRM, *The Origins of Maritime Radio*, Science Museum, London, 1972.
Poole J, *The Practical Telephone Handbook*, Pitman, London, 1918.
Portable Wireless Telegraphy, *The Marconigraph*, August 1912, p. 185.
Powell G, *The Devil's Birthday*, Macmillan, London, 1984.
Preece WH, *Signalling through Space without Wires*, Royal Institution, London, 1897.
Remarque EM, *All Quiet on the Western Front*, Mayflower, London, 1963.
Reid TR, *Microchips*, Collins, London, 1985.
Riordan M & Hoddeson L, *Crystal Fire*, Norton, New York, 1997.
Robinson SS, *Manual of Wireless Telegraphy*, Naval Institute Press, Annapolis, USA, 1913.
Robinson R, *Travellers in Time*, Macdonald, London, 1986.
Ross J, EH Armstrong — FM Pioneer, *The Broadcaster*, July 1986.
Rowe J, *Introduction to Digital Electronics*, Sungravure, Sydney, 1967.
Rowe J, *Basic Electronics*, Sungravure, Sydney, 1969.
Rowe J, *Getting into Micro-processors*, Sungravure, Sydney, 1977.
Rowe J, *Fundamentals of Solid State*, Sungravure, Sydney, 1979.
Ruby M, *F Section SOE*, Grafton, Sydney, 1988.
Ryan C, *A Bridge Too Far*, Coronet, Aylesbury, UK, 1974.
Sale A, The Colossos of Bletchley Park, *IEE Review*, March 1995.
Saunders IL, The Unidyne, *Radio Bygones*, October/November 1992, p. 21.
Scarth RN, *Mirrors by the Sea*, Hythe Civic Society, Hythe, 1995.
Schatzkin P, *The Farnsworth Chronicles*, from the Internet: www.songs.com/philo/index.html
Scott K, *Antarctica*, Halstead Press, Sydney, 1993.
Shawsmith A, Edwin Howard Armstrong, *Amateur Radio Action*, vol. 5/2, p. 29.
Shirer W, *The Rise and Fall of the Third Reich*, Pan, London, 1964.
Shurkin J, *Engines of the Mind*, Norton, New York, 1984.
Sterling GE, *The Radio Manual*, Van Nostrand, New York, 1928.
Stokes JW, *More Golden Age of Radio*, Craigs, Invercargill, NZ, 1990.
Story AT, *The Story of Wireless Telegraphy*, Hodder and Stoughton, London, 1913.
The Story of 25 Eventful Years in Pictures, Odhams, London, 1935.
Stutzman WL & Dietrich CB Jr, Moving beyond Wireless Voice Systems, *Scientific American*, April 1998, p. 80.
Swade D, *The Irascible Genius Redeemed*, Science Museum, London, 1987.
Swade D, Redeeming Charles Babbage's Mechanical Computer, *Scientific American*, February 1993, p. 86.
Swade D, Building Babbage's Dream Machine, *New Scientist*, 29 June 1991, p. 37.
Swade D, *Charles Babbage and His Calculating Engine*, Science Museum, London, 1991.
Sweeney WM, *Wireless — for Professional or Amateur Students*, EW Cole, Melbourne, 1920.
Smith T, The Origins of Morse, *Practical Wireless*, February 1986, p. 36.
Stevenson W, *A Man Called Intrepid*, Sphere Books, London, 1976.
Stokes JW, *70 Years of Radio Tubes and Valves*, Vestal Press, New York, 1982.
Taylor P, *An End to Silence*, Methuen, Sydney, 1980.
Thomas Alva Edison, *Tele-Technician*, November/December 1950.
Towler GO, Milestones in Television Pick Up Performance, International Conference on History of Television, November 1986, London, IEE Publication, No. 271, p. 64.
Tuohy F, *The Secret Corps*, John Murray, London, 1920.
Tyne GFJ, *The Saga of the Vacuum Tube*, Prompt, NJ, 1994.
Urquhart R, *Arnhem*, Pan, Sydney, 1958.
Von Clausewitz C, *On War*, Penguin, Harmondsworth, UK, 1986.
Von Rintelen, *Dark Invader*, Lovat Dickson, London, 1933.
Vyvyan RN, *Wireless Over 30 Years*, Routledge, London, 1933.

Wake N, *The White Mouse,* Macmillan, Sydney, 1985.
Wander T, Wireless Takes to the Road (Parts 1, 2 and 3), *Radio Bygones,* August/September 1989, p. 20; October/November 1989, p. 27; February/March 1990, p. 16.
War Office (UK), Signal Training — vol. III (pamphlet no. 24 — Aerials), Army Council, London, 1939.
Way P, *Codes and Ciphers,* Aldus, London, 1977.
Wedlake GEC, *SOS — The Story of Radio Communications,* Wren, Melbourne, 1973.
Welchman G, *The Hut Six Story,* Penguin, Harmondsworth, UK, 1982.
West N, *GCHQ,* Coronet, London, 1986.
West N, *MI5,* Panther, London, 1984.
Williams HS, *Practical Radio,* Funk and Wagnall, New York, 1924.
Williams N, *Australia's Radio Pioneers,* Federal Press, Sydney, 1994.
Williams N, *Basic Electronics,* Sungravure, Sydney, 1969.
Wilmot C, *The Struggle for Europe,* Fontana, London, 1952.
Winterbotham FW, *The Ultra Secret,* Futura, London, 1974.
Wright P, *Spy Catcher,* Heineman, Melbourne, 1987.
Wolff I, *In Flanders Fields,* Longmans, London, 1959.
Year Book of Wireless Telegraphy — 1914, Marconi Publishing Corporation, London, 1914.
Yeats-Brown F, *Britain at War — The Army,* Hutchinson, Melbourne, 1942.
Young P (ed.), *The Decisive Battles of World War 2,* Bison, London, 1989.
Zorkoczy P, *Information Technology,* Knowledge Industry Publication Inc., White Plains, NY, 1982.

Index